The Scientific Papers of Sir Charles Wheatstone

CHARLES WHEATSTONE

CAMBRIDGE
UNIVERSITY PRESS

CAMBRIDGE UNIVERSITY PRESS

Cambridge, New York, Melbourne, Madrid, Cape Town,
Singapore, São Paolo, Delhi, Tokyo, Mexico City

Published in the United States of America by Cambridge University Press, New York

www.cambridge.org
Information on this title: www.cambridge.org/9781108032025

This edition first published 1879
This digitally printed version 2011

ISBN 978-1-108-03202-5 Paperback

CAMBRIDGE LIBRARY COLLECTION

Books of enduring scholarly value

Technology

The focus of this series is engineering, broadly construed. It covers technological innovation from a range of periods and cultures, but centres on the technological achievements of the industrial era in the West, particularly in the nineteenth century, as understood by their contemporaries. Infrastructure is one major focus, covering the building of railways and canals, bridges and tunnels, land drainage, the laying of submarine cables, and the construction of docks and lighthouses. Other key topics include developments in industrial and manufacturing fields such as mining technology, the production of iron and steel, the use of steam power, and chemical processes such as photography and textile dyes.

The Scientific Papers of Sir Charles Wheatstone

Sir Charles Wheatstone (1802–75) was a shoemaker's son whose fascination with physics led him to become one of the most celebrated scientists and inventors of his time. Apprenticed to his uncle, a musical instrument manufacturer, Wheatstone studied the physics of sound, publishing his first scientific paper in 1823. He was the chief developer of telegraphy, inventing increasingly advanced instruments for transmitting and receiving information. Telegraphy revolutionized communication in the Victorian era, eventually making almost instantaneous global communication possible. This collection of Wheatstone's works, first published in 1879, spans his entire career and includes fully illustrated details of many of his pioneering inventions. His broad-ranging research led to numerous important advances; those in telegraphy and cryptography were still in military use as late as the Second World War. This collection is a valuable source for the history of science, and a fitting tribute to Wheatstone's 'industry and versatility'.

Cambridge University Press has long been a pioneer in the reissuing of out-of-print titles from its own backlist, producing digital reprints of books that are still sought after by scholars and students but could not be reprinted economically using traditional technology. The Cambridge Library Collection extends this activity to a wider range of books which are still of importance to researchers and professionals, either for the source material they contain, or as landmarks in the history of their academic discipline.

Drawing from the world-renowned collections in the Cambridge University Library, and guided by the advice of experts in each subject area, Cambridge University Press is using state-of-the-art scanning machines in its own Printing House to capture the content of each book selected for inclusion. The files are processed to give a consistently clear, crisp image, and the books finished to the high quality standard for which the Press is recognised around the world. The latest print-on-demand technology ensures that the books will remain available indefinitely, and that orders for single or multiple copies can quickly be supplied.

The Cambridge Library Collection will bring back to life books of enduring scholarly value (including out-of-copyright works originally issued by other publishers) across a wide range of disciplines in the humanities and social sciences and in science and technology.

THE

SCIENTIFIC PAPERS

OF

Sir CHARLES WHEATSTONE, D.C.L., F.R.S.,

ORD. BORUSS. " POUR LE MÉRITE " EQ.,

INST. FR. (ACAD. SCI.) PAR., ACAD. SCI. BEROL. CORRESP.,

SOC. REG. SCI. GOTT. ET ACAD. SCI. MONACH. SOC. EXTR.,

PROFESSOR OF EXPERIMENTAL PHILOSOPHY IN KING'S COLLEGE, LONDON.

PUBLISHED BY

THE PHYSICAL SOCIETY OF LONDON.

LONDON:

TAYLOR AND FRANCIS, RED LION COURT, FLEET STREET.

1879.

PRINTED BY TAYLOR AND FRANCIS,
RED LION COURT, FLEET STREET.

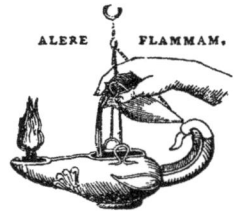

ALERE FLAMMAM.

PREFACE.

THE volume which the Physical Society now presents to its Members includes the whole of the published scientific papers of the late Sir CHARLES WHEATSTONE. Soon after his death, the Council of the Physical Society entertained the idea of paying this tribute to his memory, and a Wheatstone Committee, consisting of Prof. W. G. Adams, Dr. Atkinson, Mr. Latimer Clark, Prof. G. C. Foster, and Mr. R. Sabine, was appointed to carry it out. Sir C. Wheatstone's family placed a considerable quantity of his manuscripts and printed papers at the disposal of the Committee for the purpose of publication, and contributed generously towards the expense of the reprint.

In the arrangement of the papers subjects of the same class have been placed together, the individual papers being arranged in their respective classes in chronological order. In addition to papers read before learned Societies, the collection includes an important article which Sir Charles Wheatstone contributed to the 'London and Westminster Review' in 1837, "On Speaking Machines." The delay in the appearance of this volume has been in great measure due to the difficulty of getting original copies for the printers; and the Physical Society is indebted for

the loan of copies of certain of the papers to the Presi-
dent and Council of the Royal Society and to the Managers
of the Royal Institution.

The thanks of the Society are especially due to Prof.
Foster for reading the proofs and seeing the volume through
the press.

The unpublished papers consist, for the most part, of loose
and scattered fragments written at very different periods.
Many of them are rough notes on subjects which have afterwards been more fully considered by their author, and the
results have been already published. The Committee therefore have arrived at the conclusion that no useful purpose
would be fulfilled by the publication of the MSS. in the
incomplete state in which they were left by their author,
but would suggest their careful preservation as an evidence
of his industry and versatility. It is considered that it
would be appropriate to preserve them with the Scientific
Library and Apparatus bequeathed by him to King's College;
and Sir Charles Wheatstone's family have arranged to deposit them in the Library of the College, with which he
was so long connected.

TABLE OF CONTENTS.

NEW EXPERIMENTS ON SOUND.

Phonics, 1.—Oscillations of Phonics, 2.—Methods of observing the phenomena, 3.—Absolute magnitudes of vibrating particles, 4. —In what manner the oscillations influence the particles, 5.—Rectilineal transmission of Sound, 7.—Polarization of Sound, 9.—Nature of the Vibrations, 10.—Refraction and diffraction of Sound, 12.

EXPLANATION OF THE HARMONIC DIAGRAM.

Introductory, 14.—The Elements of Melody: § 1. Description of the Diagram, 15.—§ 2. Scales, *ib.*—§ 3. Modes, 17.—§ 4. Keys, *ib.* —§ 5. Problems on the Keys and Modes, *ib.*—§ 6. Progression of the Keys, 18.—§ 7. Transposition of Keys, 19.—§ 8. Intervals, *ib.*

DESCRIPTION OF THE KALEIDOPHONE, OR PHONIC KALEIDOSCOPE.

Its uses, 21.—Description of the apparatus, 22.—Experiments, 24. —Duplication and multiplication of objects, 28.—A new Optical experiment, *ib.*

EXPERIMENTS ON AUDITION.

§ 1. Perception by the closed ear of sounds produced internally, 30.—§ 2. The Microphone, 32.—§ 3. Production of the grave harmonic, 33.—§ 4. Acute sounds rendered inaudible, 34.—§ 5. Augmentation of acute sounds, *ib.*—§ 6. Varying intensities of certain notes, 35.

ON THE RESONANCES, OR RECIPROCATED VIBRATIONS OF COLUMNS OF AIR.

Definition of Resonance, 36.—§ 1. Reciprocation of sound of Tuning-fork, 37.—§ 2. Reciprocation of sound of Wind-instruments, *ib.*—§ 3. Sounds which a column of air will reciprocate, 38.

—§ 4. Separate audibility of any one of several simultaneous sounds, 39.—§ 5. The *Génder*, 40.—§ 6. New phenomena of Resonance, 42.—§§ 7, 8. The Jew's harp, 43.—§ 9. On simultaneous reciprocation of several sounds of a Chord, 45.—§ 10. Beats, *ib.*

On the Transmission of Musical Sounds through Solid Linear Conductors, and on their subsequent Reciprocation.

§ 1. Experiments of previous investigators, 47.—§ 2. Experiments of Wheatstone, 49.—§ 3. Laws of Resonance, 50.—§ 4. Length of Conductors, 52.—§§ 5, 6. Transmission of sound of Stringed instruments, 53.—§ 7. Of Reed Wind-instruments, 57.—§ 8. Of other Wind-instruments, 58.—§ 9. Transmission through bent Conductors, 59.—§ 10. Longitudinal and transverse undulations, 61.— § 11. Application of results, 62.

On the Figures obtained by strewing Sand on Vibrating Surfaces, commonly called Acoustic Figures.

§ 1. Statement of Chladni's results, 64.—§ 2. Modes of Vibration of a Rectangular surface, 67.—§§ 3, 4. Superposition of two similar modes of Vibration, 68.—§ 5. Superposition of four Modes, 71.— § 6. Perfect resultant figures, 74.—§ 7. Imperfect resultant figures, 76.—§ 8. Figures of Irregular plates, 77.—§ 9. Table of Sounds and Vibrations, 78.—§ 10. Theories of previous investigators, 80.— § 11. Plates of wood, 81.

An Account of some Experiments to Measure the Velocity of Electricity and the Duration of Electric Light.

§ 1. Description of apparatus, 84.—§§ 2, 3. Experiments with Revolving Mirror, 85.—§ 4. Instrument to determine velocity of transmission through conducting bodies, 89.—§ 5. Proposed modifications of instrument, 95.—§ 6. Applications of instantaneousness of Electric Light, *ib.*—§ 7. General conclusions.

An Account of several New Instruments and Processes for Determining the Constants of a Voltaic Circuit.

§ 1. Uses of Instruments, 97.—§ 2. Results of Ohm's theory, 98. § 3. New Terminology, 102.—§ 4. Measurement of force of a Current, 103.—§ 5. Description of the *Rheostat*, 105.—§ 6. Uses of the *Rheostat*, 108.—§ 7. Standard of Resistance, 110.—§ 8. Resistance Coils, *ib.*—§ 9. Measurement of resistance of interposed Bodies, 111. —§ 10. Process to ascertain the sum of the Electro-motive forces in a voltaic circuit, 112.—§ 11. Measures of Electro-motive forces, 113.—§§ 12, 13. Processes to ascertain the resistance of a Rheomotor, 118.—§ 14. Instrument for measuring the resistance of Liquids, 122.—§ 15. Measurement of forces by means of Galvanometer, 124.—§ 16. Differential resistance measurer, 126.—§ 17. Second differential arrangement, 129.—§ 18. Comparison of different degrees of Intensity, 130.—§ 19. Process to determine degrees of deviation of Needle of Galvanometer corresponding to the degrees of force; and the converse, 132.

ON THE THERMO-ELECTRIC SPARK.

Experiments of Cav. Antinori and of Prof. Linari, 134.—Verification by Wheatstone, 135.—Experiments of Prof. G. D. Botto, 136.

DESCRIPTION OF THE ELECTRO-MAGNETIC CLOCK, 138.

ENREGISTREUR ÉLECTROMAGNÉTIQUE POUR LES OBSERVATIONS MÉTÉOROLOGIQUES, 141.

NOTE SUR LE CHRONOSCOPE ÉLECTROMAGNÉTIQUE.

Invention and description of instrument, 143.—Publication of invention, 144.—Acquaintance with M. de Konstantinoff, 146.—Construction of modified Chronoscope, 147.—Publication of modified invention, 148.—Communication to M. Breguet, 149.—Instrument constructed by M. Breguet, 150.—Proposed instrument of Wheatstone, 150.—Principle had previously been employed by Young and Wheatstone, 151.

AN ACCOUNT OF SOME EXPERIMENTS MADE WITH THE SUBMARINE CABLE OF THE MEDITERRANEAN ELECTRIC TELEGRAPH.

Description of Cable, 152.—First series of Experiments, 153.—Second series, 154.—Third series, 155.

ON THE POSITION OF ALUMINUM IN THE VOLTAIC SERIES, 158.

TÉLÉGRAPHE AUTOMATIQUE ÉCRIVANT.

Outline of the system, 160.—The *Perforateur*, 161.—The *Transmetteur*, 162.—The *Récepteur*, 164.—The *Traducteur*, 165.—Advantages of the system, 165.

ON THE CIRCUMSTANCES WHICH INFLUENCE THE INDUCTIVE DISCHARGES OF SUBMARINE TELEGRAPHIC CABLES.

§ 1. Inductive action in Telegraphic Wires, 168.—§ 2. Instrument for Charging and Discharging Wires, 170.—§ 3. Description of Galvanometers, 171.—§ 4. Influence of Electro-motive force of Battery on amount of Discharge, 172.—§ 5. Influence of the length of the wire on the Inductive Discharge, 173.—§ 6. Simultaneous Discharges, 174.—§ 7. Influence of the conductivity of the Wire, 175.—§ 8. Influence of the diameter of the Wire and the thickness of the insulating coating on the amount of Discharge, 175.—§ 9. Influence of the insulating material on the amount of Discharge, 180. —§ 10. Influence of temperature on the amount of Discharge, 184. —§ 11. Influence of pressure on the Inductive Discharge and Insulation, 185.—§ 12. Influence of interposed resistances on the time of Charging and Discharging, 186.—§ 13. Discharges from one end of a wire when the other communicates with the earth, 187.—§ 14. True and apparent Discharges, 191.—§ 15. The Magnetic Rheometer. 194. —§ 16. Means of accumulating the effects of Charges and Discharges, 196.

DESCRIPTION OF THE TELEGRAPH THERMOMETER, 206.

ON A NEW TELEGRAPHIC THERMOMETER, AND ON THE APPLICATION OF THE PRINCIPLE OF ITS CONSTRUCTION TO OTHER METEOROLOGICAL INDICATORS, 208–210.

ON THE AUGMENTATION OF THE POWER OF A MAGNET BY THE REACTION THEREON OF CURRENTS INDUCED BY THE MAGNET ITSELF.

Description of Magnet employed, 211.—Effect of excitation of Magnet, by Rheomotor, ib.—Explanation of phenomena, 212.— Residual magnetism of Magnet, 213.—Effects strongest at first moment of completing circuit, ib.—Increased effect produced by diverting portion of current from Magnet, 214.—Explanation of increase, ib.—Resistances in different branches of circuit, 215.

ON A CAUSE OF ERROR IN ELECTROSCOPIC EXPERIMENTS.

Phenomena which originated inquiry, 216.—Conditions essential to complete production of phenomena, 217.—Phenomena observed in America, 219.

EXPERIMENTAL VERIFICATION OF BERNOUILLI'S THEORY OF WIND INSTRUMENTS, 220.

REMARKS ON PURKINJE'S EXPERIMENTS, 221.

ON THE PRISMATIC DECOMPOSITION OF ELECTRICAL LIGHT, 223.

CONTRIBUTIONS TO THE PHYSIOLOGY OF VISION.—ON SOME REMARKABLE, AND HITHERTO UNOBSERVED, PHENOMENA OF BINOCULAR VISION.

Part I. § 1. Dissimilarity of Retinal Images of solid object viewed

with both eyes, 225.—§ 2. Result of placing before the eye two perspective projections of an object instead of the object itself, 228.—
§ 3. The Stereoscope, 230.—§ 4. Experiments with the Stereoscope,
232.—§ 5. Effect of transposition of Figures in the Stereoscope, 234.
—§ 6. Binocular appearance of two Drawings, one presented to each
eye, identical with that of two real objects whose projections on the
retina are the same as those of the drawings, 235.—§ 7. Experiments with Symmetrical objects, 236.—§ 8. Effects of Binocular Perspective observable in smooth plates of metal and with compasses,
236.—§ 9. Explanation of correct Vision with one eye only, 238.—
§ 10. Explanation of involuntary changes in figures of solid objects,
240.—§ 11. Illusion of inversion of relief, 242.—§ 12. Double appearance of similar pictures falling on corresponding points of the
retina, 245.—§ 13. Binocular vision of images of different magnitudes, 246.—§ 14. Phenomena which are observed when pictures,
which are neither similar nor the binocular complements of each
other, are simultaneously presented to corresponding parts of the
two retinæ, 247.—§§ 15, 16. Theories as to the cause of the single
appearance of objects seen by both eyes, 249.

Part II. § 17. Alterations of apparent magnitude of Binocular
images, 260.—§ 18. Modifications of the Stereoscope, 266.—§§ 19,
20. Experiments with photographic pictures, 268.—§ 21. Combination of increased magnitude with perception of solidity, 271.—
§ 22. Conversions of relief, 273.—§ 23. Phenomena of Stereoscope
obtainable from objects themselves — the Pseudoscope, 275.—
§ 24. Conversions of relief of objects viewed immediately, 283.

On a singular Effect of the Juxtaposition of certain Colours
under particular circumstances.

Apparent motion of pattern of red and green carpet when viewed
by gaslight—less remarkable in other colours—not apparent in daylight—how accounted for, 284.

On a Means of Determining the Apparent Solar Time by the
Diurnal Changes of the Plane of Polarization at the
North Pole of the Sky.

Atmospheric Polarization, 285.—Plane of Polarization of the
North Pole, 286.—Description of the Polar Clock, 287.—Advantages over Sun-dial, 288.—Other forms of Polar Clock, 289.

Experiments on the Successive Polarization of Light, with the
Description of a new Polarizing Apparatus.

§ 1. Successive Polarization, 290.—§ 2. Dipolarization, 291.—
§ 3. Polarizing apparatus, ib.—§ 4. Experiments on successive

Polarization, 294.—§ 5. Explanation of phenomena, 296.—§ 6. Application of phenomena, 297.—§ 7. Further experiments, 298.—§ 8. Fresnel's experiment, 299.—§ 9. Means of obtaining large surfaces of Light, 300.—§ 10. Additional use of apparatus, 301.

NOTE RELATING TO M. FOUCAULT'S NEW MECHANICAL PROOF OF THE ROTATION OF THE EARTH.

Experiment of Foucault, 303.—View of Chladni, *ib.*—Is only partially true, 304.—Conclusion of Foucault, *ib.*—Confirmation by experiment, 305.

ON FESSEL'S GYROSCOPE.

Description of Gyroscope, 307.—Experiments, 308–313.

ON THE FORMATION OF POWERS FROM ARITHMETICAL PROGRESSIONS.

Triangular arrangements of Arithmetical Progressions, 314.—§ 1. Square Numbers, *ib.*—§ 2. Cube Numbers, 316.—§ 3. Higher Powers, 320.

INTERPRETATION OF AN IMPORTANT HISTORICAL DOCUMENT IN CIPHER.

Introductory, 321.—Description of Cipher, 323.—Key to Cipher, 324.—Translation of Document, 326.—Text in Cipher, 330.

INSTRUCTIONS FOR THE EMPLOYMENT OF WHEATSTONE'S CRYPTOGRAPH.

Benefits of Cryptography, 342.—Formation of Permutated Alphabet, 343.—Rendering into Cipher, 344.—Translation from Cipher, 345.—Specimen Despatch in Cipher, 347.

REED ORGAN-PIPES, SPEAKING MACHINES, ETC.

On the Vowel Sounds, and on Reed Organ-pipes.
Le Mécanisme de la Parole, suivi de la Description d'une Machine Parlante.
Tentamen Coronatum de Voce.

Speaking Machines of the ancients, 348.—Formation of the Vowels, 353.—Table of the Vowels, 354.—Inadequacy of Written Language, 355.—Willis's Researches on Vowel Sounds, 355.—Phenomena of Resonance, 360.—Multiple Resonance, *ib.*—Classification of Consonants, 361.—Table of Consonants, *ib.*—Application to Speaking Machines, 362.

ON THE VIBRATIONS OF COLUMNS OF AIR IN CYLINDRICAL AND CONICAL TUBES, 368.

LIST OF PUBLICATIONS FROM WHICH THE PAPERS HAVE BEEN EXTRACTED.

Académie des Sciences, Comptes Rendus de l', 1845, tome xx. pp. 1554–1561. *Note sur le Chronoscope Électromagnétique*, 143.

1859, tome xlviii. pp. 214–220. *Télégraphe automatique écrivant*, 160.

Annals of Philosophy, Thomson's, 1823, vol. vi. pp. 81–90. *New Experiments on Sound*, 1.

Archives de l'Électricité, par A. de la Rive, 1844, t. iv. p. 170. *Enregistreur Électromagnétique pour les Observations Météorologiques*, 141.

Athenæum, 24 March, 1832, p. 194. Report of Lecture at Royal Institution, " *On the Vibrations of Columns of Air in Cylindrical and Conical Tubes*," 368.

British Association, Report of the, 1835, pp. 11, 12. *On the Prismatic Decomposition of Electrical Light*, 223.

1835, pp. 551–553. *Remarks on Purkinje's Experiments*, 221.

1835, p. 558. *Experimental Verification of Bernouilli's Theory of Wind Instruments*, 220.

1843, pp. 128, 129. *Description of the Telegraph Thermometer*, 206.

1844, p. 10. *On a singular Effect of the Juxtaposition of certain Colours under particular circumstances*, 284.

1848, pp. 10–12. *On a Means of Determining the apparent Solar Time &c.*, 285.

1867, pp. 11–13. *On a new Telegraphic Thermometer &c.*, 208.

Committee of Enquiry into the Construction of Submarine Telegraph Cables, Report, 1861. *On the Circumstances which Influence the Inductive Discharges of Submarine Telegraph Cables*, 168.

Comptes Rendus. *See* Académie.

Explanation of the Harmonic Diagram, invented by C. Wheatstone. London. Published by C. Wheatstone, 436 Strand. *Explanation of the Harmonic Diagram*, 14.

London and Westminster Review, Nos. XI. and LIV., Oct. 1837. *Reed Organ-pipes, Speaking Machines, etc.*, 348.

Philobiblon Society, Memoirs of the. *Interpretation of an important Historical Document in Cipher*, 321.

Philosophical Magazine, 1837, vol. x. pp. 414–417. *On the Thermo-electric Spark &c.*, 134.

Philosophical Transactions. *See* Royal Society.

Quarterly Journal of Science, Literature, and Art, 1827, vol. i. *Description of the Kaleidophone*, 21.

    1827, pt. ii. *Experiments on Audition*, 30.

    1828, vol. iii. *On the Resonances, or Reciprocated Vibrations of Columns of Air*, 36.

Royal Institution, Journal of the, 1831, vol. ii. *On the Transmission of Musical Sounds through Solid Linear Conductors, &c.*, 47.

Royal Society, Philosophical Transactions of the, 1833, pp. 593–634. *On the Figures obtained by strewing Sand on Vibrating Surfaces, commonly called Acoustic Figures*, 64.

    1834. *An Account of some Experiments to measure the Velocity of Electricity, &c.*, 84.

    1838. *Contributions to the Physiology of Vision.—Part the First, &c.*, 225.

    1843. *An Account of several new Instruments and Processes for determining the Constants of a Voltaic Circuit* (The Bakerian Lecture for 1843), 97.

    1852. *Contributions to the Physiology of Vision.—Part the Second, &c.*, 260.

Royal Society, Proceedings of the, 1840, vol. iv. pp. 249, 250. *Description of the Electro-magnetic Clock*, 138.

    1851, vol. vi. pp. 65–68. *Note relating to M. Foucault's new Mechanical Proof of the Rotation of the Earth*, 303.

    1854, vol. vii. pp. 43–48. *On Fessel's Gyroscope*, 307.

Royal Society, Proceedings of the (*continued*).

1855, vol. vii. pp. 145–151. *On the Formation of Powers from Arithmetical Progression*, 314.

1855, vol. vii. pp. 328–333. *An Account of some Experiments made with the Submarine Cable of the Mediterranean Electric Telegraph*, 152.

1855, vol. vii. pp. 369, 370. *On the Position of Aluminum in the Voltaic Series*, 158.

Vol. xv. pp. 369–372. *On the Augmentation of the Power of a Magnet by the reaction thereon of Currents induced by the Magnet itself*, 211.

Vol. xviii. pp. 330–333. *On a cause of Error in Electroscopic Experiments*, 216.

Vol. xix. pp. 381–389. *Experiments on the Successive Polarization of Light, with the Description of a new Polarizing Apparatus*, 290.

LIST OF ILLUSTRATIONS AND PLATES.

Page

Harmonic Diagram.

Kaleidophone 22

Figures formed by the Kaleidophone 26, 27

Microphone 32

Génder 41

Jew's Harp 43

Conducting-wire for transmission of Sounds 56

Perforated Slip of Telegraph Paper 162

Arrangement of Wires for accumulating Charges and Discharges .. 198

———————— for accumulating Charges and Discharges separately at the same time 201

———————— for alternately Charging and Discharging.... *ib.*

———————— for producing Discharge after separation from Battery 203

———————— for communicating Charge from one wire to another .... *ib.*

———————— for obtaining Intermittent Current from accumulated Discharges..................................... 204

Acoustic Figures, Chladni's .... Plates I., II., III.

———————— of Rectangular Surfaces .... Plate IV.

———————— of Superimposed Surfaces .... Plates V.-XII.

Apparatus for determining Velocity of Electricity....Plates XIII., XIV.

Instruments for determining Constants of a Voltaic Circuit.
Plates XV., XVI.

Discharging Key; Magnetic Rheometer; Accumulating Discharger.
Plate XVII.

Optical Apparatus and Figures illustrating the Physiology of Vision.
Plates XVIII., XIX.

Stereoscopes .... Plate XX.

Polarizing Apparatus .... Plate XXI.

SCIENTIFIC PAPERS.

I. *New Experiments on Sound.*

[From Thomson's 'Annals of Philosophy,' 1823, vol. vi. pp. 81-90.]

ON THE PHONIC MOLECULAR VIBRATIONS.

BEFORE I enter on the immediate subject of this article, it may be necessary to exhibit a general view of those bodies which, being properly excited, make those sensible oscillations which have been thought to be the proximate causes of all the phenomena of sound. These bodies, to avoid many circumlocutions otherwise inevitable, I have termed Phonics.

Linear Phonics.

Transversal,	*Longitudinal,*
Making their oscillations at right angles to their axis.	Making their oscillations in the direction of their axis.
1. Capable of tension, or variable rigidity: chords or wires.	1. Columns of aeriform fluids or liquids; cylindric and prismatic rods.
2. Permanently rigid: rods, forks, rings, &c.	

Superficial Phonics.

1. Capable of tension: extended membranes.
2. Permanently rigid: laminæ, bells, vases, &c.

Solid Phonics.

1. Volumes of aeriform fluids.

B

The sensation of sound can be excited by any of these bodies when they oscillate sufficiently rapidly, either entire or divided into any number of parts in equilibrium with each other. The laws of these subdivisions differ in the various phonics according to their form and mode of connexion or insulation; and the velocities of the oscillations, or degrees of tune, depend on the form, dimensions, mode of connexion, mode of division, and elasticity of the body employed. The points of division in linear phonics are called nodes; and the boundaries of the vibrating parts of elastic surfaces are termed nodal lines. The parts at which the oscillatory portions have their greatest excursions are named centres of vibration; these are always at the greatest mean distances from the nodal points or lines.

These mechanical oscillations are not, however, themselves the immediate causes of sound; they are but the agents in producing in the bodies themselves, and in other contiguous substances, isochronous vibrations of certain particles varying in magnitude according to the degree of tune. I convinced myself of this important fact by the following simple experiments :—I took a plate of glass capable of vibrating in several different modes, and covered it with a layer of water; on causing it to vibrate by the action of a bow, a beautiful reticulated surface of vibrating particles commenced at the centres of the vibrating parts, and increased in dimensions as the excursions were made larger. When a more acute sound was produced, the centres consequently became more numerous, and the number of coexisting vibrating particles likewise increased; but their magnitudes proportionably diminished. The sounds of elastic laminæ are generally supposed to be owing to the entire oscillations of the simple parts, as shown by Chladni when, by strewing sand over the sonorous plates, he observed the particles repulsed by the vibrating parts accumulate on the nodal lines and indicate the bounds of the sensible oscillations. Did no other motions exist in the plate but these entire oscillations, the water laid on its surface would, on account of its cohesion to the glass, show no peculiar phenomena; but the appearances above described clearly demonstrate that the oscillating parts consist of a number of vibrating particles of equal magnitudes, the excursions of which are greatest at the centres of vibration, and gradually become less as they recede further from it, until they become almost null at the nodal lines.

To multiply these surfaces, and to observe whether the magnitudes of these particles vary in different media, in a glass vessel of a cylindric form I superposed three immiscible fluids of different densities—namely mercury, water, and oil. On producing the sounds corresponding with each mode of division, I observed a number of vibrating parts, agreeing with the sound and showing similar appearances to the plate, formed on the surfaces of each of the fluids ; not the least agitation appeared in the uniform parts.   I afterwards inserted this glass in another vessel of water in order to observe the vibrations of the external surface, and found the same results as in the interior, though the levels of the surfaces were different.

The most accurate method to observe these phenomena is by employing a metallic plate of small dimensions, which must be fixed horizontally in a vice at one end, and covered on its upper side with a surface of water.   On causing it to oscillate entirely by means of a bow, a regular succession of these vibrating corpuscles will appear arranged parallel to the two directions of the plate ; and if the action of the bow be rendered continuous, their absolute number might be counted with the aid of a micrometer.   Diminishing the oscillating part of the plate to one half of its length, the double octave to the preceding was heard, agreeably to the established rule that the velocities of the oscillations are inversely as the squares of the lengths : four vibrating corpuscles then occupied the space before occupied by one ; and the absolute number was double to that in the former instance ; but the absolute number of these corpuscles have no influence whatever on the degree of tune, which entirely depends on their relative magnitude in the same substance.   Theory shows us that in plates of this description alteration of breadth does not affect the degree of tune : let us therefore reduce this

half of the plate to half its breadth, and we shall find the
note remain the same; but the absolute number of the corpuscles will in this case be equal to that in the entire plate.
Let us now take two plates of equal lengths and breadths,
but one double in thickness to the other; the rule is, that
the velocities of the oscillations are as the thicknesses of the
plates; we shall, therefore, in the thicker plate see a double
number of particles to that of the other occupying the same
extent of surface. The last circumstance in which two plates
may differ is their specific rigidity; and in this respect it will
be found that two plates of exactly equal dimensions, and
covered with the same number of vibrating corpuscles of equal
magnitudes, but of different substances, differ in sound;
therefore the absolute magnitudes of the particles cannot be
assumed as a standard of tune, unless regulated by the specific rigidity.

Unassisted by any means of actual admeasurement, the
above are but the proximate results sensible to the eye;
more extended and accurate experiments are necessary to
confirm the results with mathematical certainty. As the
absolute magnitudes of these particles will, I imagine, be
hereafter a most useful element for calculation, I will here
indicate the most effectual way I am acquainted with to arrive
at this knowledge. A thick metallic slip of considerable
length and breadth, bent similarly to a tuning-fork, and fixed
at its curved part in a vice, is very easily excited by friction;
and a more considerable surface of regularly arranged vibrating particles is seen than in most other superficies : any
description of common exciter may be employed. When
this bent plate is excited by percussion, the particles, before
their disappearance, will assume an apparent rotatory motion,
on account of the force exerted and its susceptibility of continuing the vibrations. Employing a parallelopipedal rod, the
appearances of the higher modes of subdivisions are particularly neat; the entire vibrating parts between the nodes form
ellipses, and the semipart at the free end a regular half of
the same figure. It is important to remark that the crispations of the water only appear on the sides in the plane of

oscillation; the other two sides, on one of which the exciter must be applied, do not show similar appearances.

I have also rendered the phonic molecular vibrations visible when produced by the longitudinal oscillations of a column of air   The following were the means employed :—I placed the open end of the head of a flute or flagiolet on the surface of a vessel of water; and on blowing to produce the sound, I observed similar crispations to those described above, forming a circle round the end of the tube, and afterwards appearing to radiate in right lines; on the harmonics of the tube being sounded, the crispations were correspondently diminished in magnitude. These phenomena will be more evident if the tube be raised a little from the surface of the liquid, and a thin connecting film be left surrounding it; the vibrating particles will then occupy a greater space, and be more sensible.

The existence of the molecular vibrations being now completely established, it becomes a critical question in what manner the sensible oscillations induce these vibrating particles. I do not know whether what I am now going to adduce will be admitted as the right explanation; but it is certainly analogous, so far as the superficial and transversal linear oscillations are concerned. A flexible surface, covered with a coat of resinous varnish, being made to assume any curve, the cohesion of the varnish will be destroyed in certain parts, and a number of cracks will be observed more regularly disposed as the force inducing the curve has been more regularly applied; when the original position of the surface is restored the cracks will be imperceptible, but will again appear at every subsequent motion. Be this as it may, these particles are invariable concomitants of the sensible oscillations; and there is no reason to suppose otherwise than that their vibrations are isochronous with them. To avoid confusion, I have restricted the word vibrations to the motions of the more minute parts, and the term oscillations to those of the sensible divisions. We may reasonably suppose that the molecular vibrations pervade the entire substance of a phonic; their excursions, however, are not the same in all parts; and they can only be rendered visible when these excursions are large. They may be so few in number as to be entirely inaudible, as in their transmission through linear conductors; but however few, when they are properly directed they induce the mechanical divisions of sonorous bodies, each of which will give birth to numerous vibrating corpuscles whose excursions are greater, and the sound will be rendered audible. Dr. Savart has well investigated the modes of division in surfaces put in motion by communicated vibrations. All those phonics whose limited superficies preclude them from exciting in themselves a sufficient number of vibrating corpuscles, when isolated, produce scarcely any perceptible sound—as extended chords, tuning-forks, &c.; but those whose superficies or solidities are more extended, as bells, elastic laminæ, columns of air, &c., produce sufficient volume of sound without accessory means.

Loudness of sound is dependent on the excursions of the vibrations; volume, or fulness of sound, on the number of coexisting particles put in motion. Thus the tones of the Æolian harp, on account of the number of subdivisions of the strings, are remarkably beautiful and rich without possessing much power; and the sounds of an Harmonica-glass, in which a greater number of particles are excited than by any other means, are extraordinarily so, united, according to the method of excitation, with considerable intensity: their pervading nature is one of the greatest peculiarities of these sounds.

The following is a recapitulation of the various properties of sound which are attributable to modifications of the vibrating corpuscles :—

The tune		velocities of the vibrations.
The time		continuance of the vibrations.
The intensity	Depends on the	excursions of the vibrations.
The richness, or volume		number of coexisting vibrations.
The quality (timbre)		magnitudes of the vibrating corpuscles.

It has often been thought necessary to admit the existence of more minute motions than the sensible oscillations, in order to account for many phenomena in the production of sound. Perault, in his 'Essai du Bruit,' insisted on their

necessity more than any other author I have read : he imagined that the vibrations have a much greater velocity than
the oscillations which cause them ; but the experiment he adduced to prove this is far from conclusive; he mistook for
these vibrations the oscillations of the subdivisions of the
long string he employed.   Other distinguished philosophers
have had ideas of a similar nature ; and Chladni thinks their
existence necessary to account for the varieties of quality.   I,
however, conceived I was the first who had indicated these
phenomena by experiment, until a few days ago, repeating
them, together with the others which form the subject of this
paper, in the presence of Prof. Oersted, of Copenhagen, he
acquainted me with some similar experiments of his own.
Substituting a very fine powder, Lycopodion, instead of the
sand used by Chladni for showing the oscillations of elastic
plates, this eminent philosopher found the particles not only
repulsed to the nodal lines, but at the same time accumulated
in small parcels on and near the centres of vibration ; these
appearances he presumed to indicate more minute vibrations,
which were the causes of the quality of the sound.   Subsequently he confirmed his opinion by observing the crispations of water or alcohol on similar plates, and showed that
the same minute vibrations must take place in the transmitting medium, as they were equally produced in a surface
of water when the sounding plate was dipped into a mass
of this fluid.   These experiments were inserted in Lieber's
' History of Natural Philosophy,' 1813.

Rectilineal Transmission of Sound.

As the laws of the communication of the phonic vibrations
are more evident in linear conductors, I shall confine the present article to a summary of their principal phenomena.

In my first experiments on this subject, I placed a tuning-
fork, or a chord extended on a bow, on the extremity of a
glass or metallic rod five feet in length, communicating with
a sounding-board : the sound was heard as instantaneously
as when the fork was in immediate contact; and it immediately ceased when the rod was removed from the sounding-

board or the fork from the rod. From this it is evident that the vibrations, inaudible in their transmission, being multiplied by meeting with a sonorous body, become very sensibly heard. Pursuing my investigations on this subject, I have discovered means for transmitting, through rods of much greater lengths and of very inconsiderable thicknesses, the sounds of all musical instruments dependent on the vibrations of solid bodies, and of many descriptions of wind instruments. It is astonishing how all the varieties of tune, quality, and audibility, and all the combinations of harmony, are thus transmitted unimpaired, and again rendered audible by communication with an appropriate receiver. One of the practical applications of this discovery has been exhibited in London for about two years, under the appellation of " The Enchanted Lyre." So perfect was the illusion in this instance from the intense vibratory state of the reciprocating instrument, and from the interception of the sounds of the distant exciting one, that it was universally imagined to be one of the highest efforts of ingenuity in musical mechanism. The details of the extensive modifications of which this invention is susceptible, I shall reserve for a future communication; the external appearance and effects of the individual application above mentioned have been described in the principal periodical journals.

The transmission of the vibrations through any communicating medium as well as through linear conductors is attended by peculiar phenomena. Pulses are formed similar to those in longitudinal phonics; and consequently the centres of vibration and the nodes are reproduced periodically at equal distances: in this we observe an analogous disposition with regard to light. I had intended to include in this paper all the analogical facts I have observed illustratory of the identity of the causes of these two principal objects of sensation; but want of time and the danger of delay, now the subject is occupying so much the attention of the scientific world, have induced me hastily to collect the present experiments, and to defer the others for a future opportunity.

The thicknesses of conductors materially influence the

power of transmission; and there is a limit of thickness, differing for the different degrees of tune, beyond which the
vibrations will not be transmitted. The vibrations of acute
sounds can be transmitted through smaller wires than those
of grave sounds. A proof of this is easy. Attach a tuning-
fork to one end of a very small wire, and apply the other end
to the ear or a sounding-board. On striking the fork rather
hard, two coexisting sounds will be produced: that which is
more acute will be distinctly heard; but the other will not be
transmitted. If the vibrations of a tuning-fork be conducted
through a piece of brass wire of the size and thickness of a
large needle, the sound, imperfectly transmitted, will become
more audible by the pressure of the fingers on the conducting
wire; but if a steel wire of the same length and thickness be
employed, the sound will be unaltered by any pressure,
because steel has a greater specific elasticity than brass.

POLARIZATION OF SOUND.

Hitherto I have only considered the vibrations in their
rectilineal transmission; I shall now demonstrate that they
are peculiarly affected when they pass through conductors
bent in different angles. I connected a tuning-fork with one
extremity of a straight conducting-rod, the other end of which
communicated with a sounding-board: on causing the tuning-
fork to sound, the vibrations were powerfully transmitted, as
might be expected from what has already been explained;
but on gradually bending the rod, the sound progressively
decreased, and was scarcely perceptible when the angle
became a right one; as the angle was made more acute the
phenomena were produced in an inverted order: the intensity
gradually increased as it had before diminished; and when
the two parts were nearly parallel, it became as powerful as
in the rectilineal transmission. By multiplying the right
angles in a rod, the transmission of the vibrations may be
completely stopped.

To produce these phenomena, however, it is necessary that
the axis of the oscillations of the tuning-fork should be perpendicular to the plane of the movable angle; for if they

be parallel with it they will be still considerably transmitted. The following experiment will prove this. I placed a tuning-fork perpendicularly on the side of a rectilinear rod; the vibrations were therefore communicated at right angles. When the axis of the oscillations of the fork coincided with the rod, the intensity of the transmitted vibrations was at its maximum; in proportion as the axis deviated from parallelism the intensity of the transmitted vibrations diminished; and, lastly, when it became perpendicular, the intensity was at its minimum. In the second quadrant the order of the phenomena was inverted, as in the former experiment, and a second maximum of intensity took place when the axis of the oscillations had described a semicircumference and had again become parallel, but in an opposite direction. When the revolution was continued, the intensity of the transmitted vibrations was varied in a similar manner; it progressively diminished as the axis of the oscillations deviated from being parallel with the rod, became the least possible when it arrived at the perpendicular, and again augmented until it remained at its first maximum, which completed its entire revolution.

The phenomena of polarization may be observed in many corded instruments : the cords of the harp are attached at one extremity to a conductor which has the same direction as the sounding-board ; if any cord be altered from its quiescent position, so that its axis of oscillation shall be parallel with the bridge or conductor, its tone will be full; but if the oscillations be excited so that their axis shall be at right angles with the conductor, its tone will be feeble. By tuning two adjacent strings of the harp unisons with each other, the differences of force will be sensible to the eye in the oscillations of the reciprocating string according to the direction in which the other is excited.

It now remains to explain the nature of the vibrations which produce the phenomena the existence of which has been proved by the preceding experiments. The vibrations generally assume the same direction as the oscillations which induce them : in a longitudinal phonic the vibrations are parallel to its axis; in a transversal phonic they are perpendicular to this direction; a circular or an elliptic form can be also given to the vibrations by causing the oscillations to assume the same forms. Any vibrating corpuscle can induce isochronous vibrations of similar contiguous corpuscles *in the same plane*, either parallel with, or perpendicular to, the direction of the original vibrations; and the polarization of the vibrations consists in the similarity of their directions, by which they propagate themselves equally in the same plane: therefore the vibrations being transmitted through linear conductors, it is the plane in which the vibrations are made that determines their transmission or nontransmission when the direction is altered. A longitudinal or a transversal vibration may be transmitted two ways to a conductor bent at right angles: their axis may be in that direction, as to be in the same plane with the right angle, in which case the former will be transversally, or the latter longitudinally transmitted in the new direction; or their axis may be perpendicular to the plane of this new direction, under which circumstances neither can be communicated*. In explaining the polarization of light, there is no necessity to suppose that the reflecting surfaces act on the luminous vibrations by any actual attracting or repulsing force, causing them to change their axes of vibrations; the directions of the vibrations in different planes, as I have proved exist in the communication of sound, are sufficient to explain every phenomenon relative to the polarization of light.

* I have just seen a paper by M. Fresnel, entitled "Considérations Mécaniques sur la Polarisation de la Lumière," in which this eminent philosopher had previously arrived at the same conclusions with respect to light as I have proved in this communication respecting sound. The important discoveries of Dr. Thomas Young, followed by those of M. Fresnel, have recently reestablished the vibratory theory of light; and new facts are every day augmenting its probability. The new views in acoustical science which I have opened in this paper will, I presume, give additional confirmation to the opinions of these eminent philosophers; and I hope, when I resume the subject, to be enabled to account for the principal phenomena of coloration, with regard to their acoustic analogies, in a way calculated to establish the permanent validity of the theory.

Let us suppose a number of tuning-forks oscillating in different planes and communicating with one conducting-rod : if the rod be rectilinear, all the vibrations will be transmitted ; but if it be bent at right angles, they will undergo only a partial transmission ; those vibrations whose planes are perpendicular, or nearly so, to the plane of the new direction will be destroyed. The vibrations are thus completely polarized in one direction while passing through the new path, and on meeting with a new right angle they will be transmitted or not, accordingly as the plane of the angle is parallel with, or perpendicular to, the axes of the vibrations. In this point of view, the circumstances attending the phenomena are precisely the same as in the elementary experiment of Malus on the polarization of light.

Double refraction is a consequence of the laws of polarization, by which a combination of vibrations having their axes in different planes, after travelling in the same direction, are separated into two other directions, each polarized in one plane only. That this well known property of light has a correspondent in the communication of phonic vibrations I shall now demonstrate. When two tuning-forks, sounding different notes by a constant exciter, and making their oscillations perpendicularly to each other, have their vibrations transmitted at the same time through one rod, at the opposite extremity of which two other conductors are attached at right angles, and when each of these conductors is parallel with one of the axes of the oscillations of the forks, on connecting a sounding-board with either conductor, those vibrations only will be transmitted through it which are polarized in the same plane with the angle made by the two rods through which the vibrations pass ; either sound may be thus separately heard, or they may both be heard in combination by connecting both the conductors with sounding-boards.

The phenomena of diffraction regarding only the form of the surfaces, or the superficies over which the vibrations extend, are, by the conformation of the organs of hearing, not of any consequence to the perception of sound, though the same phenomena, when the chromatic vibrations are concerned, are very evident to the eye. They, however, undoubtedly take place equally in both instances, and may be well explained by the theory already laid down. Each separate vibration propagating itself in the plane of its vibrating axis, a number of vibrations in different planes, after passing through an aperture, naturally expand themselves transversely as well as rectilineally, and thereby occupy a greater space than they would were they only longitudinally transmitted.

I have still to indicate a new property of the phonic vibrations; but whether it is analogous to any of the observed phenomena of light I am yet ignorant. When the source of the vibrations is in progressive motion, the vibrations emanating from it are transmitted when the conductor is rectilineal and parallel with the original direction; and they are destroyed when the conductor is perpendicular to the direction, though the axis of vibration and the conductor being in both instances *in the same plane,* would transmit the vibrations were the phonic stationary. These circumstances are proved by the following experiments :—When a tuning-fork, placed perpendicularly to a rod, communicating at one or both extremities with sounding-boards, and caused to oscillate with its vibrating axis parallel with the rod, moves along the rod, preserving at the same time its perpendicularity and parallelism, the vibrations will not be transmitted while the movement continues; but the transmission will take place immediately after it has remained motionless. When the tuning-fork moves on the upper edge of a plane perpendicular to a sounding-board, the vibrations rectilineally transmitted will not be influenced by the progressive motion.

II. *Explanation of the Harmonic Diagram.*

[From a pamphlet entitled "An Explanation of the Harmonic Diagram,
   invented by C. Wheatstone. London. Published by C. Wheatstone,
   436, Strand."]

THE difficulty attending the acquirement of Musical Theory
has been the principal cause of the little attention paid to it
by the generality of practical students. The intention of
the *Harmonic Diagram* is to diminish this difficulty, and to
render the groundwork of the science more familiar. For,
as on a geographical chart, the relative situations and
analogies of the various places can be more readily compre-
hended than by the most accurate and diffuse independent
explanation, so on the Diagram, which is a representation of
the principles from which the Science of Music is derived,
the rules constituting the theory, from the apparent mutual
connexion of their elements, are rendered more evident than
they could be in a desultory treatise. The author does not,
however, pretend to obtrude his invention as a substitute for
either of the many excellent theoretical works now published;
but he ventures to hope it will be found a useful preliminary
and accompaniment to any publication on the subject; and,
as it is intended to convey to the practical musician the foun-
dation only of the theory, the explanation has been confined
to that kind of information more essentially necessary, and
more immediately conducive to the means of instruction.

Definitions.—A NOTE* is a musical sound at a determined

---

* It is productive of considerable inconvenience that the most useful
and elementary words employed in the theory of music should be so
vaguely and promiscuously applied. A NOTE originally signified only
the written character of a musical sound; but for want of a more appropriate and exclusive term it has been adopted in this new sense, not only
in the language of the science but also in that of literature and common
colloquial intercourse. There are many other terms similarly equivocal.
Tone is a *general* word for a musical sound, as a fine tone, a sweet tone, a

pitch or degree of tune, and is termed *graver* than one that is higher, and more *acute* than one that is lower.

An INTERVAL is the distance between two notes : the graver sound is generally considered as the base when compared with a more acute sound.

MELODY is a succession of notes.

A CHORD is the co-existence of several notes.

HARMONY is a succession of chords, or the co-existence of two or more melodies.

Part the First.

THE ELEMENTS OF MELODY.

§ 1. The circumference of the diagram represents a perfect octave. In this interval all the principles of musical variety are contained ; and all notes situated above or below a certain octave in theory are considered the same notes as those bearing similar names within it. Therefore all musical intervals may be regarded as comprised in this compass.

§ 2. *Scales.*—A *scale* is a regular progression, commencing and terminating with notes of the same name, and included within the interval of one or more octaves. The *genera* of scales regard the number of notes in the octave : of those there are three—*Diatonic, Chromatic,* and *Enharmonic**.

---

grave tone, &c. The same word is also employed to signify a particular diatonic interval. The TUNE of sound is usually understood to be its pitch, and *to tune* is to bring each of the notes of an instrument to its proper pitch ; but *to tune* is also used synonymously with *to play,* and *a tune* is the same as *a melody.*

* There is another scale, more simple in its construction than either of these ; but it has not hitherto been described in any theoretical work, and is therefore not included in this enumeration of the scales generally admitted. It consists of five notes, the succession of which forms, like the other scales, two species of intervals, which in this are the tone and the minor third. It is indicated on the diagram by the double lines ; and the diatonic scale is derived from it by dividing each of its minor thirds into a tone and a major semitone by the lines representing the 4th and 7th notes of the diatonic series. This scale possesses several remarkable peculiarities. A diatonic scale has always a determinate key ; but this progression of notes has no decided tonic, and, according to the rules of composition, it may be considered as belonging to either of five different

Diatonic scales consist of seven notes, represented by the seven lines of the movable circle : the intervals between them, viz. *tones,* of which there are five, and *major semitones,* of which there are two, are shown by the distances between these lines.   These scales are the most important, as all regular musical compositions are constructed upon them.

Chromatic scales are formed by the addition of a note in each tone of the above, subdividing it into a major and a minor semitone ; these additional notes are shown by the dotted lines.

Enharmonic scales are formed by the subdivision of each major semitone in a chromatic scale into a minor semitone

---

keys deprived of the notes which divide their respective minor thirds : thus the progression C, D, E, G, A may be regarded as the diatonic major key of C with the omission of its 4th and 7th notes, as the minor key of D without its 3rd and 6th, as the minor key of E wanting its 2nd and 5th, as the major key of G omitting its 3rd and 7th, and as the minor key of A deprived of its 2nd and 6th.   We see from hence the cause of the apparent irregularity in the terminations of melodies composed from this progression.

There are many reasons for believing this series of notes to have constituted one of the most simple and original forms of the enharmonic scale of the ancients (for the Greek names applied to the modern scales do not at all accord with their ancient significations) ; and we have undoubted evidence that the aboriginal airs of most of the European and Oriental nations are entirely constructed upon it.   The peculiar character of Scottish melody is owing to this disposition of the intervals ; and from hence it has been occasionally called the Scottish scale; but as the melodies of India, China, and Persia equally derive their character from it, the term NATIONAL SCALE may be substituted as more appropriate.

The five black keys of the pianoforte correspond with the notes of the national scale; and it may easily be proved by trial that any real Scottish air, as " Roy's Wife," "Auld Lang Syne," &c., with the omission of the unessential notes introduced by transcribers, may be performed upon them ; and in the same manner successful imitations of the Scottish style may be composed.   Melodies thus constructed owe their peculiarly pleasing effect to the simplicity of the intervals between the notes composing them ; in addition to all the consonant intervals, the minor seventh and the perfect second only are in requisition, whereas, in the diatonic scale, the imperfect second, the superfluous fourth, the imperfect fifth and the major seventh are also additionally employed.

and *diesis*, thereby adding to it seven additional notes : these are shown by small marks between the major semitones.

The note commencing and terminating a scale is termed its *tonic* or key-note.

§ 3. *Modes.*—The different dispositions of the intervals between the successive notes of a diatonic scale are called *modes.* When the intervals are situated as represented by placing the index of the upper circle to the key-note, the mode is major; it is minor as shown by placing the index of the lower circle to the key-note. In musical compositions the major mode is most important, and the minor considered in regard to it is either *relative* or *tonic.* The minor is relative when its key-note is the sixth of its major, in which case the notes (and consequently the signatures) of both scales correspond. The minor is tonic when it has the same key-note as the major scale, from which it will be seen by the Diagram to differ in having its 3rd, 6th, and 7th notes nearer the commencing note, the others remaining the same in both scales.

§ 4. *Keys.*—A scale may commence on any note; this constitutes the difference of *keys.* All keys are named from the note on which they commence and terminate; and the seven notes of the progression are in alphabetical order, two of the same letter never occurring. Those notes of a scale to which in their progressive order a sharp or a flat is affixed, to avoid repetition transfer their signs to the commencement, which forms what is termed the *signature* of the key. The signatures of all the keys which are in use are shown on the Diagram when the index points to the corresponding key-note.

§ 5. *Problems on the Keys and Modes.*—To discover the progression of any key in the major mode, place the index of the outward circle to the key-note, which will show the signature, and the lines will point out the seven notes of its scale. Example :—Let D be the key-note, to which place the index : the signature will be shown to be two sharps, and the lines will correspond with D, E, F♯, G, A B, C♯.

c

To find the progression of any key in the minor mode, place the index of the inward circle to the key-note, and the notes and signature of the key will be shown as before. Example :—Let C be the key-note; having placed to it the minor index, the lines will point out C, D, E♭, F, G, A♭, B♭, and the signature will be shown to be three flats. N.B. Those notes which keys in the minor mode ascending borrow from their relative majors are shown by a ✳ on the corresponding dotted lines.

The relative minor to any major key will be found to commence on the 6th of the latter: consequently, when any major scale is shown without altering the circle, its relative minor is likewise seen.   Example :—Let the key-note of the major be C, and its scale shown as directed above; the minor index will then point to A, its sixth note, and the two scales will be seen to be formed of the same notes and possess the same signature.

The tonic minor of any major key may be shown by placing the minor index to the major key-note.   Example :—Let C be the major key-note; by placing to it the minor index its tonic minor will be found to be C, D, E♭, F, G, A♭, B♭, and its signature will be formed of three flats.

§ 6. *Progression of the Keys.*—To explain the progression and relations of the major keys, place the index upon C, which key has neither flats nor sharps; then by removing it to the note pointed out by 5, the fifth above, the scale of the key having one note sharpened will be shown, and every removal the same degree higher will show the key with one sharp added, which additional sharp will invariably be on the 7th of its own and 4th of the preceding key.   By descending in the same manner from C (removing it to 4, the fifth below) the progression of the keys with flats will be shown, and the additional flat will be observed to occur on the 4th of its own and 7th of the preceding key.

The progression of the minor keys may be similarly demonstrated by commencing on A and substituting the minor circle for the major; each additional sharp will then occur on the 2nd of its own and 4th of the preceding key, and each additional flat on the 6th of its own key, and 2nd of the preceding.

To render these operations more evident, under the index of the key-note is shown the number and order of the sharps and flats to the corresponding key.

§ 7. *Transposition of Keys.*—Transposition or changing the key is effected by raising or lowering all the notes of a composition in an equal degree, still reserving the true intervals between the notes of the melody. The rule for the operation is as follows:—Place the index (major or minor, according to the mode) to the tonic of the key in which the piece is written, and write down the progression shown by the lines; proceed the same with the key into which you would transpose it; then by substituting the notes of the latter for those of the former, the transposition will be effected. When a note accidentally sharpened or flattened occurs, its transposition will be shown by a dotted line. The alteration of the signature shown under the index must be observed; and it is necessary to remark, that major keys can only be transposed into other major keys, and minor into other minor keys.

§ 8. *Intervals.*—Intervals derive their names from their distances between the tonic and other notes of the diatonic scale*; on the movable circle of the diagram are exhibited the characters of all those in practical use. If it be required to know the interval between any two notes, the index must be placed to the lower note, and the intervallic character will appear under the higher.

Intervals are classed by theorists into perfect, major, minor, imperfect, superfluous, and diminished. Perfect intervals

* In a former part (§ 2) of this explanation, other terms not thus derived are used to express those intervals which exist between the successive notes of the different scales. The *tone*, the larger interval of the diatonic scale, is identical with what is here termed the *perfect second*; and the *major semitone* with the *imperfect second*. The *minor semitone* and the *superfluous unison* are also the same: the first mentioned are generally employed in explaining the construction of the scales; the latter (with all those enumerated in the present article) in explaining the principles of harmony.

are those between the tonic and those notes of the scale
which are common to both major and minor keys : they are
unison, 1*, second, 2, fourth, 4, fifth, 5, octave, 8. Major intervals are those between the tonic and the notes peculiar to
the major scale : they are third, 3, sixth, 6, seventh, 7. Minor
intervals are those peculiar to the minor mode, and are respectively less than the major intervals by a minor semitone ;
they are denoted by prefixing a flat to their characters, ♭3,
♭6, ♭7†. Imperfect intervals are the perfect intervals dimi-
dished a minor semitone, and are expressed by a slur placed
over the characters—$\widehat{2}$, $\widehat{4}$, $\widehat{5}$, $\widehat{8}$. Superfluous intervals are
the perfect and major intervals increased by a minor semi-
tone ; they are denoted by placing a line across the figures,
1, 2, 3, 4, 5̶, 6̶, 7̶. Diminished intervals are the minor intervals diminished by a minor semitone ; they are expressed by
a slur placed over the minor characters, ♭$\widehat{3}$, ♭$\widehat{6}$, ♭$\widehat{7}$.

Two intervals are reciprocal when, added together, they
form an octave. And the difference or complement of an
interval to an octave is termed an inverted or reciprocal interval : thus, 4 is an inversion of 5, and *vice versâ*. Intervals
are likewise divided into consonant and dissonant ; 1, ♭3, 3,
4, 5, ♭6, 6, 8 are the consonant intervals ; all the others are
dissonant.

Simple intervals are those not exceeding that of an octave ;
compound intervals are any of the simple intervals compounded with one or more octaves, as 9th, 11th, &c., which
in every respect are analogous to the 2nd, 4th, &c.

---

* The unison cannot in strictness be termed an interval; it may be
defined as the collision of two notes of the same degree of tune.

† These characters of the major and minor intervals are, as they should
be, employed in major scales; but in minor scales the minor intervals,
being of most importance, are denoted by plain figures, and the accidental
major intervals are distinguished by prefixing a sharp.

III. *Description of the Kaleidophone, or Phonic Kalei-doscope; a new Philosophical Toy, for the Illustra-tion of several Interesting and Amusing Acoustical and Optical Phenomena.*

[From the 'Quarterly Journal of Science, Literature, and Art,' 1827, vol. i.]

THE application of the principles of science to ornamental and amusing purposes contributes, in a great degree, to render them extensively popular; for the exhibition of striking experiments induces the observer to investigate their causes with additional interest, and enables him more permanently to remember their effects. I shall not, therefore, need an apology for presenting the tyro in science with another combination of philosophy with amusement, in addition to those already extant.

But this instrument possesses higher claims to attention; for it exemplifies an interesting series of natural phenomena, and renders obvious to the common observer what has hitherto been confined to the calculations of the mathematician; it presents another proof, that however remote from common observation the operations of nature may be, the most beautiful order and symmetry prevail through all.

In the property of " creating beautiful forms," the kaleidophone resembles the celebrated invention of Dr. Brewster, from which its name is modified; but to the instrument itself and its mode of action it is almost superfluous to say there is no similarity. Previously to entering into an explanation of its construction and effects, the following brief summary may suffice to give a general idea of the nature of the experiments it is intended to illustrate.

These experiments principally consist in subjecting to ocular demonstration the orbits or paths described by the points of greatest excursion in vibrating rods, which in the most frequent cases, those of the combinations of different modes of vibration, assume the most diversified and elegant curvilinear forms *. The entire track of each orbit is rendered simultaneously visible by causing it to be delineated by a brilliantly luminous point; and the figure being completed in less time than the duration of the visual impression, the whole orbit appears as a continuous line of light. As, besides the changes which result from the combinations of the primitive with the higher modes of vibration, the figures of the orbits are affected by the form of the rod, by the extent of the excursions of the vibrations, by the mode of producing the motions, and by many other circumstances, a great variety of pleasing and regular forms is obtained. This variety is also enhanced by giving the same motions to a number of symmetrically disposed luminous points, the mutual intersections of the orbits of which produce innumerable elegant forms; and the appearances may be still more variegated by occasionally causing these points to reflect differently coloured lights.

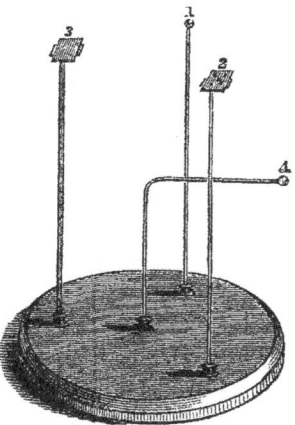

The apparatus for exhibiting these experiments consists of a circular board about nine inches in diameter, into which

* We are indebted to Dr. T. Young for the first observation of these phenomena. The following account of his experiments is quoted from the *Philosophical Transactions* for 1800 :—" Take one of the lowest strings of a square piano-forte, round which a fine silvered wire is wound in a

are perpendicularly fixed, at equal distances from the circumference and from each other, three steel rods, each about a foot in length. The first rod is cylindrical, 1-10th of an inch in diameter, and is surmounted by a spherical bead* which concentrates and reflects the light which falls upon it. The second is a similar rod, upon the upper extremity of which is placed a plate moving on a joint, so that its plane may be rendered either horizontal, oblique, or perpendicular; this plate is adapted to the reception of the objects, which consist of beads differently coloured and arranged on pieces of black card in symmetrical forms. The third is a four-sided prismatic rod, and a similar plate is attached to its extremity for the reception of the same objects.

---

spiral form; contract the light of a window, so that when the eye is placed in a proper position, the image of the light may appear small, bright, and well defined on each of the convolutions of the wire. Let the chord be now made to vibrate, and the luminous point will delineate its path like a burning coal whirled round, and will present to the eye a line of light, which, by the assistance of the microscope, may be very accurately observed. According to the different ways by which the wire is put in motion, the form of this path is no less diversified and amusing than the multifarious forms of the quiescent lines of vibrating plates discovered by Professor Chladni; and it is, indeed in one respect even more interesting, as it appears to be more within the reach of mathematical calculation to determine it."

The extremely limited extent of the excursions of a vibrating chord prevents its motions from being distinctly observed by the naked eye; but as the rods employed in the present experiments can extend their excursions to nearly two inches, and as the means employed greatly increase the intensity of the light, the phenomena are exhibited in a far more evident manner.

* The only beads well adapted for this purpose are made of extremely thin glass silvered on the interior surface, and about one sixth of an inch in diameter; they are to be obtained at the shops under the name of steel beads. The protuberances at the apertures must be removed or blackened, otherwise the reflections from them will render the images confused. To produce the coloured tracks, these beads must be coated with transparent colours, such as are ordinarily used for painting on glass; the light will then be reflected through the coloured surface; but in beads made of coloured glass, the reflection being made from the external surface, shows only white light. The bead is cemented into a small brass cup screwed to the top of the wire.

Another rod is fixed at the centre of the board; this is bent to a right angle, and is furnished with a bead similarly to the first-mentioned rod. A small nut and screw is fixed to the board near the lower end of the first rod, in order by pressing upon it to render occasionally its rigidity unequal. A hammer, softened by a leather covering, is employed to strike the rods; and a violin-bow is necessary to produce some varieties of effect.

I shall now proceed to describe the different appearances which the rods present when in action, and to give directions for the production of the different effects, following the order in which the rods have been previously mentioned.

No. 1. On causing the entire rod to vibrate, so that its lowest sound be produced*, as it is seldom that the motions

* The most simple mode of vibration of a rod vibrating transversely, when one of its extremities is fixed and the other is free, is that in which the entire rod makes its vibrations alternately on each side of the axis, which is nowhere intersected by the curve, but only touched at the fixed end. This gives the gravest sound which can be produced from the rod. In the other modes of vibration, the axis is intersected by the curve 1, 2, 3, or more times. The best means to command the production of these sounds is to touch a node of vibration lightly with the finger, and to put a vibrating part in motion by a violin-bow. In the second sound the number of vibrations is to that of the first as $5^2 : 2^2$, or as $25 : 4$; the difference of the sounds is, therefore, two octaves and an augmented fifth. Separating the first sound from the series, the number of the vibrations of all the others will be to one another as the squares of the numbers 3, 5, 7, 9, &c.; the third, in which there are three nodes, will therefore exceed the second by an octave and an augmented fourth; in the fourth, the acuteness will be augmented by nearly an octave; in the fifth, by nearly a major sixth, &c. To reduce to the same pitch all the proportions of the sounds which such a rod is capable of producing, I shall regard the sound corresponding with the most simple motion as the C one octave lower than the lowest of the piano-forte; the proportions of the sounds will then be—

Numbers of nodes........	0	1	2	3	4	5
Sounds	$\bar{C}$	$G\sharp^2$	D^4	D^5-	$B\flat^5$	F^6+
Numbers, the squares of which correspond with these sounds	(2)	$\left(\dfrac{5}{3}\right)$	5	7	9	11

of a cylindrical rod can be confined to a plane, the vibrations
will almost always be combined with a circular motion. When
the pressure on the fixed end is exerted on two opposite
points, and the rod put in motion in the direction of the
pressure, the following progression in the changes of form
will be distinctly observed :—The track will commence as a
line, and almost immediately open into an ellipse, the lesser
axis of which will gradually extend as the larger axis diminishes, until it becomes a circle; what was before the
lesser will then become the larger axis; and thus the motions
will alternate until, from their decreasing magnitudes, they
cease to be visible. In the case just described the ellipses
make a right angle with each other; but by altering the direction of the motion, so as to render it oblique to the direction of the pressure, they may be made to intersect under
any required angle; and when this angle $=0$ the motion will
be merely vibratory.

Every single sound formed by the subdivisions of the rod
will present similar appearances; but the excursions will be
smaller as the sound is higher, or, which is the same thing,
as the number of the vibrations increases.

In the most simple case of the co-existence of two sounds,
shown by putting the entire rod in motion, and producing
also a higher sound by the friction of a bow, the original
figure will appear waved or indented; and as unity is to the
number of indentations, so will the number of vibrations in
the lower sound be to the number in the higher sound. On
varying the mode of excitation, by striking the rod in different parts and with different forces, very complicated and
beautiful curvilinear forms may be obtained: some of these
are represented by the figures on p. 26.

Placing the hand on the lower part of the rod, below the
place at which it is excited, the excursions of the motions
will rapidly decrease and exhibit spiral figures.

To obtain the figures with brilliancy and distinctness, a

The possible series of sounds, regarding the fundamental as unity, will there-
fore be :—1, $6\frac{1}{4}$, $17\frac{13}{36}$, $34\frac{1}{36}$, $56\frac{1}{4}$, &c.; or, expressed in integral numbers
36, 225, 625, 1225, 2025, &c.—CHLADNI, *Traité d'Acoustique.* p. 91.

single light only should be employed, as that of the sun, a
lamp, or a candle; rays of light proceeding from several
points, as from a number of candles, or from the reflection
of the clouds, occasion the track to be broad and indistinct;
but double lights may be employed with effect, provided they
be of equal intensity and symmetrically placed; each bead
will then describe two similar figures. The appearances, in
a bright sunshine, are remarkably vivid and brilliant.

No. 2. Although very beautiful and varied forms may be
produced from the motion of a single point, yet the compound
figures which are presented by objects formed by a number
of points offer appearances still more pleasing to the eye.

An object being placed horizontally on the plate, and the rod being put in motion, the mutual intersections of the points, each describing a similar figure, present to the eye complicated, yet symmetrical figures, resembling elegant specimens of engine-turning.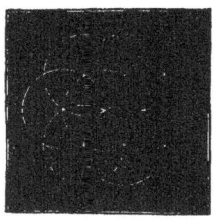

When the plate is horizontal, the figures are all in one plane; but if it be inclined or perpendicular, the curves being then made in parallel planes, gives the idea of a solid figure, and in some cases the appearances are particularly striking.

Complementary colours alone should be employed in the objects; for these harmonizing together, give greater pleasure to the eye than an injudicious combination of discordant tints: the intensities should be occasionally varied, and colourless light intermingled with the different shades.

No. 3. When this prismatic rod is put in motion, in the direction of either of its sides, the points move only rectilineally; but when the motion is applied in an oblique direction, a variety of compound curves is shown: this rod is principally employed to exhibit the optical phenomena which will be afterwards mentioned.

No. 4. When a rod is straight, the curve produced by any point describing its motion, is always in the same plane; but in a rod bent to any angle the two parts moving most frequently in different directions, curves are produced whose parts do not lie in the same plane. A few trials will soon indicate the best way of applying the motion so as to cause the two parts to vibrate in different directions.

Before the conclusion of this subject, I will avail myself of the opportunity of observing, that the application of this mode of experimenting may be extended to the delineation of every description of curvilinear and angular motion, when the amplitudes of the tracks are not too great; and by this means, it is not improbable that the experimental knowledge of many interesting principles in science may be facilitated.

On the Duplication and Multiplication of Objects.

When dark objects are substituted for luminous ones, their tracks become nearly invisible, and from the longer duration of the visual impression at the limits of vibration, the images are multiplied in proportion to the number of points at which they are retarded.   Place horizontally on the rod No. 2 a word printed or written on a piece of card; in the lowest mode of vibration, at the opposite limits of the excursions two legible images of the word will be distinctly seen, and but an indistinct shade, occasioned by the tracks of the letters, will appear in the intermediate space: the vibratory motion is imperceptible to the eye.   The images will, therefore, appear stationary in this respect; but the diminution of the excursions will cause them to approximate very slowly and gradually towards the centre: this diminution operates so gradually as to allow the images to superpose each other completely at each recurring vibration, without producing any intermingling or confusion.

On placing the object perpendicularly, the two images will appear in parallel planes, the furthest image appearing through the first apparent surface.   Small pictures have a singular effect applied in this manner.

When other sounds co-exist with the fundamental, the images are multiplied; but they become fainter as their number increases: these multiplied images are equally visible, whether the vibrations be rectilineal, elliptical, or circular.

A new Optical Experiment.

As that property of vision which occasions the apparent duration in the same places of visible images after the objects which excite them have changed their positions*, has enabled us to submit to inspection the phenomena above described, it may not be irrelevant to subjoin a description of an apparatus

---

* It has been proved by the Chevalier d'Arcy, from the only experiments approaching to accuracy which have been made on this subject, that the extent of this duration is eight thirds.   See his "Mémoire sur la durée de la sensation de la vue," Hist. de l'Acad., An. 1765.

which illustrates the transient duration of the impressions of light in a very evident manner.

At the back of a wooden frame, about six inches in height and breadth, and from one to three inches or more in depth, a circular plate of glass is placed, upon which a design is painted with transparent colours; at the front is placed, parallel to the glass, a circle of tin, covered on its exterior surface with white paper, and having the space between two adjacent radii cut out. This circle moves freely on its centre round an axis, supported by a bar in front, and is put in rapid and regular motion by the application of any mechanical principle proper for the purpose; and a catch is so placed, that when the motion ceases the aperture shall be concealed by the bar which supports the axis.

If a light be placed behind the transparent painting, and, still better, if it be concentrated by a lens, on making the circle revolve with rapidity the whole of the picture will be rendered visible at one view, although but very limited portions are successively presented to the eye.

The intensity will differ in proportion to the excess of the transmitted light above that which falls in front of the circle; it will, therefore, increase the distinctness of the picture, to darken the latter as much as possible.

IV. *Experiments on Audition.*

[From the ' Quarterly Journal of Science,' 1827, pt. ii.]

THE recent valuable experiments of Savart * and of Dr. Wollaston have added to our stock of information several important and hitherto unnoticed phenomena relating to audition; but notwithstanding the investigations of these distinguished experimentalists, and though the physiology of the ear has been an object of unceasing attention for many centuries, yet we are far from possessing a perfect knowledge of the functions of the various parts of this organ.   The description of new facts illustrative of this subject cannot, therefore, be devoid of interest; and though I do not anticipate that the observations contained in this communication will lead to any important results, their novelty may claim for them some attention from the readers of your Journal.

§ 1.

If the hand be placed so as to cover the ear, or if the entrance of the meatus auditorius be closed by the finger without pressure, the perception of external sounds will be considerably diminished, but the sounds of the voice produced internally will be greatly augmented : the pronunciation of those vowels in which the cavity of the mouth is the most closed, as *e, ou,* &c., produce the strongest effect; on articulating smartly the syllables *te* and *kew,* the sound will be painfully loud.

Placing the conducting stem of a sounding tuning-fork †

* " Recherches sur les usages de la membrane du tympan et de l'oreille externe; par M. Félix Savart." *Annales de Chimie,* tome xxvi. p. 1.

† The tuning-fork consists of a four-sided metallic rod, bent so as to form two equal and parallel branches, having a stem connected with the lower curved part of the rod, and contained within the plane of the two

on any part of the head when the ears are closed as above described, a similar augmentation of sound will be observed. When one ear remains open, the sound will always be referred to the closed ear ; but when both ears are closed, the sound will appear louder in that ear the nearer to which it is produced. If, therefore, the tuning-fork be applied above the temporal bone near either ear, it will be apparently heard by that ear to which it is adjacent ; but on removing the hand from this ear (although the fork remains in the same situation), the sound will appear to be referred immediately to the opposite ear.

In the case of the vocal articulations, the augmentation is accompanied by a reedy sound, occasioned by the strong agitations of the tympanum. When the air in the meatus is compressed against this membrane by pressing the hand *close* to the ear, or when the eustachian tube is exhausted by the means indicated by Dr. Wollaston, the reedy sound is no longer heard, and the augmentation is considerably diminished. The ringing noise which simultaneously accompanies a very intense sound proceeds from the same cause, and may be prevented by the same means. This ringing may be produced by applying the stem of a sounding tuning-fork to the hand when covering the ear, or by whistling when a hearing-trumpet is placed to the ear. As a proof that the resulting augmentation, which, when great, excites the vibrations of the tympanum, is owing to the reciprocation of the vibrations by the air contained within the closed cavity, it may be mentioned, that when the entrance of the meatus is closed by a fibrous substance, as wool, &c., no increase is obtained.

---

branches. The branches are caused to vibrate by striking one end against a hard body, whilst the stem is held in the hand. The sound produced by this instrument when insulated is very weak, and can only be distinctly heard when its branches are brought close to the ear ; but instantly its stem is connected with any surface capable of vibrating, a great augmentation of sound ensues from the communicated vibrations. The facility of its insulation and communication renders it a very convenient instrument for a variety of acoustical experiments.

If the meatus and the concha of the ear be filled with water, the sounds above mentioned will be referred to the cavity containing the water, in the same way as when it contained air and was closed by the hand; it will be indifferent whether any partition be interposed between the cavity and the external air, as the water is equally well insulated by a surface of air as by a solid body.

§ 2.

The preceding experiments have shown that sounds immediately communicated to the closed meatus externus are very greatly augmented; and it is an obvious inference, that if *external* sounds can be communicated so as to act on the cavity in a similar manner, they must receive a corresponding augmentation. The great intensity with which sound is transmitted by solid rods, at the same time that its diffusion is prevented, affords a ready means of effecting this purpose, and of constructing an instrument which, from its rendering audible the weakest sounds, may with propriety be named a Microphone.

Procure two flat pieces of plated metal, each sufficiently large to cover the external ear, to the form also of which they may be adapted; on the outside of each plate, directly opposite the meatus, rivet a rod of iron or brass wire about 16 inches in length and one eighth of an inch in diameter, and fasten the two rods together at their unfixed extremities, so as to meet in a single point. 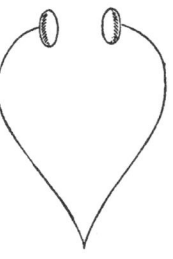 The rods must be so curved, that when the plates are applied to the ears, each rod may at one end be perpendicularly inserted into its corresponding plate and at the other end may meet before the head in the plane of the mesial line. The spring of the rods will be sufficient to fix the plates to the ears; but for greater security ribands may be attached to each rod near its insertion in the plate, and be tied behind the head.

A more simple instrument may be constructed to be applied to one ear only, by inserting a straight rod perpendicularly into a similar plate to those described above.

The Microphone is calculated only for hearing sounds when it is in immediate contact with sonorous bodies : when they are diffused by their transmission through the air, this instrument will not afford the slightest assistance.

It is not my intention in this place to detail all the various experiments which may be made with this instrument; a few will suffice to enable the experimenter to vary them at his pleasure.

1. If a bell be rung in a vessel of water, and the point of the microphone be placed in the water at different distances from the bell, the differences of intensity will be very sensible. 2. If the point of the microphone be applied to the sides of a vessel containing a boiling liquid, or if it be placed in the liquid itself, the various sounds which are rendered may be heard very distinctly.   3. The instrument affords a means of ascertaining, with considerable accuracy, the points of a sonorous body at which the intensity of vibration is the greatest or least : thus, placing its point on different parts of the sounding-board of a violin or guitar, whilst one of its strings is in vibration, the points of greatest and least vibration are easily distinguished.   4. If the stem of a sounding tuning-fork be brought in contact with any part of the microphone, and at the same time a musical sound be produced by the voice, the most uninitiated ear will be able to perceive the consonance or dissonance of the two sounds; the roughness of discords and the beatings of imperfect consonances, are thereby rendered so extremely disagreeable, and form so evident a contrast to the agreeable harmony and smoothness of two perfectly consonant sounds, that it is impossible that they can be confounded.

§ 3.

Apply the broad sides of two sounding tuning-forks, both being unisons, to the same ear; on removing one fork to the

D

opposite ear, allowing the other to remain, the sensation will be considerably augmented.

It is well known, that when two consonant sounds are heard together, a third sound results from the coincidences of their vibrations; and that this third sound, which is called the grave harmonic, is always equal to unity when the two primitive sounds are represented by the lowest integral numbers. This being premised, select two tuning-forks the sounds of which differ by any consonant interval excepting the octave: place the broad sides of their branches, while in vibration, close to one ear, in such a manner that they shall nearly touch at the acoustic axis; the resulting grave harmonic will then be strongly audible, combined with the two other sounds; place afterwards one fork to each ear, and the consonance will be heard much richer in volume, but no audible indications whatever of the third sound will be perceived.

§ 4.

Very acute sounds, such as the chirping of the *Gryllus campestris*, &c., are rendered inaudible by exhausting the air from the Eustachian tube, and thereby producing a tension of the membrane of the tympanum. The different thicknesses or tensions of this membrane may therefore occasion that diversity of the limits of audibility, with regard to the acute sounds, which Dr. Wollaston has pointed out as existing in different individuals; if so, it would be desirable to ascertain this limit in individuals in whom the tympanum is perforated, or destroyed.

§ 5.

When the auricula is brought forward, all *acute* sounds are rendered much more intense, but no sensible difference is perceived with regard to the grave sounds. The *higher* tones of glass staccados, or of an octave flute, the ticking of a watch, all kinds of sibilant sounds, &c. are thus greatly augmented: the experiment is easily tried by whistling very shrill notes. A still greater augmentation of the acute sounds is obtained, by placing the hands formed into a concave behind the ears, and by bending downwards the upper part of the auricula, so as to obtain a more complete cavity.

§ 6.

I will conclude with the following observation : I had, in consequence of a cold, a very slight pain in my left ear ; on sounding the regular notes of the piano-forte, C^3 and C^4 were much louder than the others, and the loudness was much increased by placing the hand in the manner above described to the left ear. When it was pressed close, or when the Eustachian tube was closed, the intensities of all the notes were equalized. I attribute this affection to the diminished tension of the membrana tympani, which was again increased by the operation described.

V. *On the Resonances, or Reciprocated Vibrations of Columns of Air.*

[From the 'Quarterly Journal of Science,' 1828, vol. iii.]

An elastic body may be made to assume a vibratory state in two ways:—either immediately, by any momentary impulse, which, altering the natural position of its particles, allows them afterwards to return by a succession of isochronous oscillations to their former state; or, secondarily, by means of an immediately sounding body, which causes it to reciprocate to the latter, when certain conditions, on which depends its susceptibility of vibrating in such a manner, are fulfilled. This reciprocation, to which, when the effect is referred to, the term *resonance* is applied, is effected by means of the undulations which are produced in the air, or in any fluid or solid medium, by the periodical pulses of the original vibrating body—these undulations being capable of putting in motion all bodies whose pulses are coincident with their own, and, consequently, with those of the primitive sounding body. Galileo observed that a heavy pendulum might be put in motion by the least breath of the mouth, provided the blasts be often repeated, and keep time exactly with the vibrations of the pendulum; and this remark affords a correct explanation of the phenomenon.

Some of the most obvious cases of resonance are:—the vibrations of a string when another tuned in unison with it is made to vibrate; the resounding of a drinking-glass to the sound of the voice or of a musical instrument, the reciprocated vibrations of a sounding-board communicating immediately with a vibrating string or tuning-fork, &c. In the last mentioned instance, though the string and the fork are the original vibrating bodies, the audible sound is dependent on the resonance of the sounding-board.

As all these effects are well known[*], it is unnecessary to dilate upon them here; and I may uninterruptedly proceed to the immediate object of the present paper, viz. the investigation of the laws of the resonance or phonic reciprocation of columns of air.

§ 1. If one of the branches of a vibrating tuning-fork be brought near the embouchure of a flute, the lateral apertures of which are stopped so as to render it capable of producing the same sound as the fork, then the feeble and scarcely audible sound of the fork will be augmented by the rich resonance of the column of air within the flute[†]. The sound will be found greatly to decrease by closing or opening another aperture; for the alteration of the length of the column of air in such case renders it no longer proper to reciprocate perfectly the sound of a fork. This experiment may be easily tried on a concert flute with a C tuning-fork. To insure success, it is necessary to remark, that when a flute is blown into with the mouth, the under lip partly covering the embouchure renders the sound about a semitone flatter than the sound when the embouchure is entirely uncovered; and as the latter must be unison to that of the tuning-fork, it is necessary, in most cases, to finger the flute for B when a C tuning-fork is employed.

A similar effect may be produced by substituting, for the column of air in the flute, the alterable volume cf air contained within the cavity of the mouth. I have found the sounds of tuning-forks reciprocated most intensely by placing the tongue, &c. in the position for the nasal continuous sound of *ng* (in song), and then altering the aperture of the lips until the loudest sound is obtained.

§ 2. A column of air may also reciprocate a sound originally produced by a wind instrument, as the following experiment will show. Place two concert flutes on a table,

---

[*] Biot, 'Traité de Physique,' tom. ii. p. 183. Chladni, 'Traité d'Acoustique,' pp. 222, 223.

[†] Dr. Savart has observed a similar effect by sounding a bell before a large tube enclosing an unisonant column of air ("Recherches sur les Vibrations de l'Air," Annales de Chimie, tom. xxiv.).

parallel to, and at a short distance from each other; on the one which is nearer, sound C sharp (all the lateral apertures being open), and draw out the tube of the second flute so that it shall be about a semitone flatter, to make it equivalent to the flattening of the first flute by the partial closing of the embouchure by the lip; a material difference will then be distinguished in the intensity of the tone by alternately closing and opening the first hole of the more distant instrument, thereby rendering it incapable or capable of reciprocating the original sound. That this effect is occasioned solely by the transmission of the sonorous undulations, and not by any wind actually blown into the second flute, is evident from the difference being in intensity and not in pitch*

This experiment may be varied by placing the fipple of a flageolet at a short distance from the embouchure of a flute, provided, of course, that the columns of air, both in the flageolet and the flute, be capable of producing the same note.

§ 3. A cylindric or prismatic column of air, in a tube open at both ends, may vibrate not only in its entire length, but also in any number of aliquot parts; and in all cases the number of vibrations is inversely as the length of a single vibrating part. As a column of air is capable of reciprocating every sound which, according to its different modes of vibration, it is itself capable of producing—supposing $1 = C^1$ to represent the lowest sound of the tube, it will, without any change in its length, reciprocate sounds whose relations are
$\overset{1}{C^1}, \overset{2}{C^2}, \overset{3}{G^2}, \overset{4}{C^3}, \overset{5}{E^3}, \overset{6}{G^3}, \overset{7}{B\flat^3}, \overset{8}{C^4}$, &c.

The harmonic subdivisions of a column of air in a tube

---

* Lord Bacon may be said to have anticipated this experiment in the following passage in his ' Sylva Sylvarum ':—"The experiments of sympathy may perhaps be transferred from stringed instruments to others; as, if there were two bells in unison in one steeple, to try whether striking the one would move the other, more than if it were a different chord; and so in pipes, of equal bore and sound, to try whether a light straw or feather would move in one pipe, when the other is blown in unison with it."—Art. Phonics, sect. xxi. (*On the Sympathy or Antipathy of Sounds with one another*).

closed at one end are different : a semi- vibrating part always exists near the open end; but between two nodes, or a node and the closed end, complete vibrating parts, as in an open tube, exist. The fundamental sound above mentioned, of an open tube, is given by a tube closed at one end, of one half its length; the series, corresponding with the subdivisions, compared with the above, is $\overset{1}{\mathrm{C}^1}$, $\overset{2}{\mathrm{G}^2}$, $\overset{5}{\mathrm{E}^3}$, $\overset{7}{\mathrm{B}\flat^3}$, $\overset{9}{\mathrm{D}^4}$, &c., and these sounds it can consequently reciprocate.

§ 4. Any one among several simultaneous sounds may be rendered separately audible. Thus, if two vibrating tuning-forks, differing in pitch, be held over a closed tube furnished with a movable piston, either sound may be made to predominate by altering the piston, so as to enable the column of air to reciprocate the sound required. The same result may be obtained by selecting two bottles (which may be tuned with water), each corresponding to the sound of a different tuning-fork; on bringing both tuning-forks to the mouth of each bottle alternately, in each case that sound only will be heard which is reciprocated by the unisonant bottle*.

The phenomenon of a third sound, produced by the coincidences of the vibrations of two consonant sounds, is well known. From what has been premised, it is reasonable to

---

* Had Sir Isaac Newton been acquainted with these, or with any similar facts, he might have illustrated his theory of the reflections of colours by an experimental, instead of a supposititious analogy with the reciprocation of sounds. As the passage in which this comparison is made is remarkable, I will quote it :—"If light be considered without respect to any hypothesis, I can as easily conceive that the several parts of a shining body may emit rays of different colours, and other qualities, of all which light is constituted, as that the several pipes of an organ inspired all at once, or all the variety of sounding bodies in the world together, should produce sounds of several tones and propagate them through the air, confusedly intermixed. And, if there were any natural bodies that would reflect sounds of one tone, and stifle or transmit those of another, then, as the echo of a confused aggregate of all tones would be that particular tone which the echoing body is disposed to reflect, so, since there are bodies apt to reflect rays of one colour and stifle or transmit those of another, I can as easily conceive that those bodies, when illuminated by a mixture of all colours, must appear of that colour only which they reflect."—*Philosophical Transactions*, No. 88.

suppose that if a column of air be caused to reciprocate this
third sound, or grave harmonic as it is called, and not the
two sounds which generate it, it might be heard, uncombined
with the two other sounds. But on attempting this experiment, I was unable, on the following account, to succeed.

The third sound is always unity when the ratio of the
lowest sounds is reduced to its lowest terms: thus, with
respect to a perfect fifth, 2 : 3, the third sound, 1, is an octave
below the lower sound, and the grave harmonic, 1, of a
major third, 4 : 5, is two octaves below the lower sound. We
will suppose this fundamental sound, represented by unity,
to be the C, corresponding with the sound of an open tube of
four feet, or of a closed tube of two feet: in the first case,
the column of air being capable, as explained in § 3, of
vibrating in any number of aliquot parts, not only the grave
harmonic, $= 1$, but the sounds represented by 2 : 3 and 4 : 5,
will also be reciprocated; and in the latter case, where the
subdivisions are as the arithmetical progression 1, 3, 5, 7, 9,
11, &c., one sound of each consonance will be reciprocated,
besides the grave sound. The expected result may probably
still be obtained by employing columns or volumes of air
whose subdivisions are less regular.

§ 5. Among the Javese musical instruments brought to
England by the late Sir Stamford Raffles, there is one called
the Génder, in which the resonances of unisonant columns
of air are employed to augment, I may almost say to render
audible, the sounds of vibrating metallic plates. Of these
plates there are eleven; their sounds correspond with the
notes of the diatonic scale deprived of its fourth and seventh,
and extend through two octaves. The mode of vibration of
the plates is that with two transversal nodal lines; and they
are suspended horizontally by two strings, one passed through
two holes in the one nodal line, and the other through similar
holes in the other nodal line of each plate. Under each plate
is placed an upright bamboo, containing a column of air, of
the proper length to reciprocate the lowest sound of the
plate. If the aperture of the bamboo be covered with pasteboard, and its corresponding plate be struck, a number of

acute sounds only (depending on the more numerous sub-divisions of the plate) will be heard; but on removing the pasteboard, an additional deep rich tone is produced by the resonance of the column of air within the tube.

The instrument from which the annexed drawing was taken is at present in the museum of the Honourable East-India Company; and there is another specimen in the possession of Lady Raffles.

The Génder, Musical Instrument of Java.

The same principle appears to have been employed, in more rude forms, in the construction of several Asiatic and African instruments; but I am unaware of any instrument having yet been manufactured in Europe, in which the unisonant resonances of columns of air have been made available as a means of augmenting the intensity of sounds. I shall

shortly publish an account of the several modes I have myself
devised, and practically employed, for advantageously applying this principle.

§ 6. In the experiments I have hitherto detailed, the reciprocating vibrations have always been isochronous with those
of the original sounding body ; or, in other words, the resonance and the original sound have always been unisonant.
The experiment I shall now bring forward will show that
this is not universally the case, and that there are other
phenomena of resonance which have never hitherto been investigated in their theory, or in their practical applications.
I took a tube, closed at one end by a movable piston, and
placed before its open end the branch of a vibrating tuning-
fork of the ordinary pitch, C; the length of the column of
air was six inches : on diminishing the length of the column
to three inches, the sound of the tuning-fork was no longer
reciprocated, but its octave sound (the sound of the column
when it is directly excited) was produced.   By employing a
graver tuning-fork and tubes of very small diameter, and
successively adjusting the lengths of the columns of air so as
to be one half, one third, one fourth, one fifth, &c., of the
column reciprocating the fundamental sound, the octave,
twelfth, double octave, seventeenth, &c., to that sound will
be produced. The relative numbers of the vibrations of these
sounds, considering the vibrations of the fork as unity, are
1, 2, 3, 4, 5, &c.   It therefore is evident, from experiment,
that a column of air may vibrate by reciprocation, not only
with another body whose vibrations are isochronous with its
own, but also *when the number of its own vibrations is any
multiple of those of the original sounding body.*

The converse of this law does not hold ; for when the
number of vibrations of a column of air are any sub-multiple
of those of the original sounding body, there is no resonance.
To prove this with regard to the octave, let the length of the
column of air unison to the sound of a fork be doubled, and
not the slightest trace of the octave below (*i. e.* the real
sound of the column) will be perceptible.   This negative ex-

periment must be tried with a closed tube which is incapable
of producing a harmonic octave; an open tube would resound
resonantly to the fork by its subdivision.

§ 7. On the law experimentally established in the preceding paragraph depends the explanation of the production
of sounds by the guimbarde or jew's harp.   This simple instrument consists of an elastic steel tongue, riveted at one
end to a frame of brass or iron, the form of which is represented in the annexed figure ;   the free extremity of the
tongue is bent outwards to a right angle, so as to allow the
finger easily to strike it when the instrument is placed to the
mouth, and firmly supported by the pressure of the parallel
extremities of the frame against the teeth.

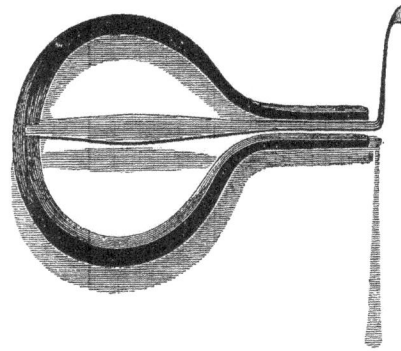

The vibrations of the tongue itself correspond with a very
low sound; but being placed before the cavity of the mouth,
the form and dimensions of which are capable of various
alterations by the motions of the tongue and the lips, when
the number of vibrations of the contained volume of air is
any multiple of the original vibrations of the tongue, a sound
is produced, corresponding to the modification of the oral
cavity.

Supposing the primitive sound of the tongue to be C^1, the
series of sounds which it can produce by multiple reciprocation will be as follows :—

Multiples of the original vibrations of the tongue.

1̊...2̊...3...4̊...5...6...7...8̊...9...10...11...12...13...14...15...16̊ etc. 32̊.

Corresponding sounds.

C¹ C² G C³ E G B♭ C⁴ D E F+ G A− B♭ B C⁵ etc. C⁶.

If the original sound be any other note, another series will arise; but the intervals of the successive sounds will always preserve the above relations. By the usual jew's harps the three first sounds of the series cannot be produced, the dimensions of the cavity of the mouth not being sufficiently large to reciprocate them.

The scale above exhibited is evidently too incomplete and too defective to allow even the most simple melodies to be performed on a single jew's harp; but the deficiencies may be supplied by employing two or more of these instruments. Mr. Eülenstein, by availing himself of the resources afforded him by the scales of sixteen jew's harps, is able to modulate through every key, and to produce effects truly original and of extreme beauty. Those who have heard only the rude twanging to which the performance of this instrument in ordinary hands is confined, can have no idea of the melodious sounds which, under Mr. Eülenstein's management, it is capable of producing.

§ 8. The following experiment will prove the accuracy of the preceding explanation, and will establish the true theory of the production of sounds by the guimbarde, beyond the possibility of doubt.

I fixed a jew's harp firmly at the two points where ordinarily it rests against the teeth, allowing sufficient space between the two supports for the tongue to vibrate freely to its greatest extent; and I tuned the tongue by applying wax to its free extremity, until it sounded C, corresponding to the sound of a closed tube four feet in length. I then placed before the tongue the open end of a closed tube, containing a column of air two feet in length and one inch in diameter, and furnished with a movable piston, by which the column could be shortened to any required length. On striking the tongue, the octave of the fundamental sound was produced, being the sound 2 of the scale in the preceding section; by shortening the column of air successively to one third, one fourth, one fifth, one sixth, one seventh, &c. &c., up to very numerous aliquot subdivisions, all the notes of the series were correctly obtained. By marking the different lengths of the piston-rod for each sound of the series, I was able to produce the notes of the scale, ascending or descending regularly, or to fix any sound at pleasure.

§ 9. No other sounds can be produced by reciprocation from columns of air but those perfectly identical with the multiples of the original vibrations of the tuning-fork, or the tongue of the jew's harp. On inquiring what takes place when the length of the column is intermediate between the lengths appropriate to reciprocate two succeeding multiples, it will be found, that though each sound of the series is heard most audibly when the column is accurately adapted to it, yet it may also be heard, unaltered in tune, though diminished in intensity, when the column is lengthened or shortened within a certain extent, which is greater for the lower sounds of the series, and less for the higher, on account of the wider intervals between the sounds in the former case.

It may now be understood in what manner a column of air is capable of reciprocating simultaneously two or three sounds of a chord. In Mr. Eülenstein's performance this effect is thus produced : suppose the perfect major chord C is required; three jew's harps, incapable of producing lower sounds than the fourth of the series, are selected; and the mouth being made to correspond with the C, the two other sounds, E and G, are likewise reciprocated, though faintly, because these sounds are nearer that to which the mouth is adapted, than any other multiples of the original vibrations of the tongues are.

§ 10. When two imperfect unisons are sounded together, the interferences of their undulations give rise to periodical alternations in the intensity of the sound; this effect is called beating; and the greater the deviation from unison is, the more rapid are the beats. From what has been above stated, the same column of air may reciprocate sounds differing in pitch; if, therefore, two vibrating tuning-forks, imperfect unisons to each other, be brought together over the embouchure of a flute, or the open end of a tube containing an appropriate column of air, the periodical recurrence of the beats will be rendered strikingly evident, slowly or rapidly succeeding each other, accordingly as the forks are more or less in tune with each other.

VI. *On the Transmission of Musical Sounds through Solid Linear Conductors, and on their subsequent Reciprocation.*

[From the Journal of the Royal Institution, 1831, vol ii.]

§ 1. The fact of the transmission of sound through solid bodies, as when a stick or a metal rod is placed with one extremity to the ear, and is struck or scratched at the other end, did not escape the observation of the ancient philosophers; but it was for a long time erroneously supposed that an aëriform medium was alone capable of receiving sonorous impressions; and in conformity with this opinion, Lord Bacon, when noticing this experiment, assumes that the sound is propagated by spirits contained within the pores of the body*. The first correct observations on this subject appear to have been made by Dr. Hooke in 1667; who made an experiment with a distended wire of sufficient length to observe that the same sound was propagated far swifter through the wire than through the air†. Professor Wunsch, of Berlin, made, in

---

* "If a rod of iron or brass be held with one end to the ear, and the other be struck upon, it makes a much greater sound than the same stroke upon the rod, when not so contiguous to the ear. By which, and other instances, it should seem that sounds do not only slide upon the surface of a smooth body, but also communicate with the spirits in the pores of the body."—*Sylva Sylvarum*, Phonics, § 3. "The pneumatical part, which is in all tangible bodies, and has some affinity with air, performs, after a sort, the office of the air. Thus the sound of an empty barrel is in part created by the air on the outside, and in part by that in the inside; for the sound will be less or greater as the barrel is more or less empty, though it communicates also with the spirit in the wood, through which it passes from the outside to the inside."—*Sylva Sylvarum*, Phonics, § 2.

† "And though some famous authors have affirmed it impossible to hear through the thinnest plate of Muscovy glass, yet I knew a way by which 'tis easy to hear one speak through a wall a yard thick. It has not yet been thoroughly examined, how far otacoustics may be improved, nor what other ways there may be of *quickening* our hearing, or *conveying*

1788, a similar experiment, substituting 1728 feet of connected wooden laths for the wire, and confirmed Dr. Hooke's results [*].   Other experiments of a similar nature were subsequently made by Herhold and Rafn[†], Hassenfratz and Gay-Lussac[‡], &c.; but the first direct observations of the actual velocity of sound through solid conductors were made by Biot, assisted at different times by Bouvard and Martin. These experiments were made on the sides of the iron conduit-pipes of Paris, through the length of 951·25 metres; and the mean result of two observations made in different ways gave 3459 metres, or 11,090 feet per second, for the velocity of sound in cast iron [§].

Previously to these last-mentioned experiments, Chladni had, in an ingenious manner, inferred the velocity of sound in different solid substances; and his results are fully confirmed by calculations from other grounds.   His method was founded on Newton's demonstration, that sound travels through a space of a given length filled with air in the same time that a column of air of the same length, contained in a tube open at both ends, makes a single vibration [‖].   His own discovery of the longitudinal vibrations of solid bodies, which are exactly analogous to the ordinary vibrations of

---

sounds through *other bodies* than the air; for, that that is not the only medium, I can assure the reader that I have, by the help of a *distended wire*, propagated the sound to a very considerable distance in an instant, or with as seemingly quick a motion as that of light—at least incomparably swifter than that which at the same time was propagated through the air,—and this not only in a straight line, or direct, but in one bended in many angles."—*Preface to Hooke's 'Micrographia.'*

 [*] Acad. Berl., deutsch. Abh. 1788, p. 87.

 [†] Reil's 'Archiv für die Physiologie,' vol. iii. no. 3, p. 178.

 [‡] 'Annales de Chimie,' tome liii. p. 64.

 [§] Mémoires de la Société d'Arcueil, tome ii. p. 403.

 [‖] A single vibration is here considered as the motion of the vibrating body between the two opposite limits of its excursion; and with this signification the expression is adopted by Chladni.   Other authors, however, regard this, with Newton and Sauveur, as a semivibration, and call an entire vibration the motion of the vibrating body from one limit of its excursion until it again arrives at the same limit.   This difference of meaning attached to the same term has given rise to several mistakes.

columns of air, enabled him to apply this proposition to solid bodies, and to establish the general law that sound is propagated through any elastic substance in the same time in which this substance makes one longitudinal vibration. In this manner he ascertained the velocities of sound in the following substances, among others: tin 7800, silver 3900, copper 12,500, glass and iron 17,500, and various woods from 11,000 to 18,000 feet in a second.

From the experiments of M. Perolle*, it would appear that the intensity with which sound is communicated through solid matters is nearly in proportion to the velocity of its transmission.

§ 2. In all the experiments above alluded to, the sounds transmitted were either mere noises, such as the blow of a hammer, or, as in Herhold and Rafn's experiment, a single musical sound, produced by striking a silver spoon attached to one end of the conducting wire; and in no case were any means employed for the subsequent augmentation of the transmitted sound. I believe that, previous to the experiments which I commenced in 1820, none had been made on the transmission of the modulated sounds of musical instruments; nor had it been shown that sonorous undulations, propagated through solid linear conductors of considerable length, were capable of exciting, in surfaces with which they were in connexion, a quantity of vibratory motion, sufficient to be powerfully audible when communicated through the air. The first experiments of this kind which I made were publicly exhibited in 1821, and notices of them are to be found in the 'Literary Gazette,' 'Ackerman's Repository,' and other periodicals of that year. On June 30, 1823, a paper of mine was read by M. Arago, at the Academy of Sciences in Paris, in which I mentioned these experiments, and a variety of others relating to the passage of sound through rectilinear and bent conductors†. I propose, in the present instance, to give a more complete detail of these ex-

* 'Journal de Physique,' tome xlix. p. 382.

† An abridgment appeared in the 'Annales de Chimie,' July 1823, and the entire paper in the 'Annals of Philosophy,' August 1823.

E

periments than I have yet published; and at the same time to add what additional facts my subsequent experience has furnished me on the same subject.

§ 3. Before proceeding any further, it will be necessary to make a few observations on the augmentation of sound which results from the connexion of a vibrating body with other bodies capable of entering into simultaneous vibration with it.   This participation of the vibrations of an original sounding body is called *resonance*, or reciprocation of sound.

Sonorous bodies are audible (the extent of their excursions being supposed equal) in proportion to the quantity of their vibrating surfaces.   Thus, a plate of glass or metal is capable of producing powerful sounds without accessory means; but the sounds of vibrating bodies of smaller dimensions, such as insulated strings or tuning-forks, are scarcely audible at a moderate distance from the ear; but the sounds of the latter are capable of considerable augmentation when communicated to surfaces, as when they are placed to a table, or to the sounding-board of a musical instrument.

There are several circumstances which influence the intensity of the resonance of a sounding-board.   The principal of these is the plane in which the vibrations of the sounding body are made with respect to the reciprocating surface. Thus, its vibrations may be so communicated as to be perpendicular, or normal to the surface, in which case the sound is the most greatly augmented; or they may be tangential to, or in the same plane with the surface, when the sound is the most feeble.   The first of these cases may be illustrated by placing a vibrating tuning-fork perpendicularly to the surface of a flat board; and the second, by placing it perpendicularly to one of the edges of the board.   In intermediate positions—viz. when the vibrations are communicated obliquely to the surface—the sound will be found to have intermediate degrees of intensity.

These facts, which the extensive investigations of Savart place in full evidence, being understood, the peculiarities of the sounding-boards of various musical instruments admit of easy explanation.

The sounding-board of the piano-forte is better disposed than that of any other stringed instrument, as the planes of the vibrations of the strings are, on account of the direction in which they are struck by the hammers, always perpendicular to its surface. The difference of intensity when a string vibrates in this way, and when it vibrates parallel to the surface, is very obvious, and may be easily tried by striking it with the finger in these two directions*. There is no other instrument now in use in which the strings make their vibrations perpendicular to the sounding-board.

In the guitar, lute, &c., the strings are also parallel to the sounding-board, but the vibrations must, for the convenience of performance, be made obliquely to it. If the sides of the instrument be of inconsiderable depth, and the back be flat, the difference of intensity between the perpendicular and oblique vibrations will be very sensible. But if the sides be deep, very little difference will be perceived, as the vibrations which are tangential to the front sounding-board are perpendicular to the sides, which thus enter readily into normal vibrations; this fact may be proved by placing the ear to the side of a guitar while a string is made to sound with its plane of vibration successively parallel and perpendicular to it. In some instruments, as the lute, mandoline, &c., the back is polygonal or curved; by this construction a considerable portion of the resonant surface enters into normal, or nearly normal, vibrations when the strings are struck obliquely to the principal sounding-board.

These laws are not so immediately applicable to the violin, and other instruments of the same class; an extensive series of experiments will yet be necessary to enable us to account for the peculiarities of their forms, their various curvatures, and the functions of that irregular conductor, resting on the

* It sometimes happens, when the impulse is oblique to the direction in which the string presses on the bridge, that its plane of vibration assumes a rotary motion; the periodical changes of intensity thus occasioned produce an effect similar to that of the beating of imperfect unisons. This phenomenon is generally erroneously attributed by tuners to a faulty string.

sounding-board at two points only, which in these instruments is called the bridge. The investigations of Savart still leave much to be desired on this head.

In no instrument are the strings perpendicular to the sounding-board; for in such case, however a string were made to vibrate, its communicated vibrations would be tangential. But they are sometimes placed obliquely, as in the harp, and then the same changes of intensity may be observed as when the strings are parallel to the board; for if the plane of their vibrations coincide with that of the inclination of the board, the communicated vibrations of the board will be oblique to its surface, and the intensity will be at its maximum; but if they be perpendicular to this plane, the communicated vibrations must be tangential to this surface, and consequently the intensity will be at its minimum.

Besides the proper adaptation of sounding-boards, there are other circumstances on which the tones of a stringed instrument materially depend. One of the most important of these is, the proper dimensions of the volume of air contained within the sides. The laws of these resonant cavities have occupied the attention of Savart; but the obvious use of the bars placed within these cavities to divide the mass of air, and thus to enable it to vibrate more readily in separate portions, seems to have escaped his notice.

§ 4. In the piano-forte, the guitar, &c., the ends of the strings are not in immediate contact with the sounding-board, but they rest on bars of wood, which are called bridges, through which the vibrations are communicated to the board. In these instruments the bridge is usually about half an inch in height, and in the violoncello does not exceed three inches. To ascertain how far the distance might be extended between the string and the sounding-board of a piano-forte without injury to the tone, I substituted a glass rod five feet in length for the bridge, and by placing at its end a string stretched on a steel bow, I found that the sound of the string was as distinctly audible as when it was immediately in contact with the board; a tuning-fork placed at the end of the rod gave the same result. These experiments, which were the first I made on the subject, and which suggested all the subsequent ones, have been repeated in the theatre of the Royal Institution on a larger scale. A series of connected deal rods, forty feet in length, was suspended so as to extend, in a straight line, obliquely from an open window of the cupola to within a short distance of the floor of the room; on the upper end of this conductor, an assistant placed the stem of a vibrating tuning-fork; when no sounding-board was placed at the lower extremity of the conductor, no sound was heard, but it became powerfully audible the instant the communication was made: this experiment was repeated with different acute and grave toned tuning-forks, employed both in combination and in succession.

Tuning-forks are the most convenient instruments for making experiments on the transmission of sound, because their vibrations are almost inaudible by themselves, and only become strongly audible when augmented by resonant surfaces.

§ 5. The vibrations of the sounding-board of any stringed instrument may be communicated in the same manner as those of a string, or of a tuning-fork, to a distant sounding-board by means of a metallic, glass, or wooden conductor; but in this case it is necessary to prevent the original sounds from being heard through the air, otherwise the communicated sounds will not be distinguishable from them. This may be effected by placing the originally vibrating, and the reciprocating instruments in different rooms, and allowing the conductors to pass through the floor or wall separating the two rooms.

In the passage of the conducting-rod or wire through these partitions care must be taken to prevent its touching their sides; for this purpose a tin tube, covered at its two ends with leather, or india-rubber, may be inserted in the partition, and the conductor be made to pass through holes in these coverings, so as not to touch the side of the tube.

A square piano-forte is a very convenient instrument to employ in these experiments. If the sound is to be transmitted upwards, nothing more is requisite than to open or remove the lid of the instrument, and to allow the conductor to rest upon the sounding-board. A metallic wire is not sufficiently rigid to support itself thus without bending; a rod of some straight-fibred wood, as lancewood or deal, is therefore better adapted for this form of the experiment; the lower end of the rod must be reduced in thickness, so as to allow it to pass between two adjacent strings; and the best place to make the contact will be found to be about a quarter of an inch from the bridge, among the middle notes, and on the side occupied by the unvibrating portions of the strings. The reciprocating instrument in the room above, may be a guitar or any other similar instrument, or a harp; in which latter case, the rod may be brought in contact with the inner surface of the belly of the instrument, through one of the apertures of the swell. These were the forms under which the experiments have been repeated at the Royal Institution.

If the sounds of the piano-forte are to be transmitted downwards, a brass wire, about the thickness of a goose-quill, will suffice for the communication, as the weight of a reciprocating instrument suspended from it below will keep it sufficiently straight. To bring the conducting wire into contact with the under surface of the sounding-board of the pianoforte, an aperture must be made in the bottom of the instrument immediately below the intended point of contact; and to ensure a perfect connexion with the sounding-board, it is advisable to furnish the wire with a shoulder just below its entrance into the aperture, and to occasion an upward pressure on this shoulder, by a piece of leather stretched on a ring (as in the insulating-tube above described) and placed at the end of a strong steel spring, the other end of which is screwed firmly to the bottom of the instrument. To assist in supporting the wire, another shoulder may be made on it, so as to rest upon the upper covering of the insulating-tube which passes through the floor; and the reciprocating instrument may be suspended by inserting the end of the wire into the sounding-board, and then securing it by a nut and screw on the opposite side. The form of the resounding instrument is a matter of choice; but, in order to obtain the freest

and loudest tones, it is requisite to have the principal vibrating surface perpendicular to the conducting-wire.   It is instructive to observe the gradual changes in the intensity and
the quality of the transmitted sounds, when the sounding-
board is made to pass through the various degrees of obliquity
from a perpendicular direction to the conductor, until it is in
the same plane with it; or, to employ Savart's language, as
the communicated vibrations change from normal to tangential; in the latter case, the sounds have a subdued, and what
is ordinarily called a metallic quality.   In the first public
experiments I made in 1821, the reciprocating instrument,
which was the representation of an ancient lyre, was so constructed as to produce tangential vibration; the tones were
consequently far inferior to what I have since been able to
produce.   The transmitted sounds are not sensibly impaired
when the wire is separated at several places, and the disunited
parts fastened together by mechanical contact.   The annexed
woodcut (p. 56) represents the divisions of the conducting-
wire, which I found it convenient to make in the original
form of the experiment, for the sake of portability and facility of removal; but, if the apparatus be intended as a fixture,
it will be easier and better to employ but one length of wire.
The wire consisted of four portions: the first part touched
the sounding-board of the piano-forte, and reached halfway to
the floor; the second passed through the insulating-tube in
the floor, and terminated when it reached the ceiling of the
room below in a hook ; a third part was suspended from this
hook by a loop; and the fourth, after identifying itself with
one of the apparent wires of the lyre, passed within the instrument, and was ultimately fixed, at its lower end, to the
point marked at the end of the dotted line on the sounding-
board; each of the disunited parts were allowed to overlap
each other at *a* and *b*, and were fastened together by means
of a clamp with a screw-nut.   The whole apparatus thus prepared may be easily removed : the clamps being unscrewed
and the resounding instrument removed, the lower wire must
be unhooked from the ceiling, the hook unscrewed, and the
middle wire withdrawn from the insulating-tube, the time

for fixing or removing the apparatus need not exceed a few
minutes.

From what has preceded, it will be obvious in what manner
two square piano-fortes or two harps may be so connected as
mutually to reciprocate each other's sounds; by such an

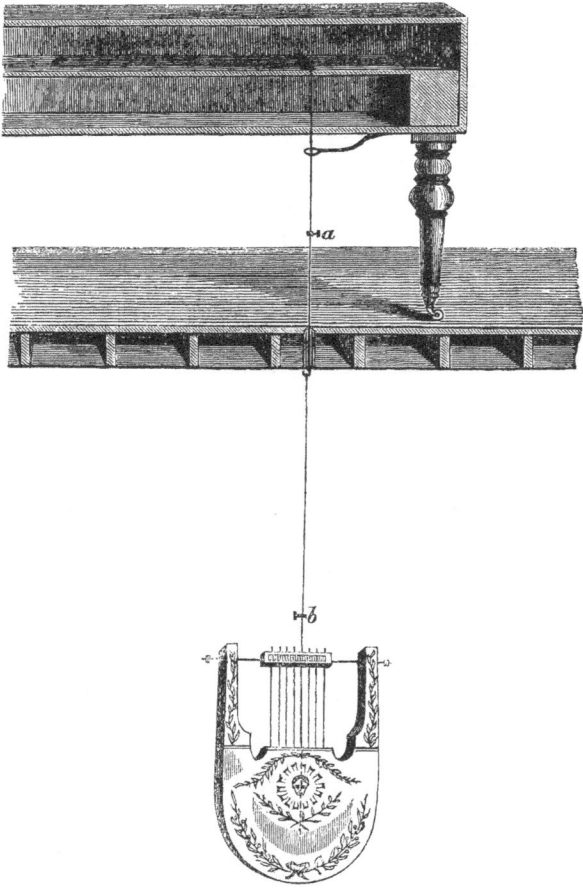

arrangement, two performers in different rooms may play a
duet together to two distinct audiences, or one may echo the
performance of the other.   If the transmission is required to

be horizontal, *i. e.* between two rooms on the same floor, cabinet piano-fortes must be employed.

The sounds of an instrument may be at the same time transmitted to more than one place : for instance, communications may be made from a square piano-forte to a resounding instrument above, and to another below; and the communication may be even continued through a series of reciprocating instruments.   If the instruments be not in adjacent rooms, but be further removed from each other, a person in the intermediate room, through which the conductor passes, will hear no sound but what is communicated by the ordinary means.   Hence it would be possible to extend a horizontal conductor through a series of rooms belonging to different houses, and (provided the instrument connected with one of its extremities be constantly played upon) to hear at pleasure the performance in any of these rooms, by merely attaching a reciprocating instrument to the conductor; on removing this instrument, the sonorous undulations would pass inaudibly to the next apartment.   These abservations will equally apply to the transmission of other musical sounds which will be hereafter noticed (§ 6, 7).

§ 6. The transmission of the sounds of those stringed instruments which produce sustained sounds, as the violin, violoncello, &c., is equally effective.   The conducting-rod may be applied either to the back or the front of the instrument; no precise directions can be given with respect to the points at which the contact should be made; but, in general, the effect has appeared to me better when the end of the conductor has not been too far removed from the situation of the sound-post.

§ 7. I have been able to effect the transmission of the sounds of reed wind-instruments through solid conductors as perfectly as that of instruments dependent on the vibrations of sounding-boards.   In the clarionet, or any other reed instrument, the column of air and the vibrating tongue (or reed) mutually influence each other in such a manner that, whether the sounds be communicated to the atmosphere from the column of air, or to a solid conductor from the vibrating tongue, the quality (timbre) of the sound undergoes no change.

To connect the conducting wire, which may be of brass, and about a tenth of an inch in diameter, with the tongue of the clarionet, the end of the wire must be bent for about a quarter of an inch, and then filed flat on both sides. This flattened end must be fastened to the fixed end of the tongue by the silk wrapping which usually fastens the tongue only, and the angle of the bend be adjusted so as to suit the position of the performer. If the sound is to be transmitted downwards, the embouchure of the clarionet must be placed in the performer's mouth in the usual way, viz. the tongue of the reed resting on the under lip; but if the sound is to be transmitted upwards, the performer must play, as some eminent masters of this instrument do, with the tongue applied to the upper lip. For the bassoon or the hautbois, it is equally convenient to the performer, whether the wire be applied to the reed above or below.

The resounding instrument may, as in the experiments above detailed, be either a harp, a piano-forte, or a guitar. It is a singular effect to hear the sounds of a wind-instrument thus reproduced by a sounding-board.

§ 8. The experiments I have made with respect to other classes of wind-instruments have not been equally successful. It is not possible to communicate the vibrations of the air to a solid conductor without an enormous loss of intensity; if, however, the intermediation of other bodies which enter readily into vibrations from the agitations of the air, be employed, the transmission may in some measure be effected. Thus, if the end of the conducting-wire be placed in the most strongly vibrating part of the column of air in a flute, there is but little perceptible transmission of sound; but if the wire touch the side of the instrument, it will more readily transmit the sounds, as the side is susceptible of entering into vibration. Even in this latter case, the sounds are scarcely audible, unless the ear be held close to the resounding instrument.

In a similar manner, the sounds of an entire orchestra

may be transmitted—viz. by connecting the end of the conductor with a properly constructed sounding-board, so placed as to resound to all the instruments. The effect of an experiment of this kind is very pleasing; the sounds, indeed, have so little intensity as scarcely to be heard at a distance from the reciprocating instrument; but on placing the ear close to it, a diminutive band is heard, in which all the instruments preserve their distinctive qualities, and the pianos and fortes, the crescendos and diminuendos, their relative contrasts. Compared with an ordinary band, heard at a distance through the air, the effect is as a landscape seen in miniature beauty through a concave lens, as compared with the same scene viewed by the ordinary vision through a murky atmosphere.

§ 9. In the preceding experiments on the transmission of sound through solid bodies, the conductors have been represented as straight; but, though sound is transmitted the more readily through straight conductors, it will yet pass, though with diminished intensity, through rods with angular and curved bendings. If a vibrating tuning-fork be placed at one end of a straight brass rod, the other end of which rests perpendicularly upon a sounding-board, the vibrations will, in accordance with what has been above stated, be powerfully transmitted. On gradually bending the rod at any part of its length, while the vibrations of the tuning-fork are kept in the same plane with the angle of the bent rod, the transmitted sound will progressively decrease in intensity, and will be very feeble when the angle has become a right one; as the bending is continued so as to make the angle between the two parts of the rod more acute, the intensity of the sound will increase in the same order in which it had before diminished; and when the two parts of the rod are nearly parallel, the sound will be nearly as loud as when the transmission was rectilineal. If, during the gradual bending of the rod, the plane of the vibrations of the tuning-fork be perpendicular to the plane of the angle made by the two parts of the rod, the same change will be observed, but in a more obvious manner than in the former case; and when the angle becomes a right one, the sound will be scarcely perceptible. At intermediate inclinations of the two planes, the gradations of intensity, occasioned by the bending of the rod, will be found to be intermediate.

The changes of intensity dependent on the variation of the angle of the two planes may be instructively shown by bending the rod permanently to a right angle, and placing, as before, the stem of a tuning-fork so as to form the prolongation of one of the parts of the rod, the other part of the rod resting on the sounding-board. On gradually turning the tuning-fork round the axis of its stem, without inclining it to the rod, the plane of the vibrations will assume every angle with respect to the plane in which the two parts of the rod is bent. During the revolution it will be observed, that when the planes coincide the intensity will be at its maximum, and when they are perpendicular to each other at its minimum : thus, supposing the sound to commence when the two planes are parallel, it will gradually diminish until they make an angle of 90° ; it will then increase through the same changes of intensity, in an inverted order, until it acquires its maximum at 180° ; it will again decrease between this and 270°, and increase until it arrives at its first position at 0° If the stem of the tuning-fork be placed perpendicularly on the side of a conducting-rod resting on a sounding-board, the same phenomena may be observed ; the stem of the tuning-fork is, in fact, a short conductor, forming a right angle with the rod.

Were it necessary for the transmission of sound that the undulations should propagate themselves only rectilinearly, it is obvious that they would not pass through a bent rod ; and, on the other hand, had they the property of diffusing themselves equally in all directions, we should not observe any differences of intensity in the experiments above noticed. These experiments lead us to conclude that sound diffuses itself in all directions, though unequally ; that it is communicated more readily in the plane in which the original vibrations are made, and with the greatest degree of intensity in the direction of these vibrations.

§ 10. In most of the experiments relating to the transmission of the sounds of musical instruments, which I have in the preceding paragraphs detailed, the conductor has been represented as receiving its impulses from a surface vibrating normally, to which it was perpendicularly attached; the communication was consequently effected by longitudinal undulations in the conducting-wire. But if, while the conductor retains its position, the surface were to vibrate tangentially or obliquely, the communication would be effected by transversal or oblique undulations.

In practice it is preferable to employ the longitudinal undulations for the purpose of transmitting musical sounds to a distance: for, firstly, the transmission is more efficacious; and, secondly, the transverse undulations have a great tendency to communicate themselves laterally from the conductor to the surrounding medium, and thereby to become audible without the assistance of a reciprocating instrument. This lateral dispersion is scarcely observable with small conductors, but is very obvious when a rod of considerable diameter is employed.

I had an opportunity of observing this fact while repeating some of the preceding experiments at the Royal Institution. A square piano-forte was placed in the apartment beneath the lecture-room; and a conductor, placed perpendicularly to its sounding-board, passed through the floor separating the two rooms, but no reciprocating sounding-board was placed at its upper end. By this arrangement, longitudinal undulations were communicated to the conductor; and, whether this was a brass wire one tenth of an inch in diameter or a square deal rod half an inch thick, the insulation appeared to be equally perfect. But it was not so when the conductor, instead of being placed on the sounding-board of a piano-forte, was made to rest on the top of the bridge of a violin, and the strings put into vibration by drawing a bow across them, communicated transverse vibrations to the conductor; it was now observed, that the metal wire insulated the sound tolerably well, but that when the wooden rod was employed, the sound communicated to the air from the entire surface of the portion of the rod above the floor, was nearly as loud as if a sounding-board were placed at its extremity.

§ 11. I have in this paper given the general results of a variety of experiments made at different and distant periods during the last ten years ; but they are far from forming so complete a course as I have been desirous of making. To extend these experiments much further would be attended with some difficulties ; but as the velocity of sound is much greater in solid substances than in air, it is not improbable that the transmission of sound through solid conductors, and its subsequent reciprocation, may hereafter be applied to many useful purposes. Sound travels through the air at the rate of 1142 feet in a second of time ; but it is communicated through iron wire, glass, cane, or deal-wood rods with the velocity of about 18,000 feet per second, so that it would travel the distance of 200 miles in less than a minute.

When sound is allowed to diffuse itself in all directions as from a centre, its intensity, according to theory, decreases as the square of the distance increases ; but if it be confined to one rectilinear direction, no diminution of intensity ought to take place. But this is on the supposition that the conducting-body possesses perfect homogeneity, and is uniform in its structure,—conditions which never obtain in our actual experiments. Could any conducting substance be rendered perfectly equal in density and elasticity, so as to allow the undulations to proceed with a uniform velocity without any reflections or interferences, it would be as easy to transmit sounds through such conductors from Aberdeen to London, as it is now to establish a communication from one chamber to another. Whether any substance can be rendered thus homogeneous and uniform remains for future philosophers to determine.

The transmission to distant places, and the multiplication of musical performances, are objects of far less importance than the conveyance of the articulations of speech. I have found by experiment that all these articulations, as well as the musical inflexions of the voice, may be perfectly, though feebly, transmitted to any of the previously described reciprocating instruments by connecting the conductor, either immediately with some part of the neck or head contiguous to the larynx, or with a sounding-board, to which the mouth of the speaker or singer is closely applied. The almost hopeless difficulty of communicating sounds produced in air with sufficient intensity to solid bodies, might induce us to despair of further success; but could articulations similar to those enounced by the human organs of speech be produced immediately in solid bodies, their transmission might be effected with any required degree of intensity. Some recent investigations lead us to hope that we are not far from effecting these desiderata; and if all the articulations were once thus obtained, the construction of a machine for the arrangement of them into syllables, words, and sentences, would demand no knowledge beyond that we already possess.

VII. *On the Figures obtained by strewing Sand on Vibrating Surfaces, commonly called Acoustic Figures.*

[From the Philosophical Transactions of the Royal Society, 1833,
pp. 593–634.]

§ 1.

HALF a century has nearly elapsed since the attention of philosophers was first called to the curious phenomena exhibited when sand is strewed on vibrating surfaces. Long before this time, Galileo had noticed that small pieces of bristle laid on the sounding-board of a musical instrument, were violently agitated on some parts of the surface, whilst on other parts they did not appear to move; and our own countryman Dr. Hook, whose sagacity in anticipating many of the discoveries of later times has been so frequently remarked, had proposed to observe the vibrations of a bell by strewing flour upon it. But to Chladni is due the sole merit of having discovered the symmetrical figures exhibited on plates of regular forms when caused to sound. His first investigations on this subject, 'Entdeckungen über die Theorie des Klanges,' were published in 1787; this work was followed by his 'Akustik' in 1802, and his 'Neue Beyträge zur Akustik,' 1817. A French translation, by himself, of his second work was published at Paris in 1809.

All the figures obtained by Chladni on square surfaces are delineated in plates I., II., III.; they are copied from the 'Neue Beyträge,' which work contains his most mature experiments; but not having been translated either into French or English, it is but little known in this country. The following are the general results deduced by Chladni from his observations respecting these figures: his works may be referred to for the details omitted, and for those concerning the vibrations of plates in general.

In all the modes of vibration of a square or rectangular plate, the figures, even if they consist of diagonal or tortuous lines, may all be referred to a certain number of nodal lines in the two directions parallel to the sides.

To establish a convenient notation for these figures, he represents the lines in the two directions by numbers separated by a vertical line. Thus, for example, 3|0 signifies the mode of vibration, in which there are three lines in one direction and none in the other; 5|2 denotes that in which there are five lines parallel to one side, and two to the other, &c.

The nodal lines, which may always be considered as having been originally straight, may curve themselves more or less; and in general the flexions of these lines, whether they adjoin each other or are separated by a straight line, mutually approach to or recede from each other. In some modes of vibration the nodal lines are never straight.

In some instances, the same mode of vibration may manifest itself in two essentially different ways, according as the flexions of the lines, or the greater number of them, are inward or outward; in the first case, the sound is usually graver than in the second. This difference is remarked in those figures where there is an entire number of flexions, as in 2|0, 3|1, 4|0, 5|3, 6|2, &c.; but never in those figures where there are $1\frac{1}{2}$, $2\frac{1}{2}$, &c., as in 3|0, 4|1, 5|0, 5|2, &c. To distinguish the first from the second figure, Chladni places a horizontal line above the numbers in the first case, and below them in the second case, thus, $\overline{4|2}$, $\underline{4|2}$.

When two or more figures having the same notation, occur in the Table, the others are to be considered as distortions of the first, occasioned by altering the fixed points, and the place at which the bow is applied.

If four plates of the same size, and upon which the same figure has been produced, be placed together so as to form a larger square, this compound figure may also be more or less accurately produced on a single larger plate. Several instances of this may be seen by reference to the Table of figures. (Plates I., II., III.)

The figures placed as exponents indicate the octaves in which the sounds occur; and the characters + and − denote that the sounds to the characters of which they are affixed are respectively sharper and flatter than the true intervals.

1	2	3	4	5	6	7	8	9	
								D##^{6}, 612	9
							B^5, 480	C##^{6}+561	8
						F#^{5}, 360 364	A^5, 432 435	C^5 −495 / 510 512 C^5	7
					C^5, 256 264	E^5, 320 324 325	G^5, 377+ 384 / G#^{5}, 390 392	B♭^{5}, 450	6
				F#^{4}, 180	B♭^{4}, 224+ 231−	C#^{5}, 275 280 / D^5, 286 288	F^5, 336 338	G#^{5}−390 392	5
			B^3−, 110+, 112	D#^{4}, 150, 153	G^4, 189 192 / G#^{4}, 196 198 200−	B^4, 240 240	D^5, 286 288 / D#^{5}, 294 299	F#^{5}, 360	4
		C^3+, 64, 65	F#^{3}, 90, 91	B^3, 119 120 / C^4, 125 126 128−	E^4, 160 162	G#^{4}+, 209 210 / A^4 216 220	C^5+, 256+ 264−	F^5 330 336 / F^5+, 343 346	3
	A^1..B^1, 27+, 28−	F#^{2}, 45	C#^{3}, 70 / D^3, 72	G#^{3}, 98, 99, 100−	C#^{4}, 135 140 / D^4, 144	F#^{4}+ 180+? 189−?	B^4, 240 242 / C^5 −245 250	E^5−, 315	2
G, 6	B, 15	B^1, 30 / C^2+, 32? 33?	B^2, −, 55, 56−	F^3, 84 / F#^{3}, 90 91	C^4, 128	F^4 F#^{4} 175? 180?			1
D−, 9− / E..F, 10+		G#, 25	G#^{2}−, 49− / G#^{2}, 50	B^3+, 81	B^3, 130 121 / C^4−, 125 126	F^4, 169	B♭^{4}, 224 225 / B♭^{4}+ 225? 231?	D^5, 289	0
1	2	3	4	5	6	7	8	9	

The preceding Table contains the relative sounds (expressed both by their musical names and the number of their vibrations) of all the modes of vibration of a square plate, experimentally ascertained by Chladni. The horizontal series of numbers denotes the lines parallel to one of the sides, and the vertical series those parallel to the other.

Having thus briefly stated the general results deduced by Chladni from his experimental researches, I shall proceed to class and analyze the phenomena; and I shall endeavour to show, that all the figures of vibrating surfaces are the resultants of very simple modes of vibration, oscillating isochronously, and superposed upon each other; the resultant figure varying with the component modes of vibration, the number of the superpositions, and the angles at which they are superposed. In this first part of the investigation I shall confine myself to the figures of square and other rectangular plates.

§ 2.

The most simple modes of vibration of a rectangular surface are those which exhibit quiescent lines parallel to one of its edges. Euler has theoretically established, that a rod or band, having both its ends unfixed, can vibrate with 2, 3, 4, 5, 6, &c. quiescent lines parallel to the ends, and that the corresponding numbers of vibrations are very nearly as the squares of the arithmetical progression 3, 5, 9, 11, &c. These conclusions are fully confirmed by experiment. He has proved, moreover, that when the same mode of vibration of different plates is compared, the number of vibrations is inversely as the square of the length of the plate, but that increase of breadth occasions no difference in the sound; and that the distance from a free end to a quiescent line is rather less than half the distance between two quiescent lines.

Fig. 1. a. Plate IV., shows the number and situations of the quiescent lines in the first four modes of vibration of this series. Fig. 1. b. and c. are profiles of the preceding, and represent the curvature of each parallel fibre perpendicular to the quiescent lines at the two opposite limits of their vibration. The quantity of motion at each point is indicated

by the corresponding ordinate of the curve, and its direction
by its situation above or below the horizontal line. It will
be convenient to distinguish these states of motion, in which
every corresponding point is moving in direct opposition;
and I shall therefore call the first, *b.* positive states of vibration, and the second, *c.* negative states of vibration. When
there is an even number of quiescent points, the positive
state of vibration may be considered as that in which the
motion at the *central* part is above the plane of equilibrium,
and the negative, that in which it is below it. If we suppose
two similar surfaces with the same number of quiescent lines
to be superposed, and both to vibrate in concurrence, *i. e.*
both either positively or negatively, they will mutually assist
each other's effects; but if they vibrate in opposing directions, they will destroy each other's motions, and the entire
surface will be at rest.

§ 3.

When the rectangular surface is equilateral, it is obvious
that it may vibrate in two different rectangular directions, so
as to give the same sound, and present the same arrangement
of quiescent lines. Now this plate may be excited at various
points where the motion of each mode of vibration is at its
maximum, in the same direction, and of equal intensity:
such being the case, there is no reason why one mode of
vibration should be produced in preference to the other; and
on calculating the effect of such coexistence, it will be found
that the resultants of these combined modes of vibration,
similar in everything but in their direction with regard to the
sides of the plate, give rise to new quiescent lines which accurately correspond with figures described by Chladni; while
the number of vibrations does not materially differ from that
of the component modes of vibration.

The principal results of the superposition of two similar
modes of vibration are these: 1st, The points where the
quiescent lines of each figure intersect each other, remain
quiescent points in the resultant figure; 2ndly, The quiescent
lines of one figure are obliterated when superposed by the

vibrating parts of the other; 3rdly, New quiescent points, which may be called points of compensation, are formed wherever the vibrations in opposite directions neutralize each other; and, lastly, At all other points the motion is as the sum of the concurring, or the difference of the opposing vibrations.

A primary figure, having an even number of quiescent lines, may be superposed two ways, and may consequently give rise to two distinct resultant figures: one, when the central vibrating parts concur; and the other, when they are in opposition; but if the number of the quiescent lines in the primary figure be uneven, there can be only one resultant figure.

The quiescent lines which thus result may be very easily ascertained. I will take as an example the first mode of vibration, having two parallel quiescent lines; this being superposed in two rectangular directions, and so that the states of vibration are opposing (Plate IV. fig. 2), it is obvious that no lines of compensation can exist in the four rectangular segments *a a a a*, as every point included within them is actuated by concurrent motions; but in all other rectangles they must necessarily be formed, as every point within them is affected by the two opposing motions, and if the two modes be of equal intensity, the compensations must occur at every point equally distant from the two rectangular quiescent lines, each appertaining to a different mode of vibration. The resultant figure will thus be found to consist of two diagonal lines perpendicular to each other, and passing through the centre of the plate.

But if the two superpositions vibrate in concurrence, the rectangles *b b b b* will be free from compensating points; but these will occur in the other rectangles, and form the figure represented (fig. 3), which also consists of diagonal lines.

In the same manner the resultant of any two similar modes of vibration with nodal lines, parallel to the sides, may be proved to consist of lines parallel to the diagonals.

§ 4.

It is not a necessary condition for the vibrations of a square plate that the primary nodal lines shall be parallel to a side; they may also be parallel to a diagonal, or to any line intermediate between a transverse and a diagonal line. In these cases the superpositions take place according to the following rule: That the axes of the superposed modes of vibration must make equal angles with a transversal line passing through the centre; for otherwise the modes of vibration would not be similar. By the axis of a primary mode of vibration, I mean a straight line passing through the centre of the plate and parallel to the quiescent lines. Considerations of the kind already employed will show that in all these instances the resultant figures consist of lines parallel to the edges of the plate, and that they are always the same in number as the nodal lines of a component mode of vibration, but differently distributed in the two directions, according as the angle of superposition varies.

The various primary modes of vibration, transverse, intermediate, and diagonal, and the angles which the quiescent lines of two similar figures make with each other when they are superposed, are represented in the first column of the general Table, Plates V. to XI.; in the second column of this Table are placed the figures resulting from their opposing superpositions, and in the third column those which arise from their concurring superpositions.

We obtain by experiment a limited number only of figures which can be considered the resultants of primary modes of vibration consisting of any given number of oblique lines; but it would seem, that as the various degrees of obliquity are infinite, so there should be an infinite number of resultant figures passing into each other by insensible gradations: by calculation this should be so, but there are causes of limitation which I shall proceed to explain.

It appears that no resultant figure is maintainable unless the greatest excursions of the external vibrating parts occur at the edges of the plate. In the concurring superpositions of eight oblique lines, this condition can only be fulfilled

when the angles they make with each other are either 90° or
143° 8'; in the first case the resultant figure consists of four
lines in each transverse direction, in the second of six lines
in one direction and two in the other. In the opposing
superpositions of the same number of lines, the condition is
fulfilled when the angles at which the lines are inclined are
118° 4' and 163° 44': the resultant figure of the former consists of five lines in one direction and three in the other, and
that of the latter of seven in one direction and one in the
other.

§ 5.

I have, in the preceding sections, described the various
modes of binary superpositions which may take place on a
square surface. But there are numerous cases in which four
superpositions may coexist, and these I shall now proceed to
take into consideration.

When the axis of a primary figure corresponds either with
a diagonal or with a transverse line, passing through the
centre of the plate, it is obvious that there can be only one
other line of equal length, which can be considered as the
axis of a similar and isochronous mode of vibration; in these
cases it is evident, therefore, that there can only be two superpositions. But in every intermediate direction of an axis
there are three other lines of equal length which constitute
axes of similar modes of vibration; and four superpositions
can therefore take place whenever the axis of a component
mode of vibration is neither a diagonal nor a transverse line.

It would be a tedious and laborious process to ascertain a
resultant figure by combining its four component modes of
vibration; but the same purpose will be effectually answered
by combining them first in pairs, as explained in the preceding section, and then combining two of these first resultants rectangularly together.

The following process affords great facility for ascertaining
the second resultant figure, which arises from the superposition of two first resultants. *a.* and *b.* (fig. 4. Plate IV.) are
the two component first resultants, the similar lines of one

being placed rectangularly to those of the other; the vibrating
parts are indicated by the letters P and N, according as the
vibrations are positive or negative.   At C the two figures are
superposed, A being represented by the continuous and B
by the dotted lines.   The surface is now subdivided into a
number of unequal rectangles, and by comparing the two
component figures together, it is easy to see which of these
rectangles are influenced by conspiring, and which by opposing motions; if the motions are found to conspire, the
letters P or N must be placed in these rectangles according
as the coexisting motions are positive or negative: if the
motions are in opposition, a mark may be made to indicate
that a quiescent line passes through this rectangle.   Wherever
a continuous line intersects a dotted line, a mark is to be
made, to indicate that a quiescent point is formed; and as in
every other part, the quiescent lines of one figure pass over
vibrating parts of the other, the boundary lines of all the
rectangles must be marked with the letters indicating the
motions of the vibrating parts they superpose.   The figure
C being thus marked, the resultant figure is easily described
by joining the fixed points by lines drawn through the rectangles shown to be actuated by opposing motions; carefully
avoiding to encroach upon the rectangles of conspiring motion
marked P or N.

    That the diagonal line is perfectly straight may be proved
in the following manner.   It must first be premised that the
rectangles included within the quiescent lines of each of the
first resultant figures have precisely the same quantity of
motion in the same relative points with respect to the surrounding sides, and that the vibrating parts at the edges and
corners of the plate must be considered respectively as exact
halves and quarters of a complete vibrating part.   This being
understood, if two similar first resultants be laid alongside
each other in the directions in which they are to be superposed, and if a diagonal line be similarly drawn through
each of them, it will be obvious that every corresponding
point of each line must possess the same intensity of motion.
If the successive segments (equal in each figure) through

which the lines simultaneously pass, be in opposite states of vibration, they will neutralize each other's effects, and a diagonal quiescent line will be formed; and if they be concurring, all the parts between the coincident or fixed points will be in motion. In the diagram (fig. 4. *c.*) one diagonal is in the first state and the other in the second.

If the number of lines in the component first resultant figures be uneven, they admit of only one mode of superposition. But first resultants having an even number of quiescent lines, admit of being superposed in two ways, according as they are vibrating in concurrence or in opposition.

It frequently occurs that entire quiescent lines superpose each other; thus, when in the component figure there is an uneven number of lines in each direction, the two rectangular central lines of each superposed figure must coalesce, and consequently they continue in both the resultant figures; examples of this are seen in the resultants of 3|1, 5|1, 5|3, 7|1, 7|3, 7|5, &c. Again, when the number of quiescent lines in one direction of the component figure is three times greater than that in the other, the number of coinciding fixed lines in each direction of the resultant figure is equal to the smallest number in the first resultant, as in 6|2, 9|3, 12|4, 15|5, &c. See the general Table, Plates V. to XI.

The following general results are obtained from constructing the second resultant figures according to the rules above given.

In superposing first resultants consisting of an even number of lines: 1st, When the number of lines in each direction of the first resultants is even, and the modes of vibration are concurring, no line passes through the centre of the second resultant figure. When the modes of vibration are opposing, two rectangular diagonal lines of compensation occur. 2ndly, When the number of lines in each direction is uneven, and the modes of vibration are concurring, there are always two perpendicular transversal lines passing through the centre; and when the modes are opposing, there are, in addition to these fixed lines, the two diagonal lines of compensation.

When the first resultants consist of an uneven number of lines, in which case there is no distinction of concurring and opposing vibrations, one diagonal line only invariably occurs.

In no case is it necessary to calculate an entire figure. When the number of nodal lines in the primary mode of vibration is even, only one quarter of the figure is required to be calculated, as it is obvious that every second resultant of this kind consists of four symmetrical, and as it were reflected, portions. But when the primary number of quiescent lines is uneven, it is necessary to calculate one half; the other half is symmetrical and inverted.

Some of the first resultants are never obtained by experiment. When the number of quiescent lines in the primary mode of vibration is uneven, either the first or the second resultant may be obtained at pleasure; thus in 3|2 if the impulses be made at a corner, where the motion of both superpositions is at its maximum, the second resultant must arise; but if they be made at the middle of a side, the first resultant only will appear, because the point of excitation is a quiescent point of the other. But in all cases where there is an *even* number of lines, it is impossible to obtain the first resultant, because each maximum of vibration equally belongs to both superpositions.

§ 6.

I have given in Plates V. to XI., a Table which shows every perfect resultant figure of a square surface when the number of quiescent lines in the primary modes of vibration does not exceed twelve. I have carefully calculated each figure by the rules laid down in the preceding sections, and have shown in the Table the successive processes of superposition. The first vertical row exhibits the two primary modes of vibration superposed at the required angles; one figure being represented by the continuous lines, and the other by the dotted lines. The second row contains the first resultants which arise from the opposing superpositions of the preceding; and the third row, those which result from their concurring superpositions. The fourth and fifth rows exhibit the perfect

second resultants which are formed, the former by the opposing and the latter by the concurring superpositions of the
first resultant which the plate has been already found com-
petent to produce.

On comparing the calculated figures with those obtained
experimentally by Chladni, the greater number are found
exactly to agree; there are, however, some differences which
it will be necessary to explain. In the first place, there is an
obvious cause of error in delineating figures from experiment,
from this circumstance,—that the sand accumulates in the
spaces where two convex curves are near and opposite to each
other, the motion being there very small, so that it is difficult to ascertain whether the curves join, or not. Secondly,
Inequalities in the plate will sometimes occasion lines which
ought to intersect each other, so as to appear separated
curves. On comparing together the figures Chladni has
marked as 6|4, $\overline{6|4}$, 7|2, 7|4, 8|3 c. &c. with those of the cal-
culated Table, they will be found to differ only in these
respects.

Another cause of difference is this : When the lines of one
component figure very nearly coincide with those of the
other, but without actually doing so, the resultant figure
may be such as would arise from their actual superposition,
instead of that which accurate calculation would give. In
Chladni's Table there are two instances of this alteration,
7|2 a. and 8|3 a.

A few of the figures delineated by Chladni are irregular
resultants formed by the superpositions of dissimilar modes
of vibration. These irregular resultants can be formed only
when the dissimilar component modes of vibration give the
same sound, and have a maximum point of vibration in common, at which they can be simultaneously excited. The
figure marked by Chladni 6|1, I find to be an irregular resul-
tant formed by the combination of 6|1 with 3|5, which both
give the same sound C^4. The irregular figure marked in his
Table 5|1 a. is a compound of 5|1 and 2|5.

The calculated figures 6|1, 7|1, 8|1, 9|1, 10|1, 10|2, and
11|1, are not to be found in Chladni's Table. The near ap-

proach of the inclined lines of their primary component figures to parallelism is the cause of the great difficulty in obtaining these figures by experiment.

The figures marked by Chladni 10|3, 9|4, 8|5, 7|6, 10|4, 9|5, 8|6, and 9|7, exceed the limits within which I have calculated the Table, and are therefore not to be found in it.

§ 7. *Imperfect resultant Figures.*

I have hitherto considered those resultant figures only which arise from the superpositions of similar modes of vibration, each exactly equal in intensity; these I shall in future call *perfect* resultant figures. But when the vibrations of the superposed modes are unequal in intensity, then a figure intermediate between the perfect resultant and one of its components is formed; these intermediate figures I shall call *imperfect* resultants. They are experimentally obtained by varying in a slight degree the places at which the plate is held or touched, from those necessary to determine the corresponding perfect resultant figure; the place at which the bow is applied remaining in both cases the same.

Fig. 6. *a. b. c. d. e.* in Chladni's Table, Plate I., represents the successive transformations of figure which take place when each component mode of vibration presents three transversal lines : *a.* and *e.* are the two components; *c.* the perfect resultant; *b.* an imperfect resultant, in which the excursions of *a.* are the greatest : and *d.* an imperfect resultant, in which *e.* has the greatest energy.

Fig. 11. *a. b.* Plate I., exhibits the transitions of the opposing superposition of two primary figures, each presenting four transversal lines; and fig. 12. *a.* and *b.* the changes of the concurring superposition of the same.

These are the principal types of the transformation of primary figures into first resultants. In each of these series of transitions there are certain points which are invariable during every change : these are, the quiescent points formed by the nodal lines of one figure intersecting those of the other, and the centres of vibration where the maxima of positive or negative vibration agree in each component mode of vibration. The points of compensation are changeable.

Figs. 29. *a.* and 30. *a.* Plate II., represent imperfect second resultants, formed by two superpositions of the first resultant figure 6|2. Fig. 29. *a.* arises from concurring, and fig. 30. *a.* from opposing superpositions. The straight lines in these imperfect figures arise from the coincidence of entire quiescent lines in each component figure, and, consequently, they remain unaltered whatever may be the relative intensities of the superposed modes of vibration. But the curved lines, which are formed of compensating points, change with the varying intensities.

§ 8. *Figures of irregular Plates.*

If the sides of the square be nearly, but not exactly, equal, the superpositions of two similar modes of vibration with transversal lines still take place; but instead of exhibiting perfect resultants, figures resembling transitional figures appear. Thus in the binary superpositon (Plate I. fig. 1.) of the figure with two transversal lines, if the sides be unequal, the crossed lines separate at their point of intersection and are converted into two curves, the summits of which recede from each other as the difference in the lengths of the sides becomes greater.

Also, if the diagonals of the square be unequal, the resultant figure (Plate II. fig. 2.) arising from two superposed modes of vibration with diagonal axes, will be modified in a similar manner.

Corresponding modifications are occasioned, through accidental differences of elasticity, &c. in the directions of the axes of the superposed modes of vibration, even when the dimensions of the plate are apparently equal.

If a plate of glass be covered on one of its sides with leaf-gold, or if a plate of ground glass be substituted for an ordinary glass plate with smooth surfaces, the figures may be obtained distinctly delineated by lines consisting of a single row of grains of sand. Experiments made in this manner, upon square plates carefully prepared, induced Professor Strehlke to conclude, after many minute measurements of these lines, that all acoustic figures are formed of hyperbolic curves, and that the quiescent lines never intersect each other. But however correct these experiments may have been, the conclusions drawn from them are unwarranted; were it possible to obtain plates of a perfectly homogeneous substance, and of accurately equal dimensions, there can be no doubt that the lines, however finely defined, would actually intersect each other.

§ 9.

I have already, § 1, given Chladni's Table of the comparative sounds, and numbers of vibrations of the figures of square surfaces obtained from experiment. In the following Table these results are arranged so as to correspond with the views taken in this paper. The numbers in the first vertical row indicate the number of parallel quiescent lines in the primary figure. In the horizontal rows the angles at which the lines of the primary modes of vibration intersect each other are shown; and the figures between brackets give the notation of the first resultant produced by their superposition; below these its number of vibrations, and the character representing its musical sound, are given; when there are two of these lower lines separated by a horizontal dash, the one above indicates the sound of the opposing superposition, and that below it the concurring superposition.

Thus it appears, that every figure of a square surface which experiment can give, may be reduced to a primary figure with parallel lines, giving the same sound: if, therefore, the analytical investigation be confined to these, many of the difficulties will disappear. Euler has investigated the subject when lines parallel to a side only are concerned; it remains to extend the inquiry to modes of vibration the nodal lines of which are perpendicular to any line passing through the centre of the surface. An analytical expression for all the sounds of a square plate may probably be obtained, which shall be a function of the number of quiescent lines, and the length of the axis of the mode of vibration.

No.							
2	90° (1|1) G, 6	180° (2|0) D—, 9					
3	126° 52' (2|1) B, 15	180° (3|0) G♯1, 25 E. F, 10+					
4	90° (2|2) A1. B1, 27+, 28—	143° 8' (3|1) B1, 30	180° (4|0) G♯2, 49— C2+, 32. ? 33. ?				
5	112° 38' (3|2) F♯2, 45	151° 56' (4|1) B2—, 55. 56—	180°, (5|0) E3+, 81 G♯2, 50				
6	90°, (3|3) C3+, 64. 65	126° 52' (4|2) C♯3, 70	157° 22' (5|1) F3, 84 D3, 72	180° (6|0) B♭3, 120. 121			
7	106° 16' (4|3) F♯3, 90. 91	138° 22' (5|2) G♯3, 98. 99. 100—,	161° 4' (6|1) C4, 128	180° (7|0) F♯4, 169 C4—, 125. 126			
8	90° (4|4) B♭3—, 110+. 112	118° 4' (5|3) B3. 119. 120	149° 8' (6|2) C♯4 135. 140	163° 44' (7|1) F♯4. F♯4 175. ? 180. ? D4. 144	180° (8|0) B♭4 224. 225		
9	102° 40' (5|4) D♯4 150. 153	126° 52' (6|3) B4 160. 162	148° 6' (7|2) F♯4+, 180+. ? 189—	165° 44' (8|1)	180° (9|0) D5 289		
10	90° (5|5) F♯4, 180	112° 38' (6|4) G4 189. 192	133° 26' (7|3) G♯4+ 209. 210	151° 56' (8|2) B4 240. 242	167° 18' (9|1) unascertained.	180° (10|0)	
11	100° 24' (6|5) B♭4, 224+, 231—	120° 30' (7|4) C5+, 256+. 264—	138° 52' (8|3) C5+, 256+. 264—	154° 56' (9|2) C5— 245. 250 E5— 315	168° 34' (10|1)	180° (11|0)	
12	90° (6|6) C5 256+. 264—	108° 56' (7|5) C♯5, 275. 280	129° 52' (8|4) D5, 286. 288 D♯5 294. 299	143° 8' (9|3) F5, 330. 336 F5+ 343, 345	157° 22' (10|2)	169° 36' (11|1)	180° (12|0)

§ 10.

Immediately after the publication of Chladni's experiments on square plates, James Bernouilli attempted to demonstrate them analytically; but his investigation was entirely unsuccessful; his conclusions were founded on erroneous data, and the results he obtained were at variance with experiment. His assumptions were these: that the primary figures consisted of 2, 3, 4, 5, 6, 7, &c. lines parallel to the side only, these being the modes of vibration of a lamina as investigated by Euler; that any two similar or dissimilar modes of vibration might superpose each other rectangularly, the nodal lines of the two components appearing together in the resultant mode of vibration; and that the sound of the resultant differed from that of either of the two components, being much higher. The figures given by experiment he considered accidental distortions of these compound figures. This theory gave no account of those figures in which a single line in one direction coexists with any number in the other, as there is no primary figure consisting of one nodal line only; and Bernouilli acknowledged his theory to be imperfect in this respect.

The failure of Bernouilli led Chladni inconsiderately to state, that "the supposition of regarding such a rigid membraniform body as a network formed by curved lines in one direction applied upon curved lines in another direction, is not conformable to nature, and will never give, either results agreeing with experiment, or an appearance of explanation of some of the most simple vibrations." That this assertion is erroneous, the considerations in the present paper have, I conceive, fully proved. The error of Bernouilli did not consist in assuming that the observed acoustic figures were formed by superposing simple modes of vibration on each other, for this has been shown to be true; but in his assumptions of the manner in which these superpositions were made, and the effects which resulted from such hypothetical superpositions.

The various mathematicians who have more recently undertaken to investigate the laws of vibrating surfaces, as Poisson, Cauchy, Mademoiselle Germain, &c., do not appear to have taken into consideration any thing resembling the theory of superposition.

Dr. Young seems to have had a correct notion of the origin of the acoustic figures; for in his Lectures, when slightly noticing Chladni's experiments, he remarks, "The vibrations of plates differ from those of rods in the same manner as the vibrations of membranes differ from those of chords, the vibrations which cause the plate to bend in different directions being combined with each other, and sometimes occasioning singular modifications."

The brothers Weber, in their excellent work the Wellen-lehre, published in 1825, have advanced a step nearer the truth than any of their predecessors. They have shown, that in a square vessel containing water or mercury, two series of stationary waves, one parallel to each side, may be made to intersect each other; and that the compound wave formed by their interferences assumes the form shown in § 3. to be the resultant of two superpositions of parallel transverse lines. Their observations are confined to those modes of undulation, analogous to the first resultants of primary modes of vibration with lines parallel to a side. Though I had advanced considerably in the present inquiry before I saw this work, yet I should be wanting in justice to these philosophers did I not here state that theirs is the merit of having shown, in the most simple case, the way in which the superpositions of modes of vibration or undulation actually do take place.

§ 11. *Plates of Wood.*

From the rules already laid down, it is obvious that the series of figures presented by a square plate of any homogeneous material ought not to be obtained on a square plate of wood, in which substance the elasticity is not the same in all directions. If a square plate of wood be prepared with its fibres parallel to one of the sides of the square, the axes of greatest and least elasticity will be disposed rectangularly, and parallel to the adjacent sides; in this case the same primary mode of vibration in the two directions will not give the same sound, although the dimensions of the vibrating parts are the same in both; consequently they cannot coexist, and the resultant figures with diagonal lines will be wanting on such a plate. But if the axes of the two component modes of vibration be equally inclined to either of the axes of elasticity, these modes of vibration will be necessarily similar and isochronous, and therefore capable of superposition: on a square plate of wood, consequently, all those first resultants which consist of any number of lines parallel to the sides intersecting each other rectangularly, may be obtained; and the same figure will be accompanied by different sounds, according as the axes of the modes of vibration are inclined to the axes of least or greatest elasticity. It is easy to foresee that none of the second resultants, which consist of four isochronous superpositions, can be obtained on such a plate.

But if the wooden plate, instead of being square, be a rectangle, the sides of which are inversely as the squares of their resistance to flexion, the two modes of vibration parallel to the sides, though differing in length, will be isochronous; and their coexistence will give rise to a resultant figure with lines parallel to the diagonal. Thus on a rectangular plate of straight-fibred deal wood, in which the proportion of the sides were as 28 to 59, I obtained the two crossing diagonal lines, corresponding to the second figure of a homogeneous square plate of glass or metal.

Savart has made a series of numerous and accurate experiments on the changes which take place in the sound, and also in the form and position of the figure of the first mode of vibration on circular plates of wood of similar dimensions, cut in different directions with respect to the three principal axes of elasticity. All the results he has obtained are in perfect accordance with the rules stated in this paper, and might have been predicted by them. He has extended his investigations to circular slices of crystals, cut in various directions with respect to their axes, and has obtained in this way much valuable information. These researches of Savart point out a new direction to our inquiries respecting the structure of bodies; and the utility of his experiments will be greatly enhanced by the knowledge we now possess of the causes on which these phenomena depend. I shall shortly return to this subject.

An Account of some Experiments to measure the Velocity of Electricity and the Duration of Electric Light.

[From the 'Philosophical Transactions,' 1834.]
[Plates XIII. & XIV.]

§ 1.

THE path of a luminous or an illuminated point in rapid motion, it is well known, appears as a continuous line, in consequence of the after duration of the visual impression. There is nothing, however, in the appearance of such a line by which the eye can determine either the direction or the velocity of the motion which generates it. It occurred to me some years since, that if the motion which described the line in these cases were to be compounded with another motion, the direction and velocity of which were known, it would be easy, from an inspection of the resultant straight or curved line, to determine the velocity and direction of the former. Following up this idea, I made a series of experiments relating to the oscillatory motions of sonorous bodies, too numerous, and not sufficiently connected with the subject of the present communication, to be detailed in this place. The satisfactory results thus obtained made me desirous to ascertain whether, by similar means, some information might not be gained respecting the direction and velocity of the electric spark : the method by which I then proposed to effect this purpose was first announced in a lecture delivered by Dr. Faraday at the Royal Institution in June 1830. My attention was again drawn to the subject at the commencement of last year, and I attempted to realize the idea in the following manner.

Fig. 1 (Plate XIII.) represents the apparatus employed, which was screwed at *a* to the spindle of a whirling machine, so that a rapid rotatory motion might be given to it. The upper and lower parts, which were all of brass except the wooden disc *b c*, were insulated from each other by a stout glass rod d e; a slip of tinfoil connected the ball h with a, and the upper ball g was capable of adjustment to various distances from the lower one h.   When the ball f was placed within striking distance of the prime conductor of an electric machine, a spark passed between them, and also between the balls g and h, which could be separated to the distance of four inches, so as to exhibit a spark of that length.   It is obvious, that if the angular motion of the balls were in any sensible proportion to the velocity of electricity, there would be a deviation between the upper and lower terminations of the line. The instrument revolving from left to right, if the motion of the spark be downwards, the deflection of the line should be as in fig. 2 (Plate XIII.); and if its motion be upwards, it should be deflected as in fig. 3 (Plate XIII.).

When the apparatus was made to revolve rapidly, the sparks passed in the same manner as when it was at rest, and no deviation of the extremities of either of the two sparks from the same vertical line was observed.   The apparatus revolved fifty times in a second, and as a difference of the twentieth part of the circumference described by the balls could have easily been observed had it existed, we may safely conclude that the spark passed jointly through the air and the metallic conductor in less time than the thousandth part of a second.

§ 2.

Having failed to observe any deflection of the spark by the means just mentioned, I found it necessary, if I would continue the inquiry, to contrive some more effectual means of prosecuting it.   It occurred to me that the motion of the reflected image of the electric spark in a plane mirror would answer all the purposes of the motion of the apparatus itself connected with the spark.   Several advantages, it was evident, would result from this substitution ; the apparent motion of the reflected image in a small moving mirror would be equal to an extensive motion of the object itself; the same mirror might be presented to any object to be examined, thus forming, with its moving machine, an independent and universally applicable instrument; and many experiments might be tried, which, without this expedient, would be difficult or impossible to perform, from the size or immobility of the apparatus.

The most convenient form of the revolving mirror is represented in fig. 4 (Plate XIII.) ; it rotates on a vertical axis, and in its motion successively assumes every vertical plane. If a luminous point, the flame of a candle for instance, be placed at any distance before this revolving mirror, the successive places of its reflected image will describe a circle, the radius of which is equal to the perpendicular distance between the luminous point and the axis of rotation. The angular velocity of the image is twice that of the mirror; the entire circle is consequently described while the mirror makes a semi-revolution ; and if the back of the mirror be also a reflecting surface, the image will describe two entire circles during one revolution of the mirror. If the motion exceed a certain rapidity, the successive images leave their impressions on the retina, and the eye, properly placed, takes in the view of a perfectly continuous line of light, being an arc of the circle described, which arc is larger in extent in proportion to the proximity of the eye to the mirror.

If now, while the mirror is in motion, the luminous point be moved in a direction parallel to the axis of rotation, the composition of the two motions of the image, the one depending on the motion of the object, the other on the motion of the mirror, will give rise to a diagonal resultant ; and if the number of rotations made by the mirror in a given time are known, the direction and velocity of the moving point may be calculated.

By screwing the axis of the mirror to a machine with multiplying wheels, I was enabled to cause it to revolve fifty times in a second. The reflected image of a luminous point, therefore, passed over half a degree in the 72,000dth part of a second, the angular velocity of the image being, as before noticed, double that of the mirror. An arc of half a degree is easily estimated by the eye, and is equal to about an inch seen at the distance of ten feet. Supposing this to be the limit of distinct observation, though perhaps a much smaller arc might be distinguished even by the unassisted eye, we might expect, when a line of electric light is placed parallel to the axis of the revolving mirror, to ascertain two things : first, the duration of the light at each point where it appears : and secondly, the time which elapses between the appearance of the light in two successive points of its path ; provided that the time, in either case, be not less than the 72,000dth part of a second.   The first would be indicated by the horizontal elongation of the reflected image, and the second by the distance between two lines drawn from the images perpendicular to the horizontal plane.   If the duration and velocity were both rendered sensible by the mirror, the reflected image would appear as a deflected band of light.

I successively presented to the mirror, sparks four inches in length drawn from the prime conductor of a powerful electrical machine ; the explosions of a charged jar ; a glass tube four feet in length, exhibiting a spiral of electric sparks passing between dots of tinfoil : an exhausted glass tube six feet in length, through which the spark passed, and produced an unbroken line of attenuated electric light ; various pictures, such as birds, stars, &c., formed of electric sparks.   But in all these cases, when the reflected images occurred within the field of view, they appeared perfectly unaltered, and precisely as they would have done had they been reflected from the mirror while at rest.

When sparks were made to follow each other quickly, several reflected images were simultaneously seen in different positions, owing to the images having been renewed before the visual impression caused by the first had disappeared. The exhausted tube being held near a prime conductor, when looked at directly, will sometimes appear to gleam with a continuous light ; but examined in the mirror, this apparent continuity is seen to be owing to a rapid succession of transient flashes.

§ 3.

For some experiments another position of the revolving mirror is preferable to that just described.   Fig. 5 (Plate XIII.) represents the reflecting surface inclined to the axis of rota-

tion, and nearly perpendicular to it.   If a luminous point be
placed anywhere in the prolongation of the axis, its images,
succesively reflected from different parts of the mirror, form
together a circle, the whole circumference of which may be
seen at once.   In this form of the experiment the angular
velocity of the image is equal to that of the mirror, and both
move in the same direction ; whereas in the former case the
image moved with double the velocity of the mirror, and
in the opposite direction.   The visual magnitude of the circle
described increases with the distance of the object and inclination of the mirror.   The flame of a candle presented to it
appears as a broad luminous ring ; the image of the sun is
converted into a magnificent fiery belt, &c.

A series of minute sparks made to pass between two points,
or between a point and the prime conductor of a machine,
presents to the eye, from the rapidity of their succession, the
appearance of a permanent star of light.   When this star is
placed in the prolongation of the axis of the revolving mirror,
the successive sparks of which it consists are reflected to the
eye each from a different part of the surface, and they are exhibited arranged at regular distances in a circle.   When the
intermissions are rapid the appearance is extremely beautiful.

The brush of light which appears on a point when presented
at some distance from the conductor, is also by this means
shown to be an intermitting action, notwithstanding its permanent appearance ; its reflected images present, however,
this remarkable peculiarity, they are elongated in the direction of the motion, proving that a brush is not so transient
as a spark, and that the emissions which constitute it last
during an interval of time measurable by the motion of the
mirror.

But this instrument is not confined to observe merely the
intermittences of electric light ; whenever a rapid succession
of alterations occurs in an object which does not change its
place, they may be separately examined by this means.   Vibrating bodies afford many instances for investigation ; one
among these is perhaps worthy to be mentioned.   A flame
of hydrogen gas burning in the open air presents a continuous

circle in the mirror; but while producing a sound within a glass tube, regular intermissions of intensity are observed, which present a chain-like appearance, and indicate alternate contractions and dilatations of the flame corresponding with the sonorous vibrations of the column of air.

§ 4.

Experiments have frequently been made with a view to determine the velocity of the transmission of electricity through conducting bodies. In all the recorded trials of this kind it was attempted to measure the interval of time supposed to occur between two discharges made at opposite extremities of the wire, which were brought near each other so that they might be seen at the same time. In one experiment, performed at Shooter's Hill in 1747 under the superintendence of Dr. Watson, the circuit was four miles in extent, two miles through wire, and two miles through the ground; but the discharges appeared, as in all similar experiments, to be perfectly simultaneous. Nor need we feel surprised at this result, when we know that the eye is unable to distinguish the succession of luminous objects which follow at the interval of the eighth or tenth of a second, from their simultaneous appearance; and that, therefore, with a circuit even of four miles extent, the velocity of a few miles per second would be the utmost observable by such means.

I determined, therefore, to repeat a similar experiment, substituting for the imperfect judgment of the eye a revolving mirror, but more rapid in its motion and accurate in its indications than any I had previously employed. The instrument I am about to describe will, unless there be some error in the estimate which I have not been able to perceive, measure beyond the millionth of a second; and this degree of minuteness may be yet far surpassed by more costly instruments and more careful observations.

But as it is only on the hypothesis of an actual transfer of a fluid from one end of the wire to the other that a difference of time between the two sparks at its opposite extremities might be expected to be observed, in order to render the proposed experiment independent of this theoretical view, I took the necessary precaution of bringing a third spark, formed by disconnecting the middle of the wire, near to and in a line with the extreme sparks. For on the supposition of the transfer of two fluids in opposite directions, the extreme sparks would be simultaneous, but the middle spark later in its occurrence; the same appearances would also accord with the theory of one electricity, if we admit that a disturbance of electric equilibrium is simultaneously propagated from each end, arising in the one case from successive additions to, and in the other from successive subtractions from, the neutral quantity in the conducting wire.

The experiment was tried at the Gallery in Adelaide Street. The insulated wire, the total length of which was half a mile, was disposed as in fig. 6 (Plate XIII.). The parallel portions of the wire were each 120 feet in length, and six inches apart, and were tied to the balustrade with silk loops six inches long. The swagging of the wire was prevented by silk cords extending across the gallery; and to keep the lengths at their proper distances apart they were tied to the cords wherever they crossed them. The ends of the wire marked 2, 3, 4, 5, were continued to the similarly marked wires of the spark-board, fig. 7 (Plate XIV.), which was so fixed against the wall beneath the gallery, that the balls between which the sparks were to pass were in the same horizontal line. The striking-distance between each spark was the tenth of an inch, and the spark-board itself was three inches and a half in diameter. The conducting wire I employed was of copper, and its thickness the fifteenth of an inch.

Fig. 8 (Plate XIV.) represents the measuring instrument with its appendages; and fig. 10 (Plate XIV.) shows in a more distinct manner some of its essential parts. A B C D is a solid board of well-baked mahogany one foot in length, and eight inches in breadth. E is a circular mirror of polished steel one inch in diameter, so fixed to the horizontal axle F G, that the axis of rotation is in the plane of the mirror. The pivots of the axle work in the uprights of the brass frame H I. Motion is communicated from the wheel K to the axle by means of a thread passing over grooves made on the circumferences of both; and a band passing over the wheel L, on the same axis with K, may be attached to the wheel of any machine capable of giving to it a rapid motion. In the experiments I have made with this instrument the train of wheels was so arranged that the axle carrying the mirror would have made 1800 revolutions while the wheel to which the motion was first communicated was turned round once, had there been no retardation to have been taken into consideration arising from the slipping of the bands. M is a small Leyden jar, the inner coating of which is to be constantly supplied, through the chain N, with electricity either positive or negative, from a machine; the bent wire proceeding from the inner coating of the jar is in immediate contact with the fixed discharger O P, and the spontaneous discharge of the jar is to be regulated by varying the distance between the two balls. The wire 1 in connexion with the outer coating of the jar, and the wire 6 attached to the knob of the brass frame, are continued to the similarly numbered wires of the spark-board. When the jar is fully charged, and the arm Q, revolving with the axle, is brought opposite the knob of the discharger, the discharge of electricity, or disturbance of electric equilibrium, passes through the entire circuit, and the three sparks appear perfectly simultaneous to the eye. When the face of the mirror is level with and turned towards the spark-board, and is so adjusted as to form an angle of 45° with the horizontal plane, the eye looking directly downwards sees the reflected images of the three sparks. The plane glass or lens R is for the purpose of preventing the eye approaching too near the mirror, and for accommodating the vision of long- or short-sighted observers. The arm Q is so placed that the circuit may be completed when the mirror is in the position just described; the other arm serves merely as a counterpoise. To obviate the inaccuracy which would result from discharges taking place when the arm is in different positions with respect to the knob of the discharger, a plate of mica, S, is interposed, having a very small horizontal slit exactly opposite the axis of the discharger; this fixes within narrow limits the occurrence of the discharge, and, with whatever rapidity the mirror moves, the sparks are generally within the field of view.

It was a point of essential importance to determine the angular velocity of the axle carrying the mirror. No confidence could be placed in the result obtained by calculating the train of wheels, as in such rapid motion many retarding causes might operate and render the calculation uncertain : it was necessary, therefore, to devise a means independent of these sources of error, and which should immediately indicate the ultimate velocity. Nothing appeared more likely to effect this purpose than to attach a small syren to the instrument, the plate of which should be carried round by the axle of the mirror. T is a small hollow box an inch in diameter, into which wind was conveyed through a tube placed to the aperture u.

On the face of this box a number of equidistant apertures were arranged in a circle, and a disc moving before it having the same number of apertures, periodically intercepted the issuing current, and produced a sound corresponding to the frequency of the impulses. It is obvious that the number of revolutions would be ascertained by dividing the number of vibrations in a second, corresponding to the sound, by the number of apertures. I at first employed ten apertures : when the motion was slow, the sound could be easily determined; but on augmenting the velocity it became inappreciable. I then reduced the number of apertures to five, but with no better success, and ultimately to two ; but the sound was then so feeble, compared with the accompanying noises, that it could not be distinctly heard.

The difficulty was at last overcome by employing the arm Q itself to produce the sound. A small slip of paper was held to it ; and as at every revolution a blow was given to the paper, its rapid recurrence gave rise to a sound the pitch of which varied with the velocity of the motion. When the machinery was put in motion with the maximum velocity I employed in my experiments, the sound G$\sharp^4$ was obtained, indicating 800 revolutions of the mirror in a second. I am not aware that anything can have interfered with the accuracy of this result;

the same sound was heard when different pieces of paper or card were used; and on moderating the velocity, the sound descended through all the degrees of the scale below it, until distinct percussions were perceived*.

Let us now consider what is the shortest duration of the electric light, and the greatest velocity of transmission through the wire, that can be detected by means of the instrument I have described. The mirror revolves 800 times in a second; and during this time the image of a stationary point would describe 1600 circles: the elongation of a spark through half a degree, a quantity obviously visible, and equal to one inch seen at the distance of ten feet, would therefore indicate that it exists the 1,152,000dth part of a second. The deviation of half a degree between the two extreme sparks, the wire being, as above stated, half a mile in length, would indicate a velocity of 576,000 miles in a second. This estimated velocity is on the supposition that the electricity passes from one end of the wire to the other: if, however, the two fluids in one theory, or the disturbances of equilibrium in the other, travel simultaneously from the two ends of the wire, the two external sparks will keep their relative positions, the middle one will be alone deflected, and the velocity measured will be only half that in the former case, viz. 288,000 miles in a second.

Repeated experiments gave the following results. In all cases, when the velocity of the mirror exceeded a certain limit, the three sparks were elongated into three parallel lines, and the lengths became greater as the velocity of the motion was increased. The greatest elongation observed was about 24°, indicating a duration of about the 24,000dth of a second. The lines did not always commence at the same places; sometimes they appeared immediately below the eye, sometimes to the right, at other times to the left, and occasionally they were out of view altogether. This indetermination, it has

* Since this paper was read, a registering apparatus has been attached to the instrument; it consists of an index, communicating with the axis by a light train of wheels, and making one complete revolution while the mirror revolves 10,000 times. The number of revolutions of the mirror indicated by this means, did not, in consequence of the increased resistance to the motion, exceed 600 in a second.

already been explained, is owing to the arm not always taking
the spark at the same distance from the discharger : several
discharges are therefore required to be made before the eye
can distinctly observe the appearances.   When the velocity
was low, the terminating points appeared to be exactly in the
same vertical line; but when the velocity was considerable,
and the mirror revolved towards the right, the lines assumed
this appearance, ⸻⸻ ; when it revolved towards
the left, they appeared thus, ⸻⸻.   In no case did
I see them thus, ⸻⸻ , or thus ⸻⸻ , as
required in the hypothesis of the actual transfer of a single
fluid.   I found it convenient to place at the side of and near
the spark-board, the flame of a taper or candle, to serve as a
guide to the eye : the lines of electric light in the mirror
were immediately above and parallel to the constant line
formed by the reflection of this flame, and thus the eye could
be more readily directed to them : it also served to keep the
focal distance of the eye properly adjusted.   The spark-
board was in all the experiments placed at the distance of ten
feet from the mirror.

The deviation between the extreme sparks and the middle
one could not, I am tolerably certain, have exceeded half a
degree.

Having obtained a considerable elongation of these sparks,
I expected also to be able to elongate the sparks or widen
the lines in some of the various arrangements of electric
light described in § 2 ; but even with the extraordinary velocity now attained, no alteration whatever could be observed
in them; they were still reflected as distinct and unaltered
as the objects themselves when directly looked at.   The elongation of the sparks at the interruptions of the wire above
noticed were no doubt owing to this circumstance,—that the
diameter of the wire was not sufficiently great to allow the
charge of the jar to pass through it except in a successive
manner.   The duration of the discharge in the cases of these
sparks appeared to be longer than the time required for the
electricity to pass through many miles of wire.

The sparks from the great magnet constructed by Mr.

Saxton, which is at the Gallery in Adelaide-street, were considerably elongated even when the mirror was moving with a comparatively low velocity.

§ 5.

For the purpose of increasing the chances of observing sparks, &c., when their appearance cannot be commanded at the moment the mirror is in the proper position to reflect them to the eye, I propose to employ a mirror with polygonal faces symmetrically placed with respect to the axis of rotation, a hexagon for instance, fig. 9 (Plate XIII.), where a, b is the moving axis, and c, d, e three of the reflecting surfaces. During one rotation of the axis, if the object be continuously luminous, six luminous arcs will be successively presented to the eye, all occupying the same position; and if the light be transient, we shall have six times the number of chances of observing its reflection than if one reflecting surface only were employed. It is true the arcs are not circular ones, but the difference is scarcely noticeable when the radius of the polygonal section is very small compared with the distance of the luminous object, which would be the case in all our experiments.

I have also proposed various modifications of parts of the instrument, § 4, to suit particular experiments, and to ensure additional accuracy in the repetition of those already made; but not having yet put these to the test of experiment, it would be premature at present to describe them.

§ 6.

The instantaneousness of the light of electricity of high tension, rendered evident by the preceding investigations, affords the means of observing rapidly changing phenomena during a single. instant of their continued action, and of making a variety of experiments relating to the motions of bodies when their successive positions follow each other too quickly to be seen under ordinary circumstances.

A few obvious instances will at present suffice. A rapidly moving wheel, or a revolving disc on which any object is painted, seems perfectly stationary when illuminated by the explosion of a charged jar. Insects on the wing appear, by the same means, fixed in the air. Vibrating strings are seen at rest in their deflected positions. A rapid succession of drops of water, appearing to the eye a continuous stream, is seen to be what it really is, not what it ordinarily appears to be, &c.

§ 7.

The preceding experiments having been directed rather to detect elongations and deviations than to measure them, I am not prepared to state the results with numerical accuracy. I shall endeavour to supply this deficiency in further investigations, but must at present content myself with stating the following general conclusions, deduced from the appearances which I have observed, though, I must allow, more accurately performed experiments are required before they can be considered as fully established. 1st, The velocity of electricity through a copper wire exceeds that of light through the planetary space. 2ndly, The disturbance of electric equilibrium in a wire communicating at its extremities with the two coatings of a charged jar, travels with equal velocity from the two ends of the wire, and occurs latest in the middle of the circuit. 3rdly, The light of electricity in a state of high tension has a less duration than the millionth part of a second. 4thly, The eye is capable of perceiving objects distinctly, which are presented to it during the same small interval of time.

By prosecuting these researches with instruments of higher power, and of greater accuracy in their indications, numerical laws may be established for a large class of phenomena, the relations of which we have had hitherto no means of observing. The relative velocities of electricity in different metallic wires; the modifications in the velocity of electricity in different states of tension when passing through the same conductor, if any such differences exist; the duration of the electric spark under different circumstances of tension and quantity, &c., will be among these objects of investigation.

An account of several new Instruments and Processes for determining the Constants of a Voltaic Circuit.

[THE BAKERIAN LECTURE for 1843.   From the Philosophical Transactions of the Royal Society, vol. 133, pp. 303–327.]

[Plates XV. & XVI.]

§ 1.

I INTEND in the present communication to give an account of various instruments and processes which I have devised and employed during several years past for the purpose of investigating the laws of electric currents.   The practical object to which my attention has been principally directed, and for which these instruments were originally constructed, was to ascertain the most advantageous conditions for the production of electric effects through circuits of great extent, in order to determine the practicability of communicating signals by means of electric currents to more considerable distances than had hitherto been attempted.   In this endeavour, guided by the theory of Ohm and assisted by the instruments I am about to describe, I have completely succeeded.   But the use of the new instruments is not limited to this especial object; they will, I trust, be found of great assistance in all inquiries relating to the laws of electric currents, and to the various and daily increasing practical applications of this wonderful agent.   An energetic source of light, of heat, of chemical action and of mechanical power, we only require to know the conditions under which its various effects may be most economically and energetically manifested, to enable us to determine whether the high expectations formed in many quarters of some of these applications are founded on reasonable hope, or on fallacious conjecture.   The theory we now possess is amply sufficient to direct us rightly in this inquiry, but experiments have not

H

yet been sufficiently multiplied to enable us to obtain, except in a few cases, the numerical values of the constants which enter into various voltaic circuits; and without this knowledge we can arrive at no accurate conclusions.

§ 2.

The instruments and processes I am about to describe being all founded on the principles established by Ohm in his theory of the voltaic circuit, and this beautiful and comprehensive theory being not yet generally understood and admitted, even by many persons engaged in original research, I could scarcely hope to make my descriptions and explanations understood without prefacing them with a short account of the principal results which have been deduced from it. It will soon be perceived how the clear ideas of electromotive forces and resistances, substituted for the vague notions of intensity and quantity which have been so long prevalent, enable us to give satisfactory explanations of most important phenomena, the laws of which have hitherto been involved in obscurity and doubt. Viewing the laws of the electric circuit from the point at which the labours of Ohm have placed us, there is scarcely any branch of experimental science in which so many and such various phenomena are expressed by formulæ of such simplicity and generality; in most of the physical sciences the facts of observation and experiment have kept pace with theoretical generalization, in this science alone they had gone on accumulating in prolific abundance without any successful attempt having been made to reduce them to mathematical expression. But this is now happily effected, and what has hitherto been mere matter of speculative conjecture is removed into the domain of positive philosophy.

By *electro-motive force* is meant the cause which in a closed circuit originates an electric current, or in an unclosed one gives rise to an electroscopic tension. By *resistance* is signified the obstacle opposed to the passage of the electric current by the bodies through which it has to pass : it is the inverse of what is usually called their conducting power.

When the activity of any portion of the circuit is increased or diminished, either by a change in the electro-motive force or in the resistance of that portion, the activity of all the other parts of the circuit increases or decreases in a corresponding degree, so that the same quantity of electricity always passes in the same instant of time through every transverse section of the circuit.

The force of the current is directly proportional to the sum of the electro-motive forces which are active in the circuit, and inversely proportional to the total resistance of all its parts, or in other words the force of the current is equal to the sum of the electro-motive forces divided by the sum of the resistances.

Let F denote the force of the current, E the electro-motive forces, and R the resistances : then

$$F = \frac{E}{R}.$$

The length of a copper wire of a given thickness, the resistance of which is equivalent to the sum of the resistances in a circuit, Ohm calls its *reduced length*, an expression which it will frequently be found convenient to employ.

If the electro-motive forces and resistances in a circuit are proportionately increased or diminished the force of the current remains the same, or $\frac{E}{R} = \frac{nE}{nR}$. Hence a single voltaic element, or a battery consisting of any number of exactly similar elements, if no additional resistance be interposed in the circuit, produces the same effect. Also a thermo-electric element and a voltaic element will produce the same effect when the greatly inferior electro-motive force of the former is compensated by a corresponding decrease in its resistance ; in a thermo-electric arrangement the resistance is in general small, because the circuit is entirely metallic, while in a voltaic element the resistance of the liquid is always considerable.

Any interposed resistance weakens the force of the current, but less so as it is smaller in proportion to the other resistances in the circuit. Hence in two circuits, both producing currents of equal force, when the same resistance is introduced, the strength of the two currents may be weakened in very different proportions. A single voltaic element, $\frac{E}{R}$, and a series consisting of any number of such elements, $\frac{nE}{nR}$, form circuits in which the currents have the same force, but very different results will be obtained according as the added resistance is great or small compared with the original resistances in the circuits; if it be small, the effects of the two circuits will remain sensibly the same; but if it be large, the resistance that weakens to a very great extent the current in the circuit of the single element produces but a trifling diminution in that of the series. This explains the necessity of employing a series to overcome considerable resistances. The same remarks will apply to the comparison of a thermo-electric with a voltaic circuit.

The following is the general formula for the force of the current in a voltaic circuit when completed by a connecting wire: the metallic plates of the voltaic elements being parallel to each other and of equal size:

$$F = \frac{nE}{\frac{nRD}{S} + \frac{rl}{s}}.$$

F is the force of the current, E the electro-motive force of a single element, n the number of elements, R the specific resistance of the liquid, D the thickness of the liquid stratum or distance of the plates, S the section of the plates in contact with the liquid, r the specific resistance of the connecting wire, l its length, s its section.

Expressed in words we have the following laws :—

The electro-motive force of a voltaic circuit varies with the number of the elements, and the nature of the metals and liquids which constitute each element, but is in no degree dependent on the dimensions of any of their parts.

The resistance of each element is directly proportional to

the distance of the plates from each other in the liquid, and to the specific resistance of the liquid, and is also inversely proportional to the surface of the plates in contact with the liquid.

The resistance of the connecting wire of the circuit is directly proportional to its length and to its specific resistance, and inversely proportional to its section.

The limits of this communication will not allow me to dwell longer on the consequences of Ohm's theory of the electric circuit; for further developments I must refer to the author's work, 'Die Galvanische Kette mathematisch bearbeitet,' Berlin 1827, a translation of which has appeared in Taylor's Scientific Memoirs, vol. ii.; to his various other memoirs published in Schweigger's 'Jahrbuch der Physik;' and to the more recent applications of the theory made by Fechner, Lenz, Jacobi, Poggendorff, Pouillet, &c.

There is, however, one class of considerations which it is indispensable I should bring forward, because upon it are founded many of the instruments and processes which I shall have occasion hereafter to mention,—I allude to the laws of the distribution of the electric current in the various parts of a circuit, when a branch conductor is placed to divert a portion of the current from a limited extent thereof.

Let λ be the reduced length of the portion of the circuit from which the current is partially diverted, λ' that of the wire which diverts the current, and L that of the undivided part of the circuit. The force of the current in each of the adjacent conductors, λ and λ', can be shown to be in the inverse ratio of their reduced lengths, and the reduced length of a single wire, which substituted for both would not alter the force of the current, to be $\dfrac{\lambda\,\lambda'}{\lambda+\lambda'}$, which we will designate by Λ.

The force of the current in the original circuit before the introduction of the branch wire will then be expressed thus:

$$F = \frac{E}{L+\lambda},$$

and the strength of the current in the three different portions of the altered circuit by the following expressions :—

In the principal or undivided portion L,

$$F_1 = \frac{E}{L + \Lambda} = \frac{E(\lambda + \lambda')}{L(\lambda + \lambda') + \lambda \lambda'}.$$

In the portion from which the current has been partially diverted, or λ,

$$F_2 = \frac{E}{L + \Lambda} \cdot \frac{\Lambda}{\lambda} = \frac{E \lambda'}{L(\lambda + \lambda') + \lambda \lambda'}.$$

In the portion which partially diverts the current, or λ',

$$F_3 = \frac{E}{L + \Lambda} \cdot \frac{\Lambda}{\lambda'} = \frac{E \lambda}{L(\lambda + \lambda') + \lambda \lambda'}.$$

§ 3.

It is seldom that any real advance is made in a scientific theory without a corresponding change in its terminology being required. Now that it is proved beyond doubt that the various sources of continued electric action differ from each other only in the amount of their electro-motive forces, modified by the resistance of the circuit of which they form part, it becomes of importance, in order to give precision to our statements and to avoid circumlocutions otherwise inevitable, to adopt general terms to express the source of a current without reference to the peculiar mode of its production ; I shall therefore employ the word *Rheomotor* to denote any apparatus which originates an electric current, whether it be a voltaic element or a voltaic battery, a thermo-electric element or a thermo-electric battery, or any other source whatever of an electric current; when speaking of a single element I shall term it a rheomotive element, and what is usually called a voltaic or thermo-electric-pile or battery I shall term a rheomotive series. I shall still use the ordinary expressions when I have to refer to the specific sources of the production of electric currents, but when I employ the general terms they must be understood to apply to all these sources indifferently.

The want of a general term to designate an instrument to measure the force of an electric current without reference to its particular construction has been long felt. I shall use the word *Rheometer* for this purpose, continuing occasionally to employ galvanometer, voltameter, &c. to distinguish the particular instruments to which these names have been applied, though perhaps the terms Magnetic, Chemical, Calorific, &c. Rheometer would be more appropriate.

This may be the proper place to explain a few other terms which I have frequent occasion to use, though not in the course of the present communication. By *Rheotome* is meant an instrument which periodically interrupts a current, and by *Rheotrope* an instrument which alternately inverts it. A *Rheoscope* is an instrument for ascertaining merely the existence of an electric current. The word *Rheostat* will be hereafter explained.

I have not introduced these terms, which will be found greatly convenient and will enable us to state general propositions much more clearly, without good authority. The word Reophore was employed by Ampère to designate the connecting wire of a voltaic apparatus, as being the carrier or transmitter of the current; and the word Rheometer, first proposed by Peclet as a synonym for galvanometer, has been generally adopted by the French writers on physics.

§ 4.

The method of obtaining the constants of a rheophoric circuit adopted by Fechner, Lenz, Pouillet, &c., in their experimental verifications of Ohm's theory, is essentially the following :—

The resistance of a circuit is determined by observing the force of the current, first without any extra interposed resistance in the circuit, and afterwards when a known resistance is added. Then

$$F = \frac{E}{R}, \text{ and } F' = \frac{E}{R+r} \therefore \frac{F}{F'} = \frac{R+r}{R},$$

from which equation the value of R all the others being

known quantities, is easily deduced: $R = \frac{F'}{F - F'} r$. The electro-motive force of a circuit is ascertained by multiplying the force of the current into the total resistance; for since $F = \frac{E}{R} \therefore E = FR$.

The principle of this method is extremely simple, but the difficulty of determining immediately the force of a current by means of a galvanometer is an obstacle to its general employment. Fechner * measured the force of the current by the number of oscillations of the needle when placed at right angles to the coils, a very tedious operation; and others have employed the deviations of the needle, the corresponding degrees of force having been previously determined by some peculiar process, or inferred from some rule depending on the particular construction of the instrument. Another impediment to the use of a galvanometer to measure the force of a current arises from the changes in the magnetic intensity of the needle, which frequently occur, especially when it has been acted upon by too strong a current.

The principle of my method is that of employing variable instead of constant resistances, bringing thereby the currents in the circuits compared to equality, and inferring from the amount of the resistance measured out between two deviations of the needle, the electro-motive forces and resistances of the circuit according to the particular conditions of the experiment. This method requires no knowledge of the forces corresponding to different deviations of the needle.

To apply this principle it is requisite to have a means of varying the interposed resistance so that it may be gradually changed within any required limits. I have contrived two instruments for effecting this purpose, one intended for circuits in which the resistance is considerable, the other for circuits where the resistance is small †.

---

* Massbestimmungen über die Galvanische Kette. Leipzig, 1831, p. 5.

† It appears that the idea of constructing an instrument of this kind had also occurred to Professor Jacobi of St. Petersburgh. When I explained to this eminent experimentalist my instruments and processes in

§ 5.

The first instrument is represented in Plate XV. fig. 1.
A. *g* is a cylinder of wood, and *h* is a cylinder of brass, both
of the same diameter, and having their axes parallel to each
other.   On the wood cylinder, a spiral groove is cut, and at

---

the beginning of August 1840, he informed me that he had himself constructed a similar instrument, which he had exhibited to the Academy of
Sciences at St. Petersburgh, though no description of it had yet been
published, and he at the same time showed me a drawing of it.   This
instrument, which he has since called an Agometer, differs in mechanical
construction from either of mine, and is less convenient to manipulate;
but its principle is the same.   In a communication which Professor
Jacobi made in the following month to the Meeting of the British Association at Glasgow, and which was published in No. 678 of the 'Athenæum,' 1840, he thus alludes to the subject:—

" Before proceeding, I may be permitted to make some remarks concerning an instrument which I laid before the Academy of Sciences in
the commencement of this year.   It is destined to regulate the galvanic
current, and is of value in many investigations of this kind.   During my
sojourn in London, Professor Wheatstone has shown me an instrument,
founded exactly on the same principles as mine, and with very insigni-
ficant modifications and differences.   Now, it is quite impossible that he
should have had the least notice of my instrument; but as it is probable
that its use may be greatly extended, I must add, that while I have only
used this instrument for regulating the force of the currents, he has
founded upon it a new method of measuring these currents and of determining the different elements or constants which enter into the analytical
expressions, and on which depends the action of any galvanic combination.   It is principally to the measure of the electro-motive force, by
these means, that Mr. Wheatstone has directed his attention; and he has
shown me, in his unpublished papers, very valuable results which he has
obtained by this method."

Professor Jacobi has since his return employed my method of determining the constants of a voltaic circuit.   The memoirs in which his
results are given were published in Poggendorff's 'Annalen der Physik,'
vol. liv. No. 2. for 1841, and vol. lxii. No. 9. for 1842.   To the latter the
learned editor, who has made most valuable researches himself in the
same path, has appended (p. 89) the following note:—"I will take this
opportunity to call to mind that I applied the same method (or at least
one identical to it in principle) before it was communicated to the author
by Mr. Wheatstone.   See the Annals, vol. lii. p. 526." I have referred
to this volume and find it was published in the latter part of 1841, while

one of its extremities a brass ring is fixed, to which is attached one of the ends of a long wire of very small diameter, which when coiled round the wood cylinder fills the entire groove, and is fixed at its other end to the remote extremity of the brass cylinder. Two springs, j and k, pressing one against the brass ring on the wood cylinder, and the other against the extremity of the brass cylinder h, are connected with two binding screws for the purpose of receiving the wires of the circuit. The moveable handle m is for turning the cylinders on their axes. When it is placed on the cylinder h and is turned to the right, the wire is uncoiled from the wood cylinder and coiled on the brass cylinder, but when it is applied to the cylinder g and is turned to the left, the reverse is effected. The coils on the wood cylinder being insulated and kept separate from each other by the groove, the current passes through the entire length of wire coiled upon that cylinder, but the coils on the brass cylinder not being insulated the current passes immediately from the point of the wire which is in contact with the cylinder to the spring k. The effective part of the length of the wire is therefore the variable portion which is on the wood cylinder.

In the instrument I usually employ the cylinders are six inches in length and $1\frac{1}{2}$ inch diameter, the threads of the screw are forty to the inch, and the wire is of brass the $\frac{1}{100}$th of an inch in diameter. I employ a very thin wire and a badly conducting metal, in order that I may introduce a greater resistance into the circuit.

A scale is placed to measure the number of coils unwound; and the fractions of a coil are determined by an index which is fixed to the axis of one of the cylinders and points to the divisions of a graduated circle.

As the principal use of this instrument is to adjust or

---

my communication to Professor Jacobi was, as above stated, made in August 1840. I may also mention, that the experimental process employed by Professor Poggendorff had no resemblance whatever to mine, and the result he sought was likewise different; the mathematical principle of the method was however in the single case he investigated undoubtedly the same.

regulate the circuit so that any constant degree of force may be obtained, I have called it a *Rheostat*.

Plate XV. fig. 1 shows the arrangement of the circuit when prepared for an experiment. B is a delicate galvanometer with an astatic needle furnished with a microscope for reading off the divisions of the circle, which greatly facilitates the observations. C is the rheomotor.

I must here digress for a moment to describe the voltaic element which I have employed in most of my rheometric researches, and which I have found to be very constant in its action, and convenient to manipulate with. It is quite unnecessary to use large elements in such investigations, for when considerable resistances are introduced in the circuits, which is most frequently the case, they produce no perceptibly greater effect than smaller ones, and in all cases the measures may be as accurately determined by employing small elements as large ones.

The voltaic element C consists of a glazed porcelain cell, two inches square and one inch and a half high, in the centre of which is placed a small porous cylinder of earthenware or wood, filled with a liquid amalgam of zinc, the space between the two cells being charged with a solution of sulphate of copper; a slip of thin sheet copper bent round, and having one of its edges cut and turned over so that the wire of the circuit may be attached to it, or that it may dip into the amalgam of another similar cell, is placed in the solution. Fig. 3, Plate XVI. represents several such elements combined to form a series. It will be seen that, in principle, this is but a slight modification of Professor Daniell's constant battery, liquid amalgam of zinc being employed, as in Mr. Kemp's first experiment, instead of amalgamated zinc bars or plates, and the acid solution being dispensed with. This arrangement is, besides being very constant in its action, extremely economical and easy to manipulate. Any negative metal may be substituted for copper provided a solution of a salt of that metal be employed as the interposed liquid.

§ 6.

The rheostat which I employ for circuits in which the resistance is comparatively small is represented at Plate XV. fig. 2. A. *a* is a cylinder of well-seasoned wood, on the surface of which a spiral groove is cut; a thick copper wire is wound round the cylinder occupying the groove, forming as it were the thread of a screw. Immediately above the cylinder and parallel with its axis is a triangular metal bar *b*, carrying a rider or slide *c*; to this rider a spring *d* is attached, which constantly presses against the spiral wire, yielding to any slight inequality. One end of the spiral wire is attached to a brass ring *e*, against which a spring *f* presses, which is connected by means of a binding screw to one end of the circuit, the other end of the circuit is held by the binding screw which is in metallic connection with the triangular metal bar. On turning the handle *h* the cylinder is caused to move on its axis in either direction, and the rider *c* guided by the wire moves along the bar, advancing or receding according as the cylinder is moved to the right or to the left; the rider coming in contact with a different point of the spiral wire, a different resistance is introduced into the circuit, consisting of that portion of the wire only which is included between the rider and the end of the wire connected with the spring *f*. The cylinder of the instrument I have constructed is $10\frac{1}{2}$ inches in length, and $3\frac{1}{4}$ inches in diameter; the wire is of copper the 16th of an inch thick, and it makes 108 coils round the cylinder. The dimensions of the instrument, and the thickness, length, and material of the wire, may be varied according to the limits of the variable resistance required to be introduced into the circuit, and the degree of accuracy with which these changes are required to be measured.

Fig. 2 (Plate XV.) represents the arrangement of a thermo-electric circuit in which this instrument is interposed. C is the thermo-electric element: B the galvanometer, which in this case must not have numerous coils of fine wire as in the preceding arrangement, for this would introduce too great a resistance into the circuit, but must consist of a single thick plate or wire making a single convolution; or, which I think is preferable, the method of diverting a portion of the current from the wire of a delicate galvanometer described in § 15. may be adopted. Any rheometer in which the resistance is small may be employed in conjunction with this form of the rheostat, instead of a thermo-electric element, as represented.

The rheostat, especially under the form last described, may be usefully employed as a regulator of a voltaic current in order to maintain for any required length of time precisely the same degree of force, or to change it in any desired proportion. Interposed in the circuit of an electro-magnetic engine, however the rheomotor may vary in its energy, the same velocity may be constantly restored by turning the cylinder of the regulator to the left or to the right, according as the velocity increases or decreases; or any different velocity, within given limits, may be obtained by adjusting the rheostat accordingly. Since the consumption of materials in a voltaic battery in which there is no local action decreases in the same proportion as the increase of the resistance in the circuit, this method of altering the velocity has an advantage which no other possesses, the effective force is always strictly proportional to the quantity of materials consumed in producing the power, a point which, if further improvements should ever render the electro-magnetic engine an available source of mechanical power, will be of considerable importance.

In volta-typing operations the advantage of using the rheostat is obvious. By varying the rheostat from time to time so as to keep the needle of a galvanometer to the same point, a current of any required degree of energy may be maintained, without any notable increase or diminution, for any length of time; and, as the nature of the deposit, when the solution from which it is made remains the same, varies only with the force of the current and the magnitude of the surface on which the metal is reduced, when once a good effect has been obtained the same circumstances may be reproduced with ease and certainty, and the effects of chance entirely eliminated.

In the operations of voltatyping, electro-gilding, &c., and in the production of Nobili's colours, the advantage of using the rheostat is obvious.

This however is not the place to dilate on this subject.

§ 7. *Standard of Resistance.*

It is of the highest importance to have a correct standard of resistance, and one that can easily be reproduced for the purpose of comparison. A copper wire of a given length and diameter might be employed, but as very small differences of diameter are attended with considerable differences in the resistances of wires, it is more convenient to assume for the unit of resistance a wire of a given length and weight, which allows small differences to be very accurately determined. I shall in all my experiments, therefore, take for the unit of resistance a copper wire one foot in length, and weighing 100 grains. The diameter of this wire is the ·071 of an inch, and it is intermediate to the numbers designated in commerce as fifteen and sixteen.

§ 8. *The Resistance Coils.*

It is frequently required to measure resistances much greater than can be effected by means of the rheostat, though the reduced length of its wire is considerable. I may wish to know, for instance, the resistance of the wire of the electro-magnets of my telegraphic apparatus, which is sometimes many hundred yards in length; or that afforded by an extensive telegraphic line, or the resistance of a certain extent of an imperfectly conducting liquid. In all these cases and a variety of others I employ another instrument, which enables me to interpose in the circuit resistances to any amount, and yet to obtain by the conjoined use of the rheostat, which serves as its fine adjustment, any required degree of accuracy. This instrument is represented at Plate XV. fig. 1. D; it consists of six coils of fine silk-covered copper wire, about the $\frac{1}{300}$th of an inch in diameter; two of these coils are fifty feet in

length, the others are respectively 100, 200, 400, and 800 feet
in length. The two ends of each coil are attached to short
thick wires fixed to the upper faces of the cylinders, which
serve to combine all the coils into one continued length; the
two wires a, b form the extremities of the coils by which they
are united to the circuit. On the upper face of each cylinder
is a double brass spring moveable round a centre, so that its
ends may rest at pleasure either on the ends of the thick
connecting wires, or may be removed from them and rest
only on the wood. In the latter position, the current of the
circuit must pass through the coil, but in the former position, the current passes through the spring, and removes the
entire resistance of the coil from the circuit. When all the
springs rest on the wires, the resistance of the whole series
of coils is removed, but by turning the springs so as to introduce different coils into the circuit, any multiple of 50 feet
up to 1600 may be brought into it.

As the measurement of these long lengths of wire cannot
be accurately depended upon, it is advisable to ascertain the
number of units of resistance in each coil, which, with the
aid of the rheostat, may be easily effected. I find the resistance of the entire 1600 feet to be equivalent to 218,880
units of resistance, or feet of the standard wire. I occasionally employ an auxiliary series of coils combined in the same
way as the preceding, consisting of six coils of the same wire,
each 500 yards in length. The reduced length of this series
is above 233 miles of the standard wire. By combining it
with the preceding, I am able to measure resistances equal
to $274\frac{1}{2}$ miles.

§ 9.

When a perfectly constant element, a galvanometer and a
rheostat are placed in a circuit as in fig. 1 (Plate XV.), the resistance of any interposed body may be ascertained in the following way. Observe the point at which the needle stands;
then remove the body, the resistance of which is to be measured, from the circuit, and, by means of the rheostat, add a
sufficient length of wire to bring the needle again to the same

point. The number of standard units corresponding to this added length will be the measure.

It is a point of importance to determine the resistance of the wire of the galvanometer employed in the experiments; to ascertain this by the above method an auxiliary galvanometer would be required, but when a second galvanometer is not at hand, recourse may be had to the following process. Take two rheomotive elements exactly equal both in electromotive force and resistance; place one of them in the circuit fig. 1 (Plate XV.), and observe accurately the deviation of the needle; then interpose also the other element and bring the needle again to the same point by means of the rheostat. The equivalent of the wire uncoiled λ, will be the measure of the resistance of the galvanometer wire g plus that of the connecting wires r. Subtracting r from λ, the resistance of g will be determined,

$$\frac{E}{R+r+g} = \frac{2E}{2R+r+g+\lambda}, \therefore g = \lambda - r.$$

The resistance of a galvanometer wire or any other interposed resistance may be still more accurately ascertained by means of the instruments described in § 16.

§ 10. *Process to ascertain the Sum of the Electro-motive Forces in a Voltaic Circuit.*

The rheostat affords a most ready means of ascertaining the sum of the electro-motive forces active in a voltaic circuit, without requiring for this purpose the aid of a rheometer graduated to indicate proportional forces, or having recourse to the tedious process of counting the oscillations of a needle, employed by Fechner in his investigations. To save time and trouble in this operation will be of great importance to the future progress of electro-chemistry, on account of the great number of experiments of this kind which yet remain to be made, and also from the fluctuations in the electro-motive forces of many circuits from chemical and other actions, which render observations requiring considerable time to make completely valueless.

The principle of my process is as follows :—In two circuits, producing equal rheometric effects, the sum of the electromotive forces divided by the sum of the resistances is a constant quantity, i. e. $\frac{E}{R}=\frac{nE}{nR}$; if E and R be proportionately increased or diminished, F will obviously remain unchanged. Knowing therefore the proportion of the resistances in two circuits producing the same effect, we are able immediately to infer that of the electro-motive forces. But as it is difficult in many cases to determine the total resistance, consisting of the partial resistances of the rheomotor itself, the galvanometer, the rheostat, &c., I have recourse to the following simple process. Increasing the resistance of the first circuit by a known quantity r, the expression becomes $\frac{E}{R+r}$; in order that the effect in the second circuit shall be rendered equal to this, it is evident that the added resistance must be multiplied by the same factor as that by which the electro-motive forces and original resistances are multiplied, for $\frac{E}{R+r}=\frac{nE}{nR+nr}$. The relations of the lengths of the added resistances r and $n\,r$, which are known immediately, give therefore those of the electro-motive forces.

Experimentally I proceed thus :—I interpose the rheostat and the galvanometer in the circuit, and then add, by means of the former, assisted if necessary by the resistance coils, a sufficient resistance to bring the needle exactly to 45°; I then ascertain the length of wire uncoiled from the brass cylinder of the regulator necessary to reduce the deviation of the needle to 40°. The number of turns is the measure of the electro-motive force, the number corresponding to that of a standard element having been previously determined.

§ 11.

I subjoin a few measures of electro-motive forces obtained by the preceding process.

1. Three elements of different sizes, consisting of copper, a solution of sulphate of copper, and a liquid amalgam of

I

zinc, were successively placed in the circuit. The number of turns of the rheostat requisite to reduce the needle from 45° to 40° were :—

Small element described in § 5 .... 30 turns.
Copper cylinder 3½ inches high and 2½ inches diameter .. 30 turns.
Copper cylinder 6 inches high and 3½ inches in diameter . 30 turns.

Hence, conformably to the theory, the magnitude of an element occasions no difference in its electro-motive force.

2. Five small elements of copper and amalgam of zinc were charged respectively with the following five solutions of copper—the sulphate, the ammonia sulphate, the acetate, the per-muriate, and the nitrate. Though the force of the current produced by each element separately was very different, owing to the different conductibility of the solutions, yet, with the exception of the nitrate, all required the same number of turns, indicating equal electro-motive forces ; the latter fluctuated between 23 and 29, owing to some disturbing action probably of the nitric acid on the mercury of the amalgam.

3. The electro-motive forces of a circuit in which 1, 2, 3, 4, 5 similar elements were successively placed, were measured.

1 element required .... 30 turns.
2 elements .... 61 turns.
3 elements .... 91 turns.
4 elements .... 120 turns.
5 elements .... 150 turns.

The electro-motive force of a circuit is therefore, as theory indicates, proportional to the number of similar elements of which it is formed, arranged in series.

4. The next experiments were made to determine the amount of the contrary electro-motive force which is introduced into a circuit when a voltameter or decomposing cell is interposed. The liquid in contact with the platinum electrodes was dilute sulphuric acid. The measure of this contrary electro-motive force is obtained by subtracting the actual number of turns from the number corresponding with the electro-motive force of the circuit when the decomposing cell is removed from it.

				Contrary electro- motive force.
3 elements with decomposing cell	21 turns	90 —	21 = 69
4 elements with decomposing cell	50 turns	120 —	50 = 70
5 elements with decomposing cell	79 turns	150 —	79 = 71
6 elements with decomposing cell	109 turns	180 —	109 = 70

Mean 70

The contrary electro-motive force may be considered there-fore in this case to be constant, and to be to that of a single standard element as 7 : 3. It is hence obvious why three such elements are necessary to decompose water in a cell with platinum electrodes of a certain size, and charged with dilute sulphuric acid. The amount of this contrary force varies with different liquids, and according to the nature of the electrodes employed : as it is not my present object to investigate this subject, but merely to give a few examples of the measures which may be obtained by the above-mentioned method, I shall not enter on the consideration of these interesting but intricate modifications.

5. The highest electro-motive force which a voltaic element consisting of two metals and one interposed liquid can manifest, is when the liquid is a solution of a salt of the negative metal, so that by the continual deposition of this metal the negative surface is kept free from the contact of heterogeneous substances which would tend to give rise to a reverse current. When, in consequence of the chemical action, any heterogeneous solid matter is deposited on, or any evolved gas adheres to, the negative surface, the electro-motive force of the element is reduced. The following measures will show the reduction in electro-motive force of a zinc and copper, and of a zinc and platinum element, by substituting dilute sulphuric acid for the metallic salt; the changes in these cases are effected by the adhesion of hydrogen to the surface of the negative metal.

Amalgam of zinc	.	Sulphate of copper.	.	Copper	.	30 urns.
Amalgam of zinc	.	Dilute sulphuric acid	.	Copper	.	20 turns.
Amalgam of zinc	.	Chloride of platinum	.	Platinum.		40 turns.
Amalgam of zinc	.	Dilute sulphuric acid .		Platinum.		27 turns.

I 2

6. The proportion of zinc in the liquid amalgam does no appear to effect the electro-motive force of the voltaic element of which it forms part; the number of turns of the rheostat remains the same although the quantity of zinc may vary very considerably. I was therefore led to think that tolerably accurate measures might be made of the comparative electro-motive forces of the alkaline and earthy bases. An element was formed of liquid amalgam of potassium, sulphate of zinc, and zinc; the potassium was in proportion to the mercury less than 2 per cent.: there was no apparent local action, and the current was remarkably constant and continuous.

The following were ascertained to be the electro-motive forces of different elements in which the positive metal was amalgam of potassium, and the negative metals respectively were zinc, copper, and platinum.

Amalgam of potassium .	Sulphate of zinc	. Zinc	. 29 turns.
Amalgam of potassium .	Sulphate of copper	. Copper	. 59 turns.
Amalgam of potassium .	Chloride of platinum .	Platinum .	69 turns.

The electro-motive force of the first combination nearly corresponds with that of zinc and copper, and when the resistance in the circuit is equivalent, produces a current having nearly the same degree of force. The third combination is one of great electro-motive energy, and when a voltameter with small electrodes is interposed in the circuit, decomposes the water in it abundantly.

It would not be difficult to submit to experiments of this kind all the alkaline and earthy bases; as the proportion in the amalgam does not seem to be of importance, they might be easily prepared by means of a voltaic battery. It would be interesting to know what rank the hypothetical base of ammonia would hold in this scale of electro-motive forces.

7. A still higher electro-motive force may be obtained by employing, in conjunction with the amalgam of potassium, a platinum plate covered with a film of peroxide of lead*

---

* A rheomotive series of ten such elements will have an electro-motive force equal to thirty-three elements of Daniell's battery, or fifty of Wollaston's arrangement in good action. Voltaic combinations, in

Such a plate is easily prepared by making it the positive electrode in a decomposing cell, charged with a solution of acetate of lead. The films thus formed exhibit, as Nobili has shown, according to their thickness, the colours of Newton's rings.

Amalgam of zinc　.　.　Dilute sulphuric acid　Peroxide of lead　68 turns.
Amalgam of potassium　Dilute sulphuric acid　Peroxide of lead　98 turns.

The following measures were obtained when peroxide of manganese was substituted for the peroxide of lead. The peroxide of manganese was deposited on a platinum plate which formed the positive electrode of a decomposing cell containing a solution of chloride of manganese.

Amalgam of zinc　.　.　Diluted sulphuric acid　Peroxide of manganese　54 turns.
Amalgam of potassium　Diluted sulphuric acid　Peroxide of manganese　84 turns.

A weak current is produced by employing a clean platinum plate in conjunction with one covered with the peroxide, in which combination the former acts the part of zinc. In this case the positive metal undergoes no chemical action, but on the negative side the peroxide is reduced by the evolved hydrogen.

8. The following measures conclusively show, that if three metals be taken in their electro-motive order, the electro-motive force of a voltaic element, formed of the two extreme metals, is equivalent to the sum of the electro-motive forces of the two elements formed of the adjacent metals.

1.

Amalgam of potassium　Sulphate of zinc　Amalgam of zinc　29 turns.
Amalgam of zinc　.　.　Sulphate of copper　Copper .　.　.　. 30 turns.
Amalgam of potassium　Sulphate of copper　Copper .　.　.　. 59 turns.

2.

Amalgam of potassium　Sulphate of zinc .　Amalgam of zinc　29 turns.
Amalgam of zinc　.　.　Chloride of platinum　Platinum .　.　. 40 turns.
Amalgam of potassium　Chloride of platinum　Platinum .　.　. 69 turns.

which peroxide of lead is substituted for the negative metal, have been experimented with by Professors Schönbein (Phil. Mag., 3rd Series, vol. xii. p. 255, March 1838) and De La Rive (Archives de l'Electricité, No. 7, April 1843).

9. I wished to compare the electro-motive force of a thermo-electric element, the two metals of which were bismuth and copper, and whose opposite joints were exposed to the fixed temperatures of 32° and 212°, with that of a standard voltaic element. As the interposition of the galvanometer greatly reduced the force of the current in the thermo-electric circuit, so that I could not advance the needle to 45°, I employed, instead, the reduction of the needle from 10° to 5°. The ratios of the measures of the electro-motive forces remain the same between whatever two points the needle is made to vary, provided they do not change during the same series of experiments.

Thermo-electric element of bismuth and copper, the temperatures of the joints being 32° and 212°	8 turns.
Standard voltaic element of amalgam of zinc, sulphate of copper, and copper	757 turns.

The relative electro-motive forces are therefore as 1 : 94·6 *

§ 12.

The resistance or reduced length of a rheomotor may be ascertained by either of the following processes :—

First Method.—Place the galvanometer and the rheostat in the circuit, and adjust the latter until the needle of the galvanometer stands at a determined point. Then divide the current which passes through the wire of the galvanometer, by placing an equal resistance by its side : the needle will recede. The reduced length, measured by the number of turns of the rheostat, required to be taken out of the circuit in order to make the needle stand at its former point, will be equal to half the total resistance of the undivided portion of the original circuit. The resistance of the galvanometer and connecting wires, and of the coils of the rheostat in the circuit before the experiment, having previously been determined, that of the rheomotor is easily obtained by subtracting the former from the total resistance measured.

* Pouillet, by a very different process, has ascertained this proportion to be as 1 : 95. See Elémens de Physique Expérimentale, 3ième ed. tom. i. p. 631.

Let E be the electro-motive force, g the resistance of the galvanometer wire, and R all the other resistances in the circuit. The force of the current acting upon the needle will be $F = \dfrac{E}{R + g}$; adding by the side of the galvanometer wire another wire having the same resistance, is equivalent to substituting for it a wire of double section, and the expression for the resistance of the circuit becomes $R + \dfrac{g}{2}$; but since, in consequence of the division of the current, only one-half its force acts upon the needle, this action may be represented by $\dfrac{\frac{1}{2}E}{R + \frac{1}{2}g}$. To render this expression equivalent to the first, the resistance R must be reduced one-half, for $\dfrac{E}{R + g} = \dfrac{\frac{1}{2}E}{\frac{1}{2}R + \frac{1}{2}g}$; the resistance taken out of the circuit to effect this reduction is obviously equal to half the resistance of the undivided portion of the original circuit,

or $\qquad \dfrac{E}{R + g} = \dfrac{\frac{1}{2}E}{R + \frac{1}{2}g - \lambda} \qquad \therefore \lambda = \dfrac{R}{2}.$

Second Method.—Bring the needle of the galvanometer, by means of the rheostat, to a determined point which we will call b. Ascertain the resistance r requisite to reduce the needle to a lower point a. Restore it to b; then place a wire to divide the current with the galvanometer, and alter this wire until the needle again stands at a. When the needle stands at b, $F = \dfrac{E}{R + g}$; when it stands at a in the first case $F' = \dfrac{E}{R + g + r}$, in the second case $F' = \dfrac{E\,r'}{R(g + r') + g\,r'}$

Equating these two expressions,

$$\frac{E}{R + g + r} = \frac{E\,r'}{R(g + r') + g\,r'} \qquad \therefore R = \frac{r\,r'}{g}.$$

and as these factors are known, R may be readily determined. The resistance of the rheomotor may be obtained from this as before.

If $r' = g$, that is if the resistance of the galvanometer wire be equal to that of the wire which diverts a portion of the current from it, then $R = r$.

Third Method.—Bring the needle to any determined point, and ascertain by means of the instrument described at § 18, what degree corresponds to one half the intensity thus indicated. Since, when the electro-motive force remains the same, the force of the current is simply inversely as the total resistance, to reduce the needle from a to $\frac{a}{2}$ a resistance exactly equal to that previously existing in the circuit must be added ; therefore the number of turns of the rheostat required to reduce the needle from a to $\frac{a}{2}$ will be the measure of the total resistance of the circuit when the needle stood at a. The total resistance being thus measured, that of the rheomotor is determined by subtracting from it the other known resistances, including that of the galvanometer.

More generally, if the forces of two currents, a and b, corresponding to two stationary positions of the needle, are known (§ 19.), the total resistance of the circuit will be $R = \frac{b\,r}{a-b}$, r being the resistance added to reduce the current from a to b. If $a = 2\,b$, then $R = r$ as before.

Fourth Method.—For this and the following process two exactly equal rheomotors must be employed ; their equality may be tested by successively interposing them in the same circuit, when one and the other should deflect the needle of the galvanometer precisely to the same degree.

Place one rheomotor in the circuit and adjust the rheostat until the needle points to any degree arbitrarily fixed upon ; then add the second element by the side of the first, and increase the reduced length of the circuit by turning the rheostat until the needle again points to the same division. The known quantity, measured by the number of turns of the rheostat, by which the reduced length of the circuit is increased, is equal to one half the resistance of a single rheomotor. By placing the second rheomotor by the side of the first, the resistance of that portion of the circuit is reduced one half; therefore, to restore the former condition of the circuit, a resistance equal to one half that of the rheomotor must be added.   For

$$\frac{E}{R+r}=\frac{E}{\frac{R}{2}+r+\lambda} \qquad \therefore \lambda=\frac{R}{2},$$

R being the resistance of the rheomotor, and r the other resistances in the first circuit.

Fifth Method.—Place both the rheomotors in series, and vary the resistance until the needle stands at any determined degree.   Then place them side by side, and increase the resistance, by turning the rheostat until the needle again stands as before.   The resistance of a single rheomotor is equal to twice the resistance required to be added, plus all the resistances in the first circuit except that of the rheomotor,

$$\frac{2\,E}{2R+r}=\frac{E}{\frac{R}{2}+r+\lambda} \qquad \therefore R=r+2\,\lambda,$$

R being the resistance of the rheomotor, r the other resistances in the first circuit, and λ the resistance added by the rheostat to make the force of the current in the second circuit equal to that in the first.

The resistance of one of the elements of the battery described in § 5, I have found to be equal to 2128 standard units.

§ 13.

The resistance of a standard rheomotor having been accurately determined by either of the processes above described, the resistance of any other rheomotor, in which the electromotive force is the same, may be obtained by a still more expeditious method.   The needle of the galvanometer being brought to a determined point when the standard rheomotor is interposed in the circuit : if this be removed, and the rheomotor to be measured be substituted in its place, the number of coils of the rheostat, added to or subtracted from the circuit, to make the current in the latter case equal to that in the former, when added to or subtracted from the re-

sistance of the standard rheomotor, will give that of the
rheomotor to be measured. If R' be greater than R,
$R' = R + r$; but if R' be less than R, $R' = R - r$. By this
simple process the resistances of voltaic elements of different
forms, magnitudes, &c. may be readily compared.

§ 14. *Instrument for Measuring the Resistance of Liquids.*

We do not at present possess any accurate measures of the
conductibilities of liquids, nor have there yet been formed
any tables which show even the real order of their conducting powers. In the experiments having this object in view
which have hitherto been made, the contrary electro-motive
force, which generally arises when the electric current passes
through a liquid capable of undergoing decomposition (§ 11,
4.), has been left entirely out of consideration, and the results
therefore have widely deviated from the truth. By the
simple instrument represented at Plate XVI. fig. 4, I have
been able to eliminate completely this source of error, and to
obtain perfectly constant results. A is a glass tube about
two inches long and half an inch in internal diameter; a portion of the tube is ground away for an inch and a quarter of
its length, so as to leave a segment of 270°; at one extremity
of this aperture is fixed a metal plug terminated by a platinum plate, and at the other end is a moveable piston, terminated also by a plate of platinum, capable of being advanced
to within a quarter of an inch from the fixed plate; the range
of its motion is thus limited to one inch, and an attached
micrometric apparatus enables any portion of this distance
to be accurately measured. To obtain the measure of the
resistance of a liquid I proceed in the following way :—I interpose in the circuit a small constant battery, consisting of
about three elements, with the rheostat, the resistance-coils,
the galvanometer, and the measuring tube just described.
The end of the piston being a quarter of an inch distant from
the fixed plate, I fill the intervening space with the liquid
the resistance of which is to be measured. I then adjust the
rheostat to bring the needle of the galvanometer to a determined point; this having been noted, I draw the piston

back through the entire remaining space of one inch, and fill the vacancy with the same liquid ; the needle will recede towards zero. I then diminish the resistance of the circuit by means of the rheostat and the resistance-coils, until the needle stands at the same point that it did when only a quarter of an inch of the liquid column was interposed. The reduced length of the wire thus taken out of the circuit will be the measure of the resistance of one inch of the liquid. The contrary electro-motive force arising from the decomposition of the liquid exists in the circuit during the whole process, and therefore does not affect the result.

The measure of the resistance of a liquid must be made immediately after it is placed in the circuit, because if a current be allowed to act upon it for any length of time the nature of the solution changes. In the case of sulphuric acid, for instance, the solution is rendered stronger by the decomposition and consequent diminution of the water, while, in the case of a metallic salt, not only is the water decomposed, but the metal is reduced, and free acid is liberated Under the conditions, however, of my experiments, the chemical action is so slow, and the time of operation is so short, that no sensible changes of this kind take place.

The resistance of liquids to the transmission of electricity is, no doubt, one of their most important physical properties. An investigation of all the circumstances which occasion changes in this property, especially if accompanied with accurate quantitative determinations, must necessarily lead to important and hitherto unobserved relations. To investigate the changes due to different degrees of dilution and temperature alone will be a task requiring considerable patience. I have made many measures of the specific resistances of different conducting liquids, by the aid of the preceding process, but as they have not been sufficiently numerous to enable any general conclusions to be drawn, and as I am at present engaged in a more extensive series of experiments in which strict attention will be paid to all the known influencing circumstances, I shall defer an account of them to a future occasion.

As bodies differ so much from each other in their specific resistances, and as the means of determining this property are so easy, it cannot be doubted that hereafter this process will be extensively employed to detect the purity of substances and to distinguish them from each other.

Another method of measuring the resistance of a conducting liquid is the following:—Prepare a circuit the electromotive force and resistance of which are known, $\frac{E}{R} = F$. Interpose the liquid which is to be the subject of experiment in a small cell with two parallel platinum electrodes; the expression for the circuit will then be $\frac{E-e}{R+x} = F$, e being the contrary electro-motive force, and x the resistance of the liquid which is to be determined. Having ascertained the value of e by the process described in § 10, subtract, by means of the rheostat and coils, a resistance which shall make the force again equal to F; the expression will then become $\frac{E-e}{R+x-\lambda} = \frac{E}{R}$, whence $x = \lambda - \frac{e}{E}R$. Therefore the resistance x of the liquid is equal to the resistance λ taken out of the circuit by the rheostat, minus the total resistance of the original circuit multiplied by the ratio $\frac{e}{E}$.

§ 15.

When a galvanometer is employed to measure the force of a current, its wire is usually interposed in the circuit. But it is impossible, in this way, to make use of the same galvanometer to measure the force of the current in circuits of different kinds. A galvanometer with numerous coils of thin wire adds a very considerable resistance to a circuit in which the electro-motive force is great and the resistance small; while, on the other hand, a galvanometer with a short thick wire will give scarcely any indication in a circuit in which the resistance is very great, though the electro-motive force may be considerable. Besides, a delicate galvanometer is incapable of indicating energetic forces.

But by the following simple means the same delicate galvanometer may be employed to measure forces of every degree of energy, and in all kinds of circuits, without introducing any inconvenient resistance into them.

If the current be caused to pass simultaneously through two paths, one being the wire of the galvanometer, and the other another wire connected with its two ends, the current will be divided in the inverse proportion of the resistances of the two paths. The action upon the needle of the galvanometer may hereby, by employing different wires to divert a portion of the current, be reduced to any degree. If the proportionate forces are known for the galvanometer without the reducing wire, they will remain equally proportionate whatever the resistance of the latter may be ; but measures made with the same instrument, with different reducing wires applied, will not be comparable unless the changed resistance of the galvanometer thus modified be taken into account.

But strictly comparable measures may be obtained, if the precaution be taken of adding, to the principal portion of the circuit, a resistance which will compensate for the diminution of resistance occasioned by placing the reducing wire. Let g be the reduced length of the galvanometer wire, and $n\,g$ that of the reducing wire : the force of the current in the principal portion of the circuit will be to that in the galvanometer wire as $1 : \dfrac{n}{n+1}$ The resistance to be added to the principal portion of the circuit, in order to maintain the current the same as when no reducing wire is added, is $\dfrac{g}{n+1}$.

When the measures of energetic currents are required to be determined by means of a delicate galvanometer, it is sufficient to attach its two ends to two points of the conducting wire[*]. The distance between these points must remain

* Professor Petrina, of Linz, has proposed (Poggendorff's 'Annalen,' vol. lxii. 1842, No. 9) a similar means of measuring and comparing electric currents of every degree of force. He interposes in the circuit a canal of mercury, the section of which is four square lines, and plunges into it, at various distances from each other, the ends of the wire of a

the same in all comparative experiments, but the absolute deviations of the needle will be greater as these points are further from each other. In the case of the circuit of a powerful electro-magnetic engine, or of a volta-typing apparatus, the diminution of resistance occasioned by connecting the galvanometer wire in the manner above described is so trifling that it would be useless to take it into account, and the compensation above alluded to is therefore unnecessary.

§ 16. *The Differential Resistance Measurer.*

The method of determining the resistance of metal wires and other conductors of electricity by means of the rheostat, described in § 9, is inapplicable when small differences are to be observed. If, for instance, a short length of wire has to be examined, its resistance is so small compared with the other resistances in the circuit, including that of the battery, that whether it be interposed or not, no change is observable in the deviation of the needle; and, even if greater lengths of the conducting substance be employed, fluctuations in the power of the battery frequently render the observation uncertain.

The differential galvanometer proposed by M. Becquerel, had it been an instrument as practically as it is theoretically perfect, would have enabled us to ascertain very minute differences of resistance with great facility. But it is almost impossible so to arrange the two coils that currents of equal energy circulating through them shall produce equal deviations of the needle in opposite directions, the consequence of which is that the standing of the needle at zero is no indication of equality in the currents. This and other defects have prevented the differential galvanometer from coming into use.

---

sensitive galvanometer. He shows that if the resistance in the galvanometer wire be very considerable, and that of the mercury in the canal be small in comparison, the force acting on the galvanometer needle will be sensibly proportional to the distance between the ends of the wire; and he has founded on this principle a ready approximative method of graduating the galvanometer.

All the advantages, however, which were expected from this instrument may be obtained, without any of its accompanying defects, by means of the simple arrangement I am about to describe, which, moreover, has the advantage of being immediately applicable to any galvanometer, instead of requiring, as in the former case, the instrument to be peculiarly constructed.

Fig. 5 (Plate XVI.) represents a board on which are placed four copper wires, Z b, Z a, C a, C b, the extremities of which are fixed to brass binding-screws. The binding-screws Z, C are for the purpose of receiving wires proceeding from the two poles of a rheomotor, and those marked a, b are for holding the ends of the wire of a galvanometer. By this arrangement a wire from each pole of the rheomotor proceeds to each end of the galvanometer wire, and if the four wires be of equal length and thickness, and of the same material, perfect equilibrium is established, so that a rheomotor, however powerful, will not produce the least deviation of the needle of the galvanometer from zero. The circuits Z b a C Z, and Z a b C Z, are in this case precisely equal, but as both currents tend to pass in opposite directions through the galvanometer, which is a common part of both circuits, no effect is produced on the needle. Currents are however established in Z b C Z, and Z a C Z, which would exist were the galvanometer entirely removed. But if a resistance be interposed in either of the four wires, the equilibrium of the galvanometer will be disturbed; if the resistance be interposed in Z b or C a, the current Z a b C Z will acquire a preponderance; if it be inserted either in Z a or C b, the opposite current, Z b a C Z, will become the most energetic. If the resistance interposed in the wire be infinite, or which is the same thing, if the wire (which we will suppose to be C b) be removed, the energy of the current passing through the galvanometer will be that of a partial current Z b a passing through one of the wires plus the galvanometer wire; the path of the diverted portion of the current being Z a. According to this disposition, the force of the original current $= \dfrac{E}{R + 2r + g}$, and that of the partial current acting on the galvanometer $= \dfrac{\mathrm{E}\,r}{\mathrm{R}\,(3\,r+g)+2\,r^2+r\,g}$;

R being the resistance of the rheomotor, r that of a single wire, and g that of the galvanometer.

The equilibrium having been disturbed by the introduction of a resistance in one of the wires, it may be restored by placing an equal resistance in either of the adjacent wires. For the purpose of interposing the measuring resistance and the resistance to be measured, the wires Z b and C b are interrupted, and binding-screws, c, d and e, f, are fixed for the reception of the ends of the wires. The equilibrium when once established is not in any degree affected by fluctuations in the energy of the rheomotor.

Fig. 6 (Plate XVI.) represents a different and, in some respects, a more convenient arrangement of the wires to produce the same result; the same reference letters are employed, and the preceding observations apply to it equally.

Slight differences in the lengths, and even in the tensions of the wires, are sufficient to disturb the equilibrium; it is therefore necessary to have an adjustment, by means of which, when two exactly equal wires are placed in C a and Z a, the equilibrium may be perfectly established. For this purpose, in the instrument, fig. 6, a piece of metal n, connected with the binding-screw b, is inlaid in the board, and another piece of metal m moves round n as a centre, whilst its free extremity always rests on the wire. According as the moveable piece of metal makes a greater angle with the fixed piece, the resistance of the path Z b is diminished; if, however, the equilibrium is disturbed because the resistance in C b is too great, the moveable piece of metal must be placed on the opposite side of the fixed piece.

No fixed dimensions can be assigned to these instruments. The boards of those I employ are fourteen inches long and four inches wide, and the wire is copper $\frac{1}{20}$th of an inch in diameter. A single voltaic element of large surface will produce a more considerable effect than a battery of small elements*. A thermo-electric arrangement, or a magneto-

---

* When a single element of Daniell's battery, 6 inches high and

electric machine may be substituted for the voltaic element
or battery; and a voltameter or any other description of
rheometer may in some cases supply the place of the galvanometer.   It is scarcely necessary to state that these instruments are not adapted to measure the resistances of substances capable of undergoing chemical changes from the
action of an electric current, on account of the contrary
electro-motive forces which arise under such circumstances *.

§ 17.

Another differential arrangement, which will be found
useful in some circumstances, may be worth mentioning; it
is much more sensible than the preceding, but as the equilibrium indicated is that between two currents generated by
independent rheomotors, instead of diverted portions of the
same current as in the instruments previously described, the
state of equilibrium will be disturbed by every fluctuation,
whether of the electro-motive force, or resistance of either of
the rheomotors; it can therefore only be safely employed
when these are perfectly constant, or when the object is not
to measure resistances, but to observe the comparative
changes in two rheomotors.

Fig. 7 (Plate XVI.) represents a circular board on which
are fixed ten binding-screws; the wires proceeding from one
of the rheomotors are to be attached to C^1 and Z^1, those from

---

$3\frac{1}{4}$ inches diameter, is employed, and two copper wires two feet long and
$\frac{1}{40}$th of an inch diameter are interposed in the instrument, an augmentation of the tenth of an inch in one occasions a deviation of 2° in the galvanometer needle.   This will suffice to show the accuracy with which
resistances may be measured by this instrument.

* Mr. Christie, in his "Experimental determination of the Laws of
Magneto-electric Induction," printed in the Philosophical Transactions
for 1833, has described a differential arrangement of which the principle
is the same as that on which the instruments described in this section
have been devised.   To Mr. Christie must, therefore, be attributed the
first idea of this useful and accurate method of measuring resistances.
Another differential arrangement, proposed also in the same memoir, is
analogous to that which forms the subject of the following section.

K

the other to C^2, Z^2, and the ends of the galvanometer wire are to be fixed to a and b. The two currents, $C^1\,a\,b\,Z^1$ and $Z^2\,a\,b\,C^2$, tend to pass through the galvanometer wire in opposite directions. When two equal wires are interposed between $e\,f$ and $e'\,f'$, if the opposing currents be equal, perfect equilibrium is established in the galvanometer wire, and the needle remains at zero. But if the force of the current in either of the rheomotors varies, or, if while the force of the two rheomotors remains constant, the slightest difference is occasioned in the resistance of either of the wires interposed between $e\,f$ or $e'\,f'$, the equilibrium in the galvanometer wire is disturbed and the needle is deflected.

§ 18.

It would greatly facilitate our quantitative investigations if we had a certain and ready means of ascertaining what degree of the galvanometric scale indicated half the intensity corresponding to any other given degree. The properties of diverted currents, established by the theory of Ohm, and fully confirmed by experiment, enables me to propose a simple method by which this object may be completely attained.

If a wire of the same length, thickness and conductibility as that of the galvanometer be placed so as to divert a portion of the current from it, it is obvious that one half of the current will pass through the galvanometer wire, and the other half through the diverting path. Though it simplifies the consideration to suppose the extra wire to have the same length, diameter and conducting power, it is easy to see that the same result follows if the two wires present the same resistance, which they do whenever $s'\,c'\,l = s\,c\,l'$. If the added wire produced no alteration in the intensity of the principal current, one half of the former force would act upon the galvanometer; but this is not the case, the addition of the wire produces the same effect as doubling the section of the galvanometer wire would do, and the total resistance of the circuit is therefore diminished. If the strength of the original current when it passes wholly through the galvanometer

$=\dfrac{E}{R+r}$ (r being the resistance of the galvanometer wire, and

R all the other resistances in the circuit), $\dfrac{E}{R+\dfrac{r}{2}}$ will be the

strength of the principal current when the extra wire is added; if now an additional resistance $=\dfrac{r}{2}$, that is to say, a wire whose resistance is equal to half that of the galvanometer wire, be added to the principal portion of the circuit, the intensity will be again $\dfrac{E}{R+\dfrac{r}{2}+\dfrac{r}{2}}$, and the force acting on the galvanometer will be exactly half what it was at first.

The construction and use of the instrument (Plate XVI. fig. 8) will now be easily understood. A is a square piece of wood, having two insulated pieces of brass, D, N, inlaid on its surface, on which are fixed the binding-screws C, Z and a; B is a circle also of wood, moveable round its centre; upon this moveable circle are fixed the insulated piece of brass F, with the binding-screw b upon it, and three springs G, H, I, the free ends of which press on the board A. A coil of wire K, the equivalent resistance to the wire of the galvanometer, measured by the process described in § 16, is connected by its two ends with the brass plate F and the spring G; and another coil, L, the resistance of which is one half that of the former, is similarly interposed between the brass plate and the spring H. A short wire immediately connects the plate F with the spring I. E is a nut or pin by which the moveable circle is moved through a small arc.

The wires proceeding from the poles of a rheomotor being connected with the binding-screws C, Z, and the ends of the galvanometer wire being attached to the screws z and b; in the position of the instrument represented in the figure, the springs G and H resting respectively on the insulated pieces of brass D and N, the principal portion of the current passes through the resistance-coil L, and the current is afterwards equally divided between the coil of the galvanometer and the

K 2

resistance-coil K.   But when the circle is moved in the direction of the arrow, the springs G, H leave the brass plates,
and rest on the wood, while the spring I is brought into contact with the plate E; both of the resistance-coils are now
thrown out of the circuit, and the current passes wholly
through the wire of the galvanometer.

It is almost unnecessary to state that this instrument can
only be used in conjunction with the galvanometer to which
its resistance-coils K and L have been adjusted.

In some cases, when an experiment has been performed
with a current of a certain degree of intensity, it is required
to repeat it with currents of other degrees of strength, the
proportions of which to the first current shall have been
accurately determined.   The instrument above described
readily affords the means of doing this.   It may thus be
ascertained whether the electro-motive force in any particular combination varies or remains constant when the
energy of the current changes.

§ 19. *Process to determine the Degrees of Deviation of the
Needle of a Galvanometer corresponding to the Degrees of
Force ; and the Converse.*

When the electro-motive force in the circuit remains constant, the force of the current is simply proportional to the
resistance or reduced length of the circuit. If, therefore, the
total resistance of the circuit, when the needle stands at 1°,
be determined, and if then, by means of the rheostat and
resistance-coils, the resistance be successively reduced to $\frac{1}{2}$,
$\frac{1}{3}$, $\frac{1}{4}$, $\frac{1}{5}$, &c., the corresponding forces of the current will be
2, 3, 4, 5, &c.   Conversely, if the reduced lengths a, b, c, d,
&c., necessary to be removed from the circuit in order to
advance the needle from each degree to the one next above
it, be successively ascertained, the forces corresponding to
these successive degrees will be

$$\frac{1}{R}, \frac{1}{R-a}, \frac{1}{R-(a+b)}, \frac{1}{R-(a+b+c)}, \&c.$$

By the above processes, the relations between the degrees of force and those of the galvanometric scale may be far more readily determined than by either of the ingenious methods of Nobili, Becquerel or Melloni. When we consider the changes to which the needle of a delicate galvanometer, especially if it be astatic, is subject from the influence of strong currents, the vicinity of magnets, and, in a less degree, from changes of temperature and in the intensity of the earth's magnetism, the importance of having an easy means of re-graduating the instrument, and of detecting the changes it has undergone, will not be esteemed too lightly.

On the Thermo-electric Spark.

[From the 'Philosophical Magazine,' 1837, vol. x. pp. 414–417.]

THE following notice of some recent experiments made in Italy, on the production of the thermo-electric spark, and on the chemical effects of the thermo-electric currents, will no doubt be acceptable to many of your readers. I shall confine myself to a simple statement and corroboration of the facts, avoiding all theoretical considerations.

The Cav. Antinori, Director of the Museum at Florence, having heard that Prof. Linari, of the University of Siena, had succeeded in obtaining the electric spark from the torpedo by means of an electro-dynamic helix and a temporary magnet, conceived that a spark might be obtained by applying the same means to the thermo-electric pile. Appealing to experiment his anticipations were fully realized. No account of the original investigations of Antinori has yet reached this country; but Prof. Linari, to whom he early communicated the results he had obtained, immediately repeated them, and published the following additional observations of his own in *L'Indicatore Sanese*, No. 50, of Dec. 13, 1836.

" 1. With an apparatus consisting of temporary magnets and electro-dynamic spirals, the wire of which was 505 feet in length, he obtained a brilliant spark from a thermo-electric pile of Nobili's construction consisting only of 25 elements, which was also observed in open daylight.

" 2. With a wire 8 feet long coiled into a simple helix, the spark constantly appeared in the dark, on breaking contact, at every interruption of the current; with a wire 15 inches long he saw it seldom, but distinctly; and with a double pile, even when the wire was only 8 inches long. In all the

above-mentioned cases the spark was observed only on breaking contact, however much the length of the wire was diminished.

" 3. The pile consisting merely of these few elements, and
within such restricted limits of temperature as those of ice
and boiling water, readily decomposed water. Short wires
were employed having oxidizable extremities: the hydrogen
was sensibly evolved at one of the poles.

" 4. A mixture of marine salt moistened with water and
of nitrate of silver being placed between two small horizontal
plates of gold, communicating respectively with the wires of
the pile, the latter after having acted on the mixture gave
evident signs of the appearance of revivified silver on the
plate which was next the antimony.

" 5. An unmagnetic needle placed within a closed helix
formed by the wire of the circuit became well magnetized by
the current.

" 6. Under the action of the same current the phænome-
non of the palpitation of mercury was distinctly observed."

The interesting nature of these experiments induced me to
attempt to verify the principal result. The thermo-electric
pile I employed consisted of 33 elements of bismuth and antimony formed into a cylindrical bundle $\frac{3}{4}$ of an inch in
diameter and $1\frac{1}{4}$ inch in length; the poles of this pile were
connected by means of two thick wires with a spiral of copper
ribbon 50 feet in length and $1\frac{1}{2}$ inch broad, the coils being
well insulated by brown paper and silk. One face of the pile
was heated by means of a red-hot iron brought within a
short distance of it, and the other face was kept cool by contact with ice. Two stout wires formed the communication
between the poles of the pile and the spiral, and the contact
was broken, when required, in a mercury cup between one
extremity of the spiral and one of these wires. Whenever
contact was thus broken a small but distinct spark was seen,
which was visible even in daylight. Professors Daniell,
Henry, and Bache assisted in the experiment, and were all
equally satisfied of the reality of the appearance.

At another trial the spark was obtained from the same

spiral connected with a small pile of 50 elements, on which occasion Dr. Faraday and Prof. Johnson were present, and verified the fact. On connecting two such piles together so that the similar poles of each were connected with the same wires, the same was seen still brighter *

I conclude, therefore, that the experiment of Antinori is a real addition to our knowledge of electrical phænomena, and though it was far from being unexpected, it supplies a link that was wanting in the chain of the experimental evidence which tends to prove that electricity, from sources however varied, is similar in its nature and effects, a conclusion rendered more than probable by the recent discoveries of Faraday. The effects thus obtained from the electric current originating in the thermo-electric pile may no doubt be easily exalted by those who have the requisite apparatus at their disposal. It is not too much to expect, seeing the effects produced by a pile of such small dimensions, that by proper combinations the effects may be exalted to equal those of an ordinary voltaic pile.

I shall close this hasty communication with a notice of some experiments on the chemical action of the thermo-electric pile made earlier by Prof. G. D. Botto, of the University of Turin. The form of the pile he employed may suggest some useful hints to those who are inclined to continue the inquiry, as it admits of the application of much higher degrees of heat than one of the ordinary construction does, though the difference of the thermo-electric relations of the two metals employed is not so considerable. Prof. Botto's experiments were published in the *Bibliothèque Universelle* for September 1832, and I am not aware that they have yet been published in any English Journal. The thermo-electric apparatus was a metallic wire, or chain, consisting of 120 pieces of platinum wire, each one inch in length, and $\frac{1}{100}$th of an inch in diameter, alternating with the same number of pieces of soft iron wire of the same dimensions. This wire was coiled as a helix round a wooden rule 18 inches

* The two piles here employed were made by Mr. Newman of Regent Street.

long, in such a manner that the joints were placed alternately at each side of the rule, being removed from the wood
at one side to the distance of four lines. Employing a spirit-
lamp of the same length as the helix, and one of Nobili's
galvanometers, a very energetic current was shown to exist;
acidulated water was decomposed, and the decomposition
was much more abundant when copper instead of platinum
poles were used; in this case hydrogen only was liberated.
The current and decomposition were augmented when the
joints were heated more highly. Better effects were obtained
with a pile of bismuth and antimony, consisting of 140 elements bound together into a parallelopiped, having for its
base a square of two inches three lines, and an inch in
height.

Description of the Electro-magnetic Clock.

[From the 'Proceedings of the Royal Society,' 1840, vol. iv. pp. 249, 250.]

THE object of the apparatus forming the subject of this communication, is stated by the author to be that of enabling a single clock to indicate exactly the same time in as many different places, distant from each other, as may be required. Thus, in an astronomical observatory, every room may be furnished with an instrument, simple in its construction, and therefore little liable to derangement, and of trifling cost, which shall indicate the time, and beat dead seconds audibly, with the same precision as the standard astronomical clock with which it is connected; thus obviating the necessity of having several clocks, and diminishing the trouble of winding up and regulating them separately. In like manner, in public offices and large establishments, one good clock will serve the purpose of indicating the precise time in every part of the building where it may be required, and an accuracy ensured which it would be difficult to obtain by independent clocks, even putting the difference of cost out of consideration. Other cases in which the invention might be advantageously employed were also mentioned. In the electro-magnetic clock, which was exhibited in action in the Apartments of the Society, all the parts employed in a clock for maintaining and regulating the power are entirely dispensed with. It consists simply of a face with its second, minute, and hour hands, and of a train of wheels which communicate motion from the arbor of the second's hand to that of the hour hand, in the same manner as in an ordinary clock train; a small electro-magnet is caused to act upon a peculiarly constructed wheel (scarcely capable of being described without a figure) placed on the second's arbor, in such manner that whenever the temporary magnetism is either produced or destroyed, the wheel, and consequently the second's hand, advances a sixtieth part of its revolution. It is obvious, then, that if an electric current can be alternately established and arrested, each resumption and cessation lasting for a second, the instrument now described, although unprovided with any internal maintaining or regulating power, would perform all the usual functions of a perfect clock. The manner in which this apparatus is applied to the clocks, so that the movements of the hands of both may be perfectly simultaneous, is the following. On the axis which carries the scape-wheel of the primary clock a small disc of brass is fixed, which is first divided on its circumference into sixty equal parts ; each alternate division is then cut out and filled with a piece of wood, so that the circumference consists of thirty regular alternations of wood and metal. An extremely light brass spring, which is screwed to a block of ivory or hard wood, and which has no connexion with the metallic parts of the clock, rests by its free end on the circumference of the disc. A copper wire is fastened to the fixed end of the spring, and proceeds to one end of the wire of the electro-magnet ; while another wire attached to the clock-frame is continued until it joins the other end of that of the same electro-magnet. A constant voltaic battery, consisting of a few elements of very small dimensions, is interposed in any part of the circuit. By this arrangement the circuit is periodically made and broken, in consequence of the spring resting for one second on a metal division, and the next second on a wooden division. The circuit may be extended to any length ; and any number of electro-magnetic instruments may be thus brought into sympathetic action with the standard clock. It is only necessary to observe, that the force of the battery and the proportion between the resistances of the electro-magnetic coils and those of the other parts of the circuit, must, in order to produce the maximum effect with the least expenditure of power, be varied to suit each particular case.

In the concluding part of the paper the author points out several other and very different methods of effecting the same purpose; and in particular one in which Faraday's magneto-electric currents are employed, instead of the current produced by a voltaic battery; he also describes a modification of the sympathetic instrument, calculated to enable it to act at great distances with a weaker electric current than if it were constructed on the plan first described.

Enregistreur électromagnétique pour les Observations Météorologiques.

[From the 'Archives de l'Électricité' par A. de la Rive, 1844, t. iv. pp. 170–172.]

L'ENREGISTREUR électromagnétique construit pour l'observatoire de l'Association britannique est presque terminé. Il enregistre les indications du baromètre, du thermomètre et du psychromètre pour toutes les demi-heures de jour et de nuit, et en imprime les résultats en duplicata, sous forme figurée sur un feuille de papier. Il n'exige aucune attention pendant toute une semaine, pendant laquelle il enregistre 1008 observations. Cinq minutes suffisent pour préparer la machine à noter une autre semaine d'observations, c'est-à-dire qu'il faut simplement remonter le mouvement de pendule, fournir aux cylindres une autre feuille de papier, et recharger le petit élément voltaïque. L'échelle de chaque instrument est divisée en 150 parties ; celle du baromètre embrasse trois pouces ; celle du thermomètre renferme tous les degrés de température entre −5° et +95° F., et le psychromètre a la même étendue.

La machine consiste essentiellement en deux parties distinctes. La première est une horloge régulatrice à laquelle sont attachés tous les mouvements nécessaires récurrents. La seconde est un mécanisme ayant une force motrice indépendante, qui est mise en action, à des intervalles de temps irréguliers, par le contact de fils plongeant dans le mercure de l'appareil, ainsi qu'il sera expliqué plus loin. Les principaux mouvements réguliers récurrents qui sont liés à l'horloge sont au nombre de deux : 1° les flotteurs sont élevés graduellement et régulièrement dans les tubes de l'instrument pendant cinq minutes, et descendent ensuite en une minute ; 2° une roue à types, portant à sa circonférence 15 figures ou caractères, avance seulement d'un pas lorsque la première a complété une révolution. La révolution de la seconde roue à types ne s'effectue qu'au bout de six minutes, c'est-à-dire dans le même temps qu'emploient les flotteurs pour monter et descendre. Ainsi, chaque division successive de l'échelle d'un instrument correspond à un nombre différent présenté par les deux roues à types, la même division correspondant toujours au même nombre. Les deux blancs se présentent lors du retour des flotteurs, occupent une minute, et pendant ce temps il n'y a pas d'observation enregistrée. Le point du contact entre le flotteur et le mercure dans un instrument a évidemment lieu dans une position différente des roues à types, suivant que le mercure est à une élévation ou à une autre; si, par conséquent, les types ou figures s'impriment en ce moment, on conçoit qu'on aura ainsi enregistré la hauteur du mercure. Ce but est atteint de la manière suivante. L'extrémité d'un fil conducteur est en communication avec le mercure du tube de l'instrument, et l'autre extrémité avec la boîte en laiton de l'horloge, laquelle est à son tour en communication métallique avec le flotteur. On a mis dans l'étendue de ce circuit un aimant électrique et un élément voltaïque unique très-petit. L'aimant électrique est placé de manière à agir sur une petite armature de fer doux qui est liée avec la détente du seconde mouvement. Tant que le flotteur est dans le mercure, l'armature se trouve attirée; mais au moment où le flotteur abandonne le mercure, l'attraction cesse, et les pièces de la détente permettent à un marteau de frapper les types et de les imprimer sur le cylindre au moyen d'un papier noir à copier. L'armature, en conséquence reste sans être attirée jusqu'à ce que le flotteur descende. Immédiatement avant qu'il remonte, une pièce de mécanisme qui communique avec le mouvement de l'horloge amène l'armature en contact avec l'aimant, lequel reste en cet état par suite du rétablissement du circuit jusqu'a ce que le contact se trouve de nouveau rompu.

Note sur le Chronoscope électromagnétique.

[From the ' Comptes Rendus de l'Académie des Sciences,' 1845, tome xx. pp. 1554–1561].

JE vois dans les *Comptes rendus de l'Académie des Sciences*, que dans la séance du 20 janvier, il fut lu une communication de M. Breguet, dans laquelle il attribue à M. le capitaine de Konstantinoff, et à lui-même, l'invention du chronoscope électromagnétique, instrument que j'avais moi-même inventé et confectionné plusieurs années auparavant, dans le but de mesurer les mouvements rapides, et surtout la vitesse des projectiles.

Ce fut au commencement de 1840 que j'inventai cet instrument. Mon chronoscope se composait alors d'un mouvement d'horlogerie faisant agir une aiguille indicatrice, qui marchait ou s'arrêtait suivant qu'un électro-aimant agissait sur une pièce de fer doux, l'attirant lorsqu'un courant traversait l'hélice de l'aimant, et l'abandonnant à lui-même lorsque le courant venait à cesser, comme dans mon télégraphe électromagnétique, dont cette invention peut être considérée comme une des dérivations. La durée du courant était ainsi mesurée par l'étendue du cercle parcouru par l'aiguille du chronoscope.

Une relation était établie entre la durée du courant et celle du mouvement du projectile par les moyens suivants : un anneau en bois embrassait l'embouchure d'un canon chargé, et un fil métallique tendu reliait deux côtés opposés de cet anneau isolant, passant ainsi devant la bouche du canon. A une distance convenable, était établi un but disposé de telle facon que le moindre mouvement qu'on lui imprimait établissait un contact permanent entre un petit ressort en métal et un autre pièce de métal. Une des extrémités du fil métallique de l'électro-aimant était attachée à

un des pôles d'une petite batterie voltaïque ; à l'autre extrémité de l'électro-aimant étaient attachés deux fils métalliques
dont l'un communiquait avec le petit ressort du but, et
l'autre à l'une des extrémités du fil métallique tendu devant
la bouche du canon ; de l'autre extrémité de la batterie voltaïque partaient aussi deux fils métalliques, dont l'un abou-
tissait à la pièce métallique fixée sur le but, et l'autre, à
l'extrémité opposée du fil métallique passant devant l'embouchure du canon.   Ainsi, antérieurement à l'explosion du
canon, il se trouvait établi, entre le canon et le but, un
circuit non interrompu en fil métallique, et dont le fil métallique en travers de la bouche du canon faisait partie.   Une
fois le but frappé par le boulet, le second circuit était complété ; mais durant le passage du projectile à travers l'air,
et pendant ce temps seulement, les deux circuits étaient in-
terrompus, et la durée de cette interruption etait indiquée
par le chronoscope.

J'avais déjà démontré par mon télégraphe électromagnétique que, lorsqu'ils sont convenablement disposés, les
aimants peuvent être amenés à agir avec une batterie très-
faible, quand bien même les fils métalliques décriraient un
circuit de plusieurs milles.   Par conséquent, le canon, le but
et le chronoscope peuvent être placés à des distances quel-
conques demandées les uns des autres.   En raison de la
grande rapidité avec laquelle l'électricité se propage, comme
l'ont prouvé mes expériences publiées dans les *Philosophical
Transactions* de 1834, aucune erreur sensible ne peut résulter
de sa transmission successive.

Pendant une visite que je fis à Bruxelles, au mois de sep-
tembre 1840, je décrivis cet appareil à mon ami M. Quetelet,
qui en donna connaissance, le 7 octobre, à l'Académie des
Sciences de cette ville, communication mentionnée dans le
Bulletin de cette séance.

Dans une visite que je fis postérieurement à Paris (mai
1841), j'expliquai cet appareil et j'en montrai les dessins à
plusieurs membres de l'Académie des Sciences de Paris qui
vinrent me voir au Collége de France, où, grâce à l'oblige-
ance de M. Regnault, j'eus l'occasion de répéter devant eux

plusieurs de mes expériences électromagnétiques. Parmi les personnes présentes était M. Pouillet, qui me demanda l'autorisation de faire copier mes dessins, ce à quoi je consentis volontiers. J'appris de lui, en décembre dernier, que ces dessins étaient encore en sa possession.

A mon retour en Angleterre, mon ami le Capitaine Chapman *, de l'artillerie royale, convaincu de l'utilité de cet instrument, était très-désireux qu'il fût introduit dans la pratique de l'artillerie à Woolwich, et se donna beaucoup de peine pour y parvenir. Nous eûmes une entrevue à ce sujet avec feu Lord Vivian, alors maître général de l'*Ordnance*, et le 17 juillet 1841, j'expliquai à l'Institut de l'artillerie royale la construction de l'instrument et ses diverses applications. Vingt-deux officiers assistèrent à cette séance, dans le compte rendu de laquelle (compte rendu dont je possède une copie), il est dit que mon chronoscope " indiquait $\frac{1}{7300}$ de seconde," et que mon objet était " de montrer son application aux usages pratiques de l'artillerie," c'est-à-dire de déterminer le temps employé par un projectile à franchir les différentes sections de son parcours, ainsi que sa vitesse initiale. Dans la même séance, je montrai " un chronoscope destiné à mesurer la vitesse des éclairs, tels que ceux produits par l'ignition de la poudre." Cet instrument, le seul que M. Breguet m'attribue, n'avait cependant rien de commun avec les courants électriques, comme il le suppose ; c'était simplement une série de roues portant sur des axes trois légers disques en papier, d'environ 1 pouce de diamètre chacun. Les temps de leurs révolutions respectives étant comme 1, 10 et 100, le disque dont le mouvement était le plus rapide faisait 200 révolutions par seconde ; sur chaque disque était tracé un rayon : lorsqu'ils étaient éclairés par une étincelle électrique, tous ces rayons paraissaient en

---

* Depuis longtemps j'entretenais une correspondance avec le capitaine Chapman à ce sujet. Dans une de ses Lettres, du 27 août 1840, après m'avoir fait part de ses vues sur la manière de conduire ses expériences, il dit: "Nous obtiendrons ainsi la vitesse du projectile à chacune des sections de sa trajectoire, et j'ose croire que nous arriverons à une connaissance de l'effet de la gravitation sur le projectile, beaucoup plus satisfaisante que tout ce qu'on a obtenu jusqu'aujourd'hui."

L

repos, en raison de la durée excessivement petite de cette espèce de lumière (comme il est expliqué dans mon Mémoire : *De la vitesse de l'électricité et de la durée de la lumière électrique*, publié dans les *Philosophical Transactions* de 1834) ; mais lorsqu'ils étaient illuminés par un éclair d'une durée de la deux-centième partie d'une seconde, le troisième disque paraissait uniformément teinté pendant que le second disque montrait un secteur ombré de 36 degrés.   Quand l'éclair ne durait que la deux-millième partie d'une seconde, un secteur semblable paraissait sur le troisième disque.

Pour plusieurs raisons, les expériences avec mon chronoscope électromagnétique ne furent pas poursuivies à Woolwich. En 1842, je fis la connaissance de M. de Konstantinoff, capitaine dans l'artillerie de la garde impériale de S. M. l'Empereur de Russie, et attaché à l'état-major du général de Winspaer ; il prit beaucoup d'intérêt à cette affaire, exprima un vif désir d'avoir un appareil complet, afin d'entreprendre lui-même, à son retour en Russie, une série d'expériences telles que celles que j'avais en vue.   Comme je n'avais pas moi-même le temps de poursuivre ces expériences, et comme personne en Angleterre, plus habile ou mieux placé pour cela, ne montrait le désir de les poursuivre, je cédais volontiers à sa demande, dans l'espoir que quelques résultats importants pour la science pourraient être obtenus. La seule condition que je mis à mon consentement, était que M. de Konstantinoff ne publierait aucune description de l'appareil, jusqu'au moment où moi-même je l'aurais faite. L'instrument que je fournis à M. de Konstantinoff, et qui lui fut adressé à Paris, en janvier 1843, était autrement construit que celui précédemment décrit, quoique essentiellement le même en principe.

J'avais trouvé par expérience que lorsqu'une pièce de fer doux avait été attirée par un électro-aimant, et que le courant venait ensuite à cesser, bien que le fer parût retomber immédiatement, son contact était maintenu pendant un temps qui, plusieurs fois, équivalait à une fraction considérable de seconde.   La durée de cette adhérence augmentait avec l'énergie du courant voltaïque, et avec la faiblesse du ressort à réactions. Pour la réduire à un minimum, il était nécessaire d'employer un courant très-faible et d'augmenter la résistance du circuit jusqu'à ce que la force d'attraction de l'aimant fût réduite au point de ne surpasser que d'une très-faible quantité la force de réaction du ressort; mais alors l'aimant n'avait plus la force suffisante pour attirer le fer lorsque le projectile frappait le but. Cependant je surmontai cette difficulté de la manière suivante : j'arrangeai les fils métalliques du circuit de telle sorte, qu'avant que le boulet ne fût lancé par le canon, le courant d'un seul élément de très-petites dimensions, et réduit au degré convenable au moyen d'un rhéostat * aussi interposé dans le circuit, agissait sur l'électro-aimant; mais lorsque le boulet arrivait au but, six éléments, sans la résistance du rhéostat, agissaient simultanément sur l'aimant. Mais, même avec ces précautions, qui sont efficaces jusqu'à un certain degré, il y a encore du temps de perdu durant l'attraction du fer par l'aimant, aussi bien que pendant son adhérence après que le courant a cessé. La différence de ces deux erreurs rendrait des approximations telles que $\frac{1}{500}$ ou $\frac{1}{1000}$ de seconde tout à fait incertaines. Toutefois, l'erreur provenant de cette source peut se réduire facilement à moins de $\frac{1}{60}$ ou de $\frac{1}{100}$ de seconde, et dans mon opinion, un chronoscope qui divise la seconde en soixante parties, et qu'on peut prouver ne donner jamais lieu à une erreur dépassant une seule de ces divisions, est préférable à un instrument offrant des divisions plus avancées et qui donnerait lieu à des erreurs embrassant bon nombre de ces divisions. Guidé par ces expériences, je fus en mesure de construire un chronoscope très-simple et très-efficace. Un échappement très-simple était mis en mouvement par un poids suspendu à l'extrémité d'un bout de fil enroulé dans une hélice creusée sur un cylindre fixé sur l'axe d'une roue à échappement. Sur cet axe était aussi

* Une explication de cet instrument se trouve dans ma *Description de plusieurs instruments et procédés pour déterminer les constants d'un circuit voltaïque*, publiée dans les *Philosophical Transactions* de 1843, 2ᵉ part., et traduite dans les *Annales de Chimie et de Physique*. Cet instrument est représenté par la *fig.* 1 des dessins qui accompagnent ce Mémoire.

adaptée une aiguille qui, conséquemment, avançait d'une division à chaque échappement. Quand il était nécessaire de prolonger le temps de l'expérience, la roue à échappement et le cylindre étaient établis sur des axes différents, et leur engrenage s'opérait au moyen d'une roue et d'un pignon ; dans ce cas, deux aiguilles étaient employées. Au moyen de cette construction, on évite l'accélération du mouvement qui aurait eu lieu s'il n'y avait pas eu d'échappement, et l'index franchit chaque division dans un même temps. Le poids était disposé de manière à pouvoir se régler, et la valeur d'une seule division était obtenue en divisant le temps de la chute entière par le nombre des divisions franchies dans cet intervalle ; mais des méthodes encore plus exactes peuvent être employées.

Au moyen de cet instrument, j'ai mesuré le temps mis par une balle de pistolet à parcourir différentes portées, avec des charges différentes de poudre. La répétition de ces expériences donna lieu à des résultats passablement constants, présentant rarement une différence de plus d'une division du chronoscope *. Je mesurai aussi la chute d'une balle, de différentes hauteurs, et la loi des vitesses accélérées fut obtenue avec une rigueur mathématique. Avec l'appareil dont je me servis pour cette dernière expérience, je pouvais mesurer la chute d'une balle, de la hauteur d'un pouce. Il serait difficile, sans le secours des dessins, de donner une idée des diverses dispositions que j'ai adoptées pour rendre l'instrument applicable à différentes séries d'expériences ; mais je puis mentionner que parmi d'autres applications, je me propose de l'employer pour mesurer la vitesse du son à travers l'air, l'eau et à travers les massifs de roc, avec une approximation qu'on n'a jamais obtenue jusqu'à présent.

Indépendamment de l'instrument que je fournis à M. de Konstantinoff, en avril 1843, le professeur Christie en fit déposer un au cabinet de Physique de l'Académie militaire de Woolwich, et, vers le même temps, un autre fut fait pour

---

* Ces expériences, dans lesquelles je fus assisté par Sir James South et M. Purday, célèbre armurier, eurent lieu en octobre 1842, dans les terrains attenant à l'observatoire de Camden Hill.

M. R. Addams, qui s'en est constamment servi depuis dans ses cours à l'*United Service Museum*, et ailleurs.

Je mentionnerai une modification de l'appareil qui est importante pour certaines séries d'expériences : au lieu de rompre la continuité du circuit et de la reconstituer ensuite comme nous l'avons dit jusqu'ici, l'électro-aimant est maintainu en équilibre au moyen de deux courants égaux et opposés ; en interrompant le premier circuit, l'équilibre est détruit, et en interrompant le second, le courant occasionné par la destruction de l'équilibre cesse.   Le second circuit est rompu par une balle traversant un cadre sur lequel est tendu un fil métallique très-fin disposé en lignes parallèles et très-serrées, et formant partie du circuit.   Cette disposition fournit les moyens d'employer un chronoscope totalement différent du premier.   Deux pendules, dont l'un à demi-secondes, et l'autre d'un mouvement un peu plus accéléré, sont maintenus chacun aux extrémités de leurs arcs d'oscillation par un électro-aimant.   Quand la balle s'échappe du fusil, l'un des pendules est libéré, et quand il rompt le fil métallique du cadre, l'autre pendule est aussi libéré.   On compte alors le nombre d'oscillations d'un des pendules jusqu'à ce que le mouvement des deux pendules coïncide, et, d'après ce fait, on détermine aisément le temps qui sépare les commencements des premières oscillations des deux pendules.

Les instruments que je construisis réellement n'avaient d'autre objet que d'indiquer le temps écoulé entre le mouvement initial et le mouvement final d'une balle parcourant sa trajectoire.   M. de Konstantinoff désirait un instrument mesurant les temps correspondant aux divisions successives de la trajectoire.   Bien que je pensasse alors, et que je sois encore de l'avis qu'il est préférable de les déterminer au moyen de décharges successives, j'imaginai un appareil à cet effet, mais je n'en entrepris pas la construction, en raison de son prix plus élevé et de sa plus grande complexité, bien qu'il fût l'objet de fréquentes conversations entre nous.   C'était afin de réaliser ces idées que M. de Konstantinoff, après son départ d'Angleterre et pendant son séjour à Paris, s'adressa subséquemment à M. Breguet, afin de profiter de l'habileté et de l'ingéniosité bien connues de cet ingénieur. Je suis parfaitement persuadé que M. de Konstantinoff n'eut jamais l'intention de s'attribuer cette invention, et que c'est entièrement sans son approbation, et à son insu, que M. Breguet vient de le faire*.

Quant à l'instrument décrit par M. Breguet, je le considère comme beaucoup moins exact, beaucoup plus compliqué et plus coûteux qu'aucun de ceux que j'ai précédemment inventés. Quand il est réduit uniquement à déterminer les mouvements initial et final d'une balle, l'instrument de M. Breguet est muni de cinq électro-aimants, chacun avec son mécanisme, tandis que le mien atteint le même résultat avec un seul électro-aimant; et lorsque les différentes divisions d'une même trajectoire doivent être étudiées, M. Breguet propose un aimant complémentaire et fait d'autres additions à chacune des partitions que doit traverser la balle. Si M. Breguet avait été mieux informé des moyens par lesquels je devais obtenir une suite de mesures successives correspondant à une même trajectoire, il aurait trouvé que ce qu'il propose d'obtenir, même avec une douzaine d'électro-aimants, le serait d'une manière plus efficace au moyen d'un seul. Voici quel était mon plan.

Un cylindre exécute un mouvement de rotation autour d'une vis, de façon à avancer d'un quart de pouce par révolution; à une des extrémités du cylindre est adaptée une roue dentée d'un diamètre un peu plus grand que celui du cylindre, et qui s'engrène avec une pignon dont la longueur est égale à la portion totale d'axe que doit franchir le cylindre dans ses révolutions successives; ce pignon communique avec des rouages mis en mouvement par un poids suspendu à l'extrémité d'un fil qui tourne autour d'un cylindre, et le rouage est muni d'un régulateur qui en égalise le mouvement; un crayon, adapté à l'extrémité d'un petit électro-

---

* Je joins ici un extrait d'un écrit que me donna M. de Konstantinoff, avant de quitter Londres:

"M. Wheatstone ayant eu la complaisance de me faire confectionner un appareil complet, de son invention, pour mesurer les chutes des corps et les vitesses initiales des projectiles, je m'engage, &c."

aimant, est amené en contact avec le cylindre et y trace une hélice qui est interrompue chaque fois que le courant cesse. J'empruntai l'idée de la partie chronoscopique de cet appareil d'un instrument destiné à mesurer de trés-petits intervalles de temps, inventé par feu le docteur Young, et qui est décrit et dessiné dans son *Cours de Philosophie naturelle*. On comprend aisément, d'après ce que j'ai déjà rapporté, de quelle manière le commencement et la fin du mouvement d'un projectile sont indiqués par cet instrument. Les périodes intermédiaires sont enregistrées de la manière suivante : aux points voulus sur la ligne de passage du projectile, on établit des cadres fermés par des réseaux en fil métallique ; le projectile rompt les fils métalliques en traversant les cadres ; on emploie autant de batteries voltaïques qu'il y a de paires de cadres dont les fils métalliques communiquent avec les pôles de ces batteries voltaïques et avec le fil métallique de l'électro-aimant, de telle façon que le courant électrique traverse l'hélice en fil métallique de l'électro-aimant, ou cesse de la parcourir, suivant que l'équilibre est alternativement détruit ou rétabli par la rupture successive des fils métalliques des cadres. Pour obtenir ce résultat, il est nécessaire que la résistance des différents fils métalliques soit convenablement proportionnée.

Pour conclure, j'ajouterai que l'application de mon télégraphe électromagnétique, en vue d'enregistrer à distance le nombre des révolutions d'une machine ou de tous autres mouvements périodiques, a été exécutée par moi. sous des formes très-variées, depuis plusieurs années. Un appareil pour cet objet, enregistrant jusqu'à dix mille, se voit dans le cabinet de Physique de King's-College depuis 1840, et fut montré à M. de Konstantinoff pendant son séjour à Londres.

An Account of some Experiments made with the Submarine Cable of the Mediterranean Electric Telegraph.

[From the 'Proceedings of the Royal Society,' 1855, vol. vii. pp. 328–333.]

THE following results were obtained between May 24 and June 8 in last year, with the telegraphic cable manufactured by Messrs. Kuper and Co. of East Greenwich, for the purpose of being laid across the Mediterranean sea, from Spezia on the coast of Italy to the island of Corsica. The manufacturers, in conjunction with Mr. Thomson the engineer of the undertaking, kindly afforded me every facility in carrying on the experiments. The short time that elapsed between the opportunity presenting itself and the shipping of the cable for its destination, prevented me from determining with sufficient accuracy some points of importance, respecting which I was only able to make preliminary experiments; but the following, which I was able to effect with the means at hand, may possess sufficient interest to be made public. They present perhaps nothing theoretically new, but I am not aware that experimental verifications of some of these points have been made before. I assume that the reader is acquainted with the experiments of Dr. Faraday described in the Philosophical Magazine, N. S., vol. vii. p. 197.

The cable was 110 miles in length, and contained six copper wires, one sixteenth of an inch in diameter, each separately insulated in a covering of gutta percha one tenth of an inch in thickness. The whole was surrounded by twelve thick iron wires twisted spirally around it, forming a complete metallic envelope one third of an inch in thickness. A section of the cable presented the six wires arranged in a circle of half an inch diameter, and one fifth of an inch from the internal surface of the iron envelope.

The cable was coiled in a dry well in the yard, and one of

its ends was brought into the manufactory. The wires were numbered 1, 2, 3, 4, 5, 6, and the ends in the well were indicated by an accent; the ends 1'2, 2'3, 3'4, 4'5, 5'6 were connected by supplementary wires, so that the electric current might be passed in the same direction through all the six wires joined to a single length, or through any lesser number of them, the connexions being made at pleasure in the experimenting room.

The rheomotor employed was an insulated voltaic battery consisting of twelve troughs, each of twelve elements, which had been several weeks in action.

First Series.

The following experiments show that the iron envelope of the compound conductor gives rise to the same phenomena of induction which occur when the insulated wire is immersed in water, as in Dr. Faraday's experiments.

Exp. 1. One end of the entire length, 660 miles, was brought in connexion with one of the poles of the battery, the other end remaining insulated. The wire became charged with negative electricity when its end touched the zinc pole, and with positive electricity when it communicated with the copper pole. A current, indicated by a galvanometer placed near the battery, existed as long as the charge was going on, and ceased when it arrived at its maximum. [The feeble current attributed to imperfect insulation, which continues as long as the contact with the battery remains, is here left out of consideration.] When the wire was charged, and the discharge effected by a wire communicating with the earth, the current produced was in the same direction, whether the discharge was made near the battery or at the opposite end; *i. e.* the current in both cases proceeded from the wire to the earth in the same direction.

Exp. 2. On bringing one end of the wire in contact with one of the poles of the battery, the other pole having no communication with the earth, the wire remained uncharged. A very slight and scarcely perceptible tremor was observed in the galvanometer needle interposed between the battery and the wire.

Exp. 3. To each of the poles of the battery was attached a wire 220 miles in length, and similar galvanometers were interposed between the two wires (the remote extremities of which remained insulated) and the battery. So long as one wire alone was connected with the battery no charge was communicated to it, but on connecting the other wire with the opposite pole both wires were instantaneously charged, as the strong deflection of both needles rendered evident. On bringing the free end of one of the wires in communication with the earth it alone was discharged, the other wire remaining fully charged.

Second Series.

Exp. 4. One pole of the battery was connected with the earth, and the other with 660 miles of wire, which had an earth communication at its opposite end; three galvanometers were interposed in the course of the conductor; the first near the battery, the second in the middle of the wire, *i. e.* 330 miles from each extremity, and the third at the remote end near the communication with the earth. When the connexion of the battery with the wire was completed, the galvanometers were successively acted upon in the order of their distances from the battery, as in the experiments recorded by Dr. Faraday. When the earth connexion at the remote extremity of the wire, on the contrary, was completed, the disturbance of equilibrium commenced at this end, and the galvanometers successively acted in the reverse order, *i. e.* the galvanometer which was the most distant from the battery was the first impelled into motion. In the latter case, before the completion of the circuit the needles of the galvanometers had assumed constant deflections to a limited extent, owing to a feeble current arising from the uniform dispersion of the static electricity along the wire.

Exp. 5. The two extremities of the 660 miles of wire were brought into connexion with the opposite poles of the battery.

When one of the ends previously disconnected from the battery was united therewith, the galvanometers at the extremities of the wire, and consequently which were at equal distances from the poles of the battery, were immediately and simultaneously acted upon, while that which was in the middle of the wire was subsequently caused to move. When the wire disconnected in the middle instead of near one of the poles of the battery was again united, the middle galvanometer, which was the most remote from the battery, was the first acted upon, and those near the poles subsequently.

The comparison of the two above-mentioned experiments show that the earth must not be regarded simply as a conductor, which many suppose to be the case. Since in the first experiment there were not many yards' distance between the two earth terminations, did the extent of ground between them act only as a conductor, the two galvanometers at the extremities of the wire should have acted simultaneously, as in the second experiment, and as would have been the case had a short wire united the two extremities which proceeded to the earth.

Third Series.

Exp. 6. One pole of the battery was connected with the earth, and the opposite pole with one extremity of the 660 miles of wire, the other end remaining insulated: a delicate galvanometer was interposed near the battery. Notwithstanding there was no circuit formed, the needle showed a constant deflection of $33\frac{1}{2}°$; the feeble current thus rendered evident is not so much to be attributed to imperfect insulation, as to the uniform and continual dispersion of the static electricity with which the wire is charged throughout its entire length, in the same manner as would take place in any other charged body placed in an insulating medium. The strength of the current thus occasioned appears to be nearly, if not exactly, proportional to the length of the wire added, as the following table will show : the first column indicates the number of miles of wire subjoined beyond the galvanometer, and the second the corresponding deflections of the needle :—

miles.		
0	....	$\overset{\circ}{0}$
110	....	$6\frac{1}{2}$
220	....	12
330	....	18
440	....	$23\frac{1}{2}$
550	....	28
660	....	31

Exp. 7. One end of the 660 miles of wire was now allowed
to remain constantly in contact with one of the poles of the
battery; but the galvanometer was successively shifted to
different distances from the battery. The strength of the
current was now shown to be inversely as the distance of the
galvanometer from the battery, becoming null at its extremity, as shown in the following table. The first column
shows the distance from the battery at which the galvanometer was placed, and the second column the corresponding
deflection of the needle.

miles.		
Near the battery	....	$33\frac{1}{2}$
110	....	31
220	....	25
330	....	15
440	....	12
550	....	5
660	....	0

The deflections of the needle of the galvanometer employed in these experiments were, when they did not surpass
36°, very nearly comparable with the force of the current.
This I ascertained in the following way. I took six cells of
the small constant battery described in my paper " On new
Instruments and Processes for determining the Constants of
a Voltaic Circuit," printed in the *Philosophical Transactions*
for 1843, and placed in the circuit formed of the 660 miles
of wire, the earth, and the galvanometer, successively 1, 2, 3,
4, 5, and 6 cells. Leaving out of consideration the resistances in the cells themselves and in the earth, which were
very inconsiderable in comparison with that in the long wire,
the force of the current should be approximately proportionate to the number of the elements; and since the deflec-

tions of the needle nearly indicated this proportionality, as the following table will show, it may be assumed that the force of the current, when the deflection of the needle did not surpass 36°, nearly corresponded with the angular deviation.

cell.		
1	....	$\overset{\circ}{6}$
2	....	14
3	....	19
4	....	28
5	....	32
6	....	36

From the preceding experiments (6. and 7.) it seems to result, that whatever length of wire is connected with the battery, if a galvanometer is placed at the further extremity of the wire and a constant length added to the other termination of the galvanometer, its indication remains always nearly the same. Thus the galvanometer indicated $6\frac{1}{2}°$ when it was placed close to the battery and 110 miles of wire were subjoined beyond it; and 5° when 550 miles were interposed between the battery and galvanometer, the same length, 110 miles, being subjoined. In like manner, when 220 miles were added beyond the galvanometer placed near the battery, the indication was 12°; precisely the same as when 440 miles were interposed and 220 added. So also when 330 miles were added, the deviation of the galvanometer was 18°; and 15° when 330 miles were interposed and 330 added. I have no doubt that the correspondence would have been closer had it not been for the fluctuations of the battery.

It would appear from this, that whatever be the length of wire attached to the insulated pole of a battery, it becomes charged to the same degree of tension throughout its entire extent; so that another insulated wire brought into connexion with its free extremity exhibits precisely the same phenomena, in kind and measure, as when it is brought into immediate connexion with a pole of the battery. Some important practical consequences flow from this conclusion, which I will not develop at present, as I have not yet had an opportunity of submitting them to the test of experiment.

On the position of Aluminum in the Voltaic Series.

[From the 'Proceedings of the Royal Society,' 1855, vol. vii. pp. 369, 370.]

HAVING, through the kindness of Dr. Hofmann, been permitted to examine a specimen of aluminum prepared by M. Claire-Deville, I availed myself of the opportunity to ascertain one of the physical properties of this extraordinary metal, which it does not appear has been yet determined, viz. its order in the voltaic series. The following are the results of my experiments.

Solution of potass acts more energetically and with a greater evolution of hydrogen gas upon aluminum than it does on zinc, cadmium or tin. In this liquid aluminum is negative to zinc, and positive to cadmium, tin, lead, iron, copper and platina. Employed as the positive metal, the most steady and energetic current is obtained when it is opposed to copper as the negative metal; all the other metals negative to it which were tried became rapidly polarized, whether above or below copper in the series.

In a solution of hydrochloric acid aluminum is negative to zinc and cadmium, and positive to all the other metals above named. With this liquid also copper opposed to it as the negative metal gave the strongest and most constant current.

Nitric and sulphuric acids are known not to act chemically in any sensible manner on aluminum. With the former acid diluted as the exciting liquid aluminum is negative to zinc, cadmium, tin, lead and iron. The current with zinc is strong; with the other metals very weak, and it is probable that their apparent negative condition is the result of polarization. When aluminum is immersed in dilute sulphuric acid, this metal appears negative to zinc, cadmium, tin and iron, but with lead, on which sulphuric acid has no action, the current is insensible. In both these liquids copper and platina are negative to aluminum, and notwithstanding the apparent absence of chemical action on the latter metal, weak currents are produced.

It is rather remarkable that a metal, the atomic number of which is so small, and the specific gravity of which is so low, should occupy such a position in the electromotive scale as to be more negative than zinc in the series.

Télégraphe automatique écrivant.

[From the 'Comptes Rendus de l'Académie des Sciences,' 1859, tome
xlviii. pp. 214-220.]

J'ai l'honneur de soumettre à l'Académie un nouveau télégraphe automatique écrivant, lequel, je pense, présente des avantages que l'on n'a pas encore obtenus jusqu'ici. Avec l'appareil déposé actuellement sur le bureau, on peut imprimer 500 lettres par minute. Les bandes trouées de papier qui déterminent l'ordre et la succession des courants électriques par un mécanisme analogue à celui des métiers à la Jacquard, sont préparées de telle sorte, que les groupes de points qui constituent les différentes lettres sont distinctement separés, ce qui rend impossible la confusion à laquelle donne lieu fréquemment aujourd'hui la succession continue de lettres adjacentes ; et l'impression de la dépêche en points tracés à l'encre à écrire n'ajoute rien au poids des organes de l'appareil ou n'oppose aucune résistance à la puissance motrice des électro-aimants.

Mon invention consiste essentiellement dans une nouvelle combinaison de mécanismes ou appareils, ayant pour objet la transmission à travers un circuit télégraphique de messages ou dépêches préparées à l'avance et qu'il s'agit de signaler ou écrire à une station éloignée. De longues bandes ou rubans de papier sont percées, par une machine construite à cet effet, d'ouvertures ou trous groupés de manière à représenter les lettres de l'alphabet ou d'autres caractères conventionnels. La bande ainsi préparée est placée dans un appareil associé à un rhéomoteur ou source quelconque de puissance électrique, lequel appareil mis en action fait mouvoir longitudinalement la bande de papier, et la fait agir sur deux pointes, de telle manière que si l'une des pointes est soulevée, le courant est transmis au circuit télégraphique dans une

certaine direction, tandis que si c'est l'autre pointe qui est
soulevée, le courant est transmis dans la direction opposée.
Les soulèvements et les abaissements des pointes sont gou-
vernés ou déterminés par les trous du papier et les intervalles pleins qui les séparent. Ces courants qui se suivent
ainsi, tantôt dans une direction, tantôt dans la direction opposée, agissent sur un appareil écrivant ou imprimant à la
station distante, de manière à lui faire produire des marques
correspondantes sur un ruban ou bande de papier mue par
un mécanisme approprié.

Je vais maintenant décrire plus en détail les diverses parties
ou organes de ce système télégraphique, en faisant remarquer
à l'avance que chacun de ces organes a son individualité et
son originalité propres, et peut être appliqué aux appareils
déjà existants.

Le premier appareil est un *perforateur,* instrument destiné
à percer de trous les bandes de papier dans l'ordre voulu
pour former la dépêche. La bande de papier passe dans une
rainure servant à la guider. Sur le fond de la rainure on a
ménagé une ouverture assez large pour permettre le mouvement de va-et-vient du bord supérieur d'un châssis portant
trois emportepièce ou poinçons dont les extrémités sont
placées sur une même ligne transversale ou perpendiculaire à
la longueur de la bande. Chacun de ces poinçons peut séparément se soulever par l'action du doigt sur une touche qui
lui correspond. La pression du doigt sur la touche, en outre
du soulèvement du poinçon correspondant, soulèvement qui
a pour objet de percer le papier, produit successivement deux
mouvements différents : premièrement elle soulève une pince
qui fixe le papier dans la position qu'il occupe ; secondement
elle fait avancer le châssis qui porte les trois poinçons. Dans
ce mouvement en avant, celui des poinçons qui a été soulevé,
entraîne la bande de papier et la déplace de la quantité voulue.
Pendant le mouvement de retour de la touche qui a lieu
lorsque le doigt a cessé de la presser, la pince fixe d'abord le
papier, puis le châssis retombe dans sa position normale.
Les deux touches et les deux poinçons extérieurs servent à
percer les trous qui, groupés ensemble, représentent les

M

lettres ou autres caractères. La touche et la poinçon du milieu servent à faire les trous plus petits qui marquent les intervalles de séparation de deux lettres ou caractères consécutifs. Les perforations de la bande se dessinent donc de la manière suivante :

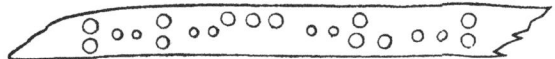

Par une addition très-simple, on rend le perforateur apte à transmettre de nouveau à une station plus éloignée une dépêche qui vient d'être reçue imprimée, sans qu'il soit aucunement nécessaire de la traduire, sans même qu'on ait besoin de savoir ce qu'elle exprime ou signifie. On fait passer la bande imprimée qui vient d'être reçue entre deux rouleaux dont l'un reçoit le mouvement d'une vis tournée à la main de manière à faire passer successivement les caractères de la dépêche sous les yeux de l'opérateur. On agit avec la main droite sur les touches du perforateur, tandis qu'on fait tourner la vis de la main gauche ; à mesure que les caractères se présentent successivement à la vue, on abaisse les touches correspondantes aux points dont les lettres sont composées. C'est une opération toute machinale et qui n'exige presque aucun effort d'intelligence. Il n'y a, en réalité, rien de changé à l'alphabet actuellement en usage ; on peut convenir, en effet, que les points d'un côté de la bande représenteront les points ou marques courtes, et les points de l'autre côté de la bande, les traits ou marques longues de l'alphabet actuel, l'ordre de succession des marques restant d'ailleurs ce qu'il est : seulement, dans mon système, les lettres occupent un espace moins long et sont, par conséquent, lues plus facilement.

Le second appareil est le *transmetteur*, dont la fonction est de recevoir les bandes de papier préalablement percées par le perforateur, et de transmettre les courants produits par une pile voltaïque ou tout autre rhéomoteur dans l'ordre et la direction déterminés par les trous faits dans le papier. Cette transmission s'opère par un mécanisme assez semblable à

celui par lequel le perforateur exerce ses fonctions. Un excentrique produit et règle la récurrence ou succession de
trois mouvements : 1° le mouvement de va-et-vient d'un petit
châssis qui contient une coulisse avec rainure destinée à
recevoir la bande de papier et à la faire avancer pendant le
déplacement en avant du châssis; 2° le soulèvement et l'abais-
sement d'un ressort qui maintient la bande de papier fixe
pendant le mouvement en arrière du châssis et lui permet de
le suivre dans son mouvement en avant; 3° l'élévation ou
soulèvement simultané de trois pointes ou fils métalliques
placés parallèlement les uns aux autres, reposant par l'une
de leurs extrémités sur l'axe de l'excentrique et pénétrant
par leur autre extrémité libre dans des trous percés dans la
rainure de la coulisse. Les trois fils ne sont pas fixés à l'axe
de l'excentrique, mais chacun d'eux est appuyé contre lui par
un ressort poussant de bas en haut, de sorte que si l'on
exerce un léger effort sur l'extrémité libre de l'un quelconque
des fils, le fil que l'on presse peut s'abaisser indépendamment
des autres. Si la bande de papier n'est pas insérée ou mise
en place, et si l'on fait mouvoir l'excentrique, une pointe
attachée à chacun des deux fils extérieurs passe, durant
chaque mouvement en avant et en arrière du châssis, du contact avec un ressort au contact avec un autre ressort; par le
moyen de ces contacts et d'isolements convenablement mé-
nagés, tout est disposé de telle sorte, que pendant qu'un des
fils est abaissé, l'autre restant soulevé, le courant passe de la
source électrique dans le circuit télégraphique suivant une
certaine direction, tandis qu'il passe dans la direction contraire si le fil d'abord soulevé est maintenant abaissé, et ré-
ciproquement; le courant sera interrompu ou cessera de
passer si les deux fils sont à la fois soulevés ou abaissés. Si
maintenant la bande de papier préparée est placée dans la
rainure et entraînée en avant, quel que soit celui des deux
fils qui entre dans un des trous de la rangée ou série qui lui
correspond, le courant passe dans une direction; et quand
l'extrémité de l'autre fil entrera à son tour dans un des trous
de la seconde rangée ou série, le courant passera dans la
direction opposée; par ce moyen les courants sont amenés à

M 2

se succéder l'un à l'autre automatiquement dans l'ordre et la direction voulus pour produire toute espèce de signaux. Le fil du milieu agit simplement comme guide du papier pendant la cessation des courants. La roue qui fait marcher l'excentrique peut être tournée à la main ou par une force motrice quelconque. Lorsque le mouvement des transmetteurs sera effectué par des machines, un ou deux aides suffiront pour en surveiller un nombre quelconque, pour transmettre un nombre égal de dépêches. Ce transmetteur n'exige qu'un seul fil télégraphique ; on peut d'ailleurs, si l'on veut, réduire la dépêche à une seule rangée de trous ; dans ce dernier cas, le perforateur pourrait n'avoir que deux touches et le transmetteur deux tiges ou fils au lieu de trois.

Le troisième appareil est le *récepteur* qui produit à la station d'arrivée, sur une bande de papier, des marques ou points noirs correspondants dans leur arrangement régulier aux trous du papier percé. Les plumes ou styles sont soulevées ou abaissées par leur liaison avec les parties mobiles des électro-aimants. Elles sont indépendantes l'une de l'autre dans leur action et tellement disposées, que si le courant passe à travers le fil inducteur de l'électro-aimant dans une direction, une des plumes est abaissée, et que lorsque le courant passe en sens contraire, c'est l'autre qui est abaissée. Lorsque le courant cesse, de légers ressorts ou mieux de petits aimants ramènent les plumes à leur position ou à leur élévation normale. L'encre est fournie aux plumes de la manière suivante, par un réservoir de 3 millimètres environ de hauteur, d'une longueur et d'une largeur convenables, fait d'une pièce de métal doré à la intérieur pour prévenir l'action corrosive de l'encre qu'on y verse : le fond de ce réservoir est percé de deux trous assez petits pour que l'action capillaire empêche l'encre de couler par leurs ouvertures ; les extrémités des plumes sont placées immédiatement au-dessus de ces petits trous ; elles y pénètrent lorsque l'action des électro-aimants les abaisse, emportant avec elles une charge d'encre suffisante pour imprimer des marques ou points très-visibles à la surface du papier qui passe sous elles. Le mouvement de progression en avant du papier est produit et réglé par un mécanisme semblable à celui des récepteurs des autres télégraphes imprimants.

Le quatrième appareil est un instrument que j'appelle *traducteur*. Son objet est de traduire le signal télégraphique formé d'une succession ou ensemble de points ou de marques conventionnelles en caractères alphabétiques ordinaires. Dans le système que j'ai adopté et qui limite à quatre le nombre des points entrant dans un signal, on dispose de trente caractères distincts. Le traducteur montre au dehors neuf touches, dont huit sont disposées sur deux rangées parallèles, quatre dans chaque rangée ; la neuvième touche est placée séparément. La partie principale du mécanisme est une roue portant à sa circonférence trente types, placés à des distances égales, représentant les lettres et autres caractères de l'alphabet. Un second mécanisme est tellement disposé et uni au premier, que si l'on abaisse les touches de la rangée supérieure, la roue s'avance de 1, 2, 4 ou 8 pas ou lettres ; que si l'on abaisse de la même manière les touches de la rangée inférieure, la roue avance respectivement de 2, 4, 8 ou 16 pas ou lettres. Par cette disposition, lorsque les touches sont abaissées successivement dans l'ordre suivant lequel les points sont imprimés sur le papier, c'est-à-dire si l'on abaisse la première touche pour un point, la première et la seconde pour deux points, etc., et que l'on choisisse les touches de la rangée supérieure ou de la rangée inférieure suivant que le point est sur la ligne supérieure ou sur la ligne inférieure de la bande de papier imprimée, la roue à types ou lettres sera amenée dans la position convenable pour montrer la lettre correspondante à la succession ou ensemble de points tracés sur la bande. Le neuvième touche, lorsqu'elle est abaissée, agit pour imprimer le type ou la lettre sur la bande, pour la faire avancer de manière à offrir une place fraîche ou libre à la roue à types, et pour ramener la roue à types à sa position première.

Je termine par quelques remarques sur les avantages que présente ce nouveau système. Quelle que soit la dextérité pratique que puisse acquérir un opérateur agissant par sa volonté, le résultat obtenu par lui sera toujours très-inférieur à celui qui sera donné par un procédé automatique qui n'est limité que par la vitesse que l'on peut imprimer aux mouvements du transmetteur. Dans l'état actuel de la construction de mon appareil, on peut transmettre à des distances moyennes cinq fois plus de signaux qu'on ne peut en envoyer aujourd'hui ; pour des distances très-considérables, et dans des conducteurs soumis à des influences inductrices, la vitesse de transmission sera nécessairement limitée par la tendance que des courants très-courts, ou qui se succèdent avec une grande rapidité, ont à s'unir ou à se fondre l'un dans l'autre.

Mais alors même que le procédé automatique ne l'emporterait pas sur le mode d'expédition à la main au point de vue de la vitesse d'impression ou de transmission des dépêches, il n'en serait pas moins vrai qu'il possède des avantages incontestables. Actuellement, pour que le travail d'une ligne télégraphique soit profitable, il est nécessaire que l'opérateur arrive à manipuler aussi rapidement que le permet l'exactitude de la transmission de la dépêche ; il faut beaucoup d'intelligence ou d'adresse pour devenir maître dans ce genre de manipulations ; il faut en outre que la langue dans laquelle la dépêche est écrite soit tout à fait familière à celui qui l'expédie ; car s'il avait à envoyer une dépêche écrite dans une langue inconnue ou en chiffres, il serait forcé de procéder avec précaution et avec lenteur.

Dans mon nouveau système, au contraire, les dépêches préparées sont transmises avec la même rapidité dans quelque langue alphabétique ou chiffrée qu'elles soient écrites ; et comme les bandes trouées peuvent être préparées à loisir, comme aussi elles peuvent être soumises à la révision d'un correcteur, on se trouve dans des conditions d'exactitude que le système de transmission volontaire à la main ne fournira jamais. S'il faut plusieurs aides pour préparer les dépêches que pourra expédier une seule ligne télégraphique constamment en activité, leur temps, au point de vue économique, a beaucoup moins de valeur ou coûtera moins que le temps employé à transmettre un message à la main. Un autre avantage du nouveau système est que la même dépêche preparée peut être transmise par un nombre quelconque de lignes distinctes, sinon simultanément, du moins par une succession si rapide, qu'elle équivaut à la simultanéité. En outre et sans aucun travail additionnel, la même dépêche peut être transmise une seconde fois, si cela est nécessaire; et les dépêches relatives à un service courant, journalier ou périodique, peuvent être gardées pour servir à une transmission nouvelle quand le besoin s'en fera sentir.

Si le système de transmission automatique était généralement adopté, il serait plus naturel que les dépêches fussent préparées dans le bureau qui commande leur expédition, d'autant plus que les appareils à l'aide desquels on les prépare sont très-portatifs et très-peu coûteux. Les opérations dans le bureau télégraphique se borneraient dans ce cas à faire passer les bandes trouées à travers le transmetteur d'une station et à recevoir à l'autre station la dépêche imprimée. La traduction comme la préparation de la dépêche resterait du ressort du bureau de l'administration qu'elle concerne.

Dans le cas actuel, il ne s'agit pas de substituer à un genre d'habileté acquise un autre genre également difficile à acquérir, ce qui condamnerait l'universalité des employés à un travail long et pénible. La grande dextérité pratique exigée aujourd'hui ne sera plus nécessaire, puisque les principales et les plus laborieuses opérations sont entièrement automatiques ou mécaniques, il n'y aura que fort peu de chose à apprendre; il y aura plutôt quelque chose à oublier.

On the Circumstances which influence the Inductive Discharges of Submarine Telegraphic Cables.

[From a Report of the Joint Committee appointed by the Lords of the Committee of Privy Council for Trade and the Atlantic Telegraph Company, to inquire into the construction of Submarine Telegraph Cables.   Eyre & Spottiswoode, 1861.]

[Plate XVII.]

1.

THE introduction, in late years, of submarine telegraphic cables, and the substitution of subterranean lines of considerable lengths for the aërial lines almost exclusively employed formerly, have brought into evidence certain conditions of electrical charge which so materially influence the rapidity and frequency of the signals transmitted as to make it an object of the highest importance that they should be strictly investigated, in order that it may be ascertained whether it be possible to obviate the evil effects arising therefrom, or in any degree to alleviate them.

These conditions could not fail to come under the observation of persons employed in telegraphic operations, as soon as such lines were constructed; but before they were made the subject of investigation by Mr. Werner Siemens * and Dr. Faraday †, only vague notions respecting their cause prevailed.

When a metallic wire is enveloped by a coating of some insulating substance, as gutta-percha or india-rubber, and is then surrounded by water or damp earth, the system becomes exactly analogous to a Leyden jar or coated pane; the insulating covering represents the glass, the copper wire the inner metallic coating, and the water or moist earth the external coating.   The electricity with which the wire is charged, by

* Annales de Chimie et de Physique, 3ième série, tome xxix. 1850.
† Proceedings of the Royal Institution, January 20, 1854, " On Electric Induction—Associated cases of current and static effects."

bringing the pole of an active battery in contact with it, acts by induction on the opposite electricity of the surrounding medium, which in its turn re-acts on the electricity of the wire, drawing more from the source, and a considerable accumulation is thereby occasioned, which is greater in proportion to the thinness of the insulating covering.

One mile of copper wire, one sixteenth of an inch in diameter, presents a surface of 85 95 square feet, and receives the same charge from a source of the same tension as a Leyden jar having an equal number of square feet of tinfoil coating. There is, however, one material difference between the two cases. Though both are discharged in a time inappreciably minute to the senses, the discharge from the wire occupies a comparatively much longer interval than that from the coatings of the jar.

A wire on insulating supports, in the open air, when it is unconnected with the earth, receives also a charge, but very much smaller in amount, the inductive action of distant surrounding bodies exerting but little influence upon it.

Although certain general conditions of this inductive action in telegraphic wires have been made the subject of experiment, and are now well known, our knowledge in regard to this subject still remains very limited.

It was therefore thought desirable that a series of experiments should be instituted, under circumstances as closely resembling those which take place in actual telegraphic lines as possible, in order to determine the amount of inductive discharge in wires of very considerable lengths, according to variations in the lengths of the wires, their diameters, the material and thickness of the insulating substance, and in the temperature and pressure of the medium by which they are surrounded.

For the purpose of these experiments the following coated wires were procured :—

Sixteen miles of copper wire, 1/16 of an inch in diameter, with a coating of gutta-percha 3/32 of an inch in thickness.

Nine wires, each one mile in length, of various diameters coated with gutta-percha of different thicknesses. Duplicates were provided of five of these wires.

Fifteen wires, each one mile in length, and 1/16 of an inch in diameter, coated with different insulating materials, prepared by different manufacturers, as hereafter described.

In all, 45 miles of insulated wire.

These lines were formed into large coils, and placed in a tank filled with water. The ends of each mile were brought into the operating-room, for the purpose of being connected with the battery, the instruments, and the earth communications, as occasion required.

The battery, which was on Daniell's principle, consisted of 512 pairs of plates; and during the course of the experiments it was so arranged that either 1, 2, 4, 8, 16, 32, 64, 128, 256, or 512 elements might be brought into action at pleasure. Every care was taken to have the battery well insulated.

2. *Instrument for charging and discharging Wires.*

The charging and discharging of the wires were effected by the simple apparatus represented by fig. 1 (Plate XVII.).

A B is a slab of ebonite*; C, D, E, three pieces of brass, furnished with binding-screws 1, 2, 3; to the middle piece is attached a double lever key, by means of which it may be brought at will into metallic communication with either of the external pieces. The part of the key to be touched by the finger is also made of ebonite. F, G, H, are three insulated pieces of brass, each having two binding-screws, 4 7, 5 8, and 6 9.

A short wire from one pole of the battery, the other being connected with the earth, is to be attached to the binding-screw 7; the insulated wire, whose charge or discharge is to be measured, to 8; and 9 is to be connected by a wire to the earth.

To ascertain the amount of *charge* communicated to the insulated wire, a galvanometer must be inserted between 1 and 4. On bringing the key into contact with C, the charge of the battery is communicated to the wire, and the amount is indicated by the galvanometer.

* Ebonite is an indurated preparation of india-rubber, and is an excellent insulator.

To measure the *discharge* from the insulated wire, the galvanometer must be placed between 3 and 6. The key is first brought into contact with C, by which the wire is charged; it is then brought into contact with E, which operation discharges the wire through the galvanometer into the earth.

If the galvanometer be interposed between 2 and 5, the currents arising both from the charge and discharge are successively indicated by the galvanometer.

The binding-screws 1 4, 2 5, 3 6, when not connected by a galvanometer must be united by short wires.

3.

Galvanometers of very different constructions were employed at various times in these experiments, which lasted several months. As the tension of the battery was subject to considerable variations at different times from a variety of causes, and as the needles of the galvanometers were occasionally affected by powerful currents sent through them, which occasioned changes in the directive power, no comparison can be safely made between experiments with the same instruments made at distant times. Every precaution, however, was taken to ensure the perfect comparability of the experiments of each group made at the same time.

The galvanometers employed were as follows :—

A galvanometer with a pair of astatic needles suspended by a silk fibre without torsion. Length of wire 80 yards; diameter of wire 1-40th of an inch .... A.

A similar galvanometer, the weight of the needles only being different .... B.

A galvanometer with very long wire, the resistance of which is stated by its constructor (Mr. Henley) to be equal to five miles of the ordinary telegraphic wire .... C.

Another galvanometer, with very long wire, having a much greater resistance..................................... D.

A galvanometer with needles not astatic suspended by a silk fibre, having 30,500 coils of wire 1-500 of an inch in diameter .... E.

A torsion galvanometer with astatic needles suspended by a glass thread. Length of wire 130 yards, diameter of wire 1-40 of an inch .... F.

The currents produced during the charging or discharging of wires, even of many miles in length, are of such extremely short duration, that they cannot affect the needle of a galvanometer in a permanent manner; we are therefore obliged to take as their measure the instantaneous maximum deflection of the needle which they occasion.

The forces which produce these deflections are, as in the similar case of the balistic pendulum, generally assumed to be as the chord of the arc through which the needle passes. The arcs themselves, when they do not surpass 30° or 40°, may, for these experiments, which do not admit of the utmost accuracy, be considered as sufficient approximations.

4. *Influence of the Electro-motive Force of the Battery on the amount of Discharge.*

A variety of experiments were made with lines varying from 1 to 16 miles in length, and with batteries varying from 16 to 512 elements; they all led to the same result, that the amount of discharge is proportionate to the electro-motive force or tension of the battery. I will record one series of experiments, in which the galvanometer A was employed, in proof of this :—

Number of Elements.	One Mile.	Eight Miles.	Sixteen Miles.	Peltier's Electrometer.
16	° ..	° 2·5	° 5	° ..
32	..	5	10	..
64	..	10	20	14
128	2·5	20	41	28
256	5	41	88	..
512	10	88	..	..

It will be found that the chords of these angles are proportionate to the electro-motive forces, or number of elements of the battery.

The fifth column gives the tensions of the battery ascertained by Peltier's electronometer; the high tensions are not given, because beyond 30° the forces are no longer in proportion to the angles.

5. *Influence of the Length of the Wire on the Inductive Discharge.*

In the following experiments the discharge was taken successively from 1, 2, 3, and 4 miles of copper wire, 2/32 of an inch in diameter, covered with gutta-percha 3/32 of an inch in thickness. A battery of 504 elements was employed to charge the wires, and two short-wire galvanometers (A and B) were used for obtaining the measures. The numbers in the first and second columns were obtained with the same galvanometer, but the tension of the battery was different in the two cases: the galvanometer B had a more sensitive needle.

	Galvanometer A.		Galvanometer B.
1 mile	7·5	7	11
2 miles	15	14	22·5
3 miles	22·5	21	33·3
4 miles	30	28	43

The discharges from one and two miles were also compared when the wires had different diameters and different thicknesses of insulating covering.

Diameter of copper wire.	Thickness of gutta-percha.	One mile.	Two miles.
2/32	3/32	8	16
2/32	6/32	6	12
2/32	12/32	5	9·5
4/32	6/32	8·5	17
8/32	6/32	13	25·5

From these experiments it is evident that the discharge is directly as the length of the wire.

In the galvanometers employed when the deviations do not exceed 30°, the angles may, without sensible error, be substituted for the chords.

When galvanometers with very long wires are employed, equally correct results are not obtained; they give for ascending degrees numbers considerably in excess, as the following table will show. A battery of 72 elements was employed with the galvanometer C, and one of 144 elements with the galvanometer D.

	Galvanometer C.	Galvanometer D.
1 mile	24, 24, 23	10, 10, 10, 11
2 miles	49, 49, 52, 49	23, 24, 23,
3 miles	89, 89	35, 36½, 38
4 miles	134, 134, 132	52, 52, 50

Assuming that the discharges are proportionate to the lengths of the wire, and that the chords of the angles accurately measure these discharges, the angles should have been :—

	Galvanometer C.	Galvanometer D.
1 mile	24	11
2 miles	49	22
3 miles	77	33
4 miles	110	45

6. *Simultaneous Discharges.*

The following experiments show that when a discharge is effected simultaneously from a number of wires united at their discharging ends, the effect on a galvanometer is the same as when the wires are all united together to form one continuous length. No doubt when four miles of wire are simultaneously discharged, the discharge is effected in one fourth of the time that it is from the same wires united continuously; but as the intensity is at the same time increased four times, the total quantity, which is the intensity multiplied by the time, remains the same, and produces the same deflection of the needle of the galvanometer.

The battery of 504 elements and the galvanometer A were employed for these experiments.

	In continuous length.	Side by side.
1 mile	7·5	7·5
2 miles	15	15
3 miles	22·5	22
4 miles	30	30

7. *Influence of the Conductivity of the Wire.*

An iron wire, one mile in length, 4/32 of an inch in diameter, and with a covering of gutta-percha 3/32 of an inch in thickness, was compared, with respect to the amount of its discharge, with a copper wire of the same diameter, and similarly coated. Employing the same battery force and the same galvanometer, the discharge was in all cases found to be the same in both lines. It appears from the particulars furnished by the Gutta-percha Company, that the weights of the gutta-percha coatings on the two wires were not precisely the same, that on the iron wire being 150 lbs. to the mile, while that on the copper wire was only 134 lbs. to the mile; the difference of thickness hereby occasioned would not, however, materially influence the result.

Iron affords eight times greater resistance to the passage of the voltaic current than copper does. It has been proposed in the construction of submarine cables to substitute for the copper conducting wire, an iron wire, increasing its thickness so as to render its conductivity equal to the former; but though the force of the current would remain the same, the induction would increase with the diameter of the wire, and a positive disadvantage would ensue.

The conductivity of the metal, other circumstances remaining the same, therefore does not seem to influence the amount of the discharge.

For the same reason, care should be taken to procure for submarine telegraphic lines copper wire of the highest conductivity. It will be seen in Dr. Matthiessen's report that the copper wires of commerce vary extremely, from 100 the standard to 14·24. By employing a bad conductor of the same diameter, the resistance of the circuit is increased, while the induction remains the same.

8. *Influence of the Diameter of the Wire and the Thickness of the Insulating Coating on the Amount of Discharge.*

In order to ascertain in what manner the diameter of the wire and the thickness of the insulating coating affects the amount of discharge, the Gutta-percha Company was requested to prepare nine experimental lines of copper wire covered with gutta-percha, each one mile in length, the wires having various diameters, and the gutta-percha covering various thicknesses, as shown in the following table. The figures in the horizontal row indicate the former, and those in the vertical row the latter; the unit is the thirty-secondth of an inch.

When the lines were manufactured, the quantities of gutta-percha in the various coverings were found not to be so exact as they ought to have been. To enable corrections to be made, I have given in the following table the weights of the gutta-percha coverings given by the company, and also the relative weights corresponding to the thicknesses intended, the line 2/3 being taken for the point of comparison. The upper figures represent the calculated weights, those below them the weights given by the company, and the lowest their differences positive or negative. The actual weights were determined in the following manner:—The coils of coated wire were weighed, and the weight of the copper wire having been previously calculated, it was deducted from the gross, and the remainder given as the weight of the gutta-percha. "It is almost impossible," says the manager of the company, "to put gutta-percha on the wires with sufficient accuracy to ensure two miles having precisely the same weight; copper wire will vary too in the same length."

In this table the weights are given in pounds :—

COPPER WIRE.

		2	4	8
	3	100 100 — 0	140 134 — 6	220 159 — 61
Gutta-percha coating.	6	320 338 — −18	400 408 — −8	560 579 — −19
	12	1100 1120 — −20	1280 1300 — −20	1600 1636 — −36

The table which follows gives the ratios of the two numbers, or the calculated divided by the actual weights. The greatest discrepancy is in the line 8/3.

	2	4	8
3	1·000	1·044	1·382
6	·946	·980	·967
12	·982	·984	·978

To control in some degree the weights given by the Company, a six-inch length of each of the lines was cut off and their gutta-percha coatings were accurately weighed. The following table gives these weights in grains, compared with the relative weights corresponding with the calculated thicknesses. The upper figures represent the calculated weights; as before, 2/3 is taken as the point of comparison.

	2	4	8
3	66	92·4	145·2
	66	93·	143·
	0	− ·6	2·2
6	211·2	264·	369·6
	229·	265·	385·
	−17·8	−1	−15·4
12	739·2	844·8	1056
	712·	878·	1090
	27·2	−33·2	−34

The next table shows the ratios of the calculated and actual weights given in the above table.

	2	4	8
3	1·000	·993	1·053
6	·922	·996	·960
12	1·038	962	·968

It will here be seen that, in the case of 8/3, the actual weight more nearly corresponds with the calculated weight

N

than that rendered by the company. This fact, added to the results of the experiments hereafter recorded, inclines me to think that there is some error in the statement of the gross weight of this wire given by the company; but on this point I have not been able to obtain any satisfactory information.

Many series of experiments were made at various times to measure the discharges from these nine lines. Battery-powers from 32 to 512 elements were employed, and the measures were made with galvanometers of very different constructions. The following tables exhibit some of these results. Those on the left-hand side give the angular deflections of the needles, and those on the right-hand side the corresponding chords of these angles. The arrangement of these tables allows the influence of the diameter of the wire and of the thickness of the gutta-percha coating to be separately seen by a simple inspection.

FIRST SERIES.
Battery of 512 elements—Galvanometer A.

	2	4	8		2	4	8
3	7·2	12·0	15·7	3	62	104	136
6	5·0	7·9	12·0	6	43	69	104
12	3·8	5·0	7·9	12	32	43	69

SECOND SERIES.
Battery of 512 elements—Galvanometer F.

	2	4	8		2	4	8
3	11·2	18·0	24·0	3	97	156	207
6	7·8	11·8	18·0	6	67	102	156
12	5·8	8·2	12·0	12	50	71	104

THIRD SERIES.
Battery of 128 elements—Galvanometer D.

	2	4	8		2	4	8
3	15·9	25·0	31·0	3	138	216	267
6	11·2	16·0	25·0	6	97	139	216
12	9·0	12·0	16·6	12	78	104	144

FOURTH SERIES.
Battery of 128 elements—Galvanometer E.

	2	4	8		2	4	8
3	25·5	40·0	53·0	3	220	342	446
6	18·7	27·0	41·0	6	162	253	350
12	14·0	19·5	27·0	12	121	169	233

FIFTH SERIES.
Battery of 64 elements—Galvanometer C.

	2	4	8		2	4	8
3	13	22	30	3	113	190	258
6	8·5	13	22	6	74	113	190
12	6	9	13·5	12	52	78	117

SIXTH SERIES.
Battery of 512 elements—Galvanometer B, with charged needles.

	2	4	8		2	4	8
3	9	14	18·5	3	78	121	160
6	6	9	14·5	6	52	78	126
12	5	6·5	9·5	12	43	61	82

Comparing all these groups of experiments together, we shall not be far from the truth if we assume, as an empirical law, that the amount of discharge from wires of different diameters, with coverings of various thicknesses of the same insulating material, is directly as the square root of the semi-diameter of the wire, and inversely as the square root of the thickness of the insulating envelope. At any rate, this law is approximatively true for every proportion that can be practically employed in telegraphic operations.

If we look more minutely to the results, it appears that the discharge increases a little more rapidly than the square root of the radius of the wire, and decreases a little more rapidly than the square root of the thickness of the gutta-percha; and that when the wire and gutta-percha increase in the same proportion, the discharge does not remain exactly the same, but increases slightly as the total diameter increases; but these differences may have arisen from variations in the diameter of the wires and thickness of the insulating material in the manufacture of the wires.

There is one practical consequence of some importance to be inferred from these results. It is seen that by increasing the diameter of the wire and the thickness of the insulating covering in the same proportion, the amount of inductive discharge remains the same. According to Ohm's law, the force of the current in a voltaic circuit, when the length of the circuit remains the same, and the resistance of the battery is so inconsiderable as to be disregarded when compared with that of the metallic portion of the circuit, increases as the square of the diameter of the wire. We have therefore a means of increasing the strength of the current without augmenting the effects of induction. If it be found inconvenient in practice to increase both the wire and insulating covering proportionably, greater advantage will be obtained by increasing the diameter of the wire than the thickness of the gutta percha; for while the latter remains the same, the inductive discharge increases only as the square root of the diameter of the wire, while the force of the current increases as the square of the diameter; whereas, if the diameter of the wire remains the same, and the thickness of the gutta-percha be made to vary, the strength of the current will remain the same, while the induction will only decrease as the square root of the thickness. If the resistance of the battery be considerable, the advantage arising from the thickness of the wire will not be so great.

9. *On the Influence of the Insulating Material on the Amount of Discharge.*

Gutta-percha is the insulating material which has been hitherto almost exclusively employed in the manufacture of submarine and subterranean telegraphic lines. Wires coated with india-rubber have, however, in some places been brought into use, and recently numerous patents have been obtained for compound materials, into which either gutta-percha or india-rubber largely enters, or for interposing other substances between the wires and the gutta-percha or india-rubber coating, or between different coatings of these materials.

To ascertain the comparative advantages and disadvantages of these various materials with respect to their inductive and insulating properties, the manufacturers of them were requested by the Committee to furnish, for the purposes of experiment, wires one mile in length and one-sixteenth of an inch diameter, coated according to their respective processes. The following is a list of the manufacturers who responded to the application, and upon whose coated wires the comparative experiments were made :—

	Thickness of coating.
1. The Gutta-percha Company:	
An improved preparation of gutta-percha, in which collodion is said to be one of the constituents	3/32
Standard gutta-percha	3/32
Do., with coating of double thickness	6/32
Ten coatings of gutta-percha alternating with 10 coatings of Chatterton's compound	6/32
2. Messrs. Silver and Co.:	
Masticated india-rubber, cut into narrow slips, which are coiled spirally round the wire, and are cemented by a heat not exceeding that of boiling water .... B	6·25/32
Do.. C	3·86/32
3. Messrs. Hall and Wells:	
Three coatings of pure india-rubber covered externally with vulcanized india-rubber thread, with cotton thread interposed between the wire and the coating ...... A	3/32
One coating next the cotton-covered wire of pure india-rubber, three coatings of manufactured india-rubber, and one externally of vulcanized india-rubber thread .. B	3/32
4. Mr. Leonard Wray:	
Wray's compound, consisting of india-rubber, shell-lac, powdered flint, and gutta-percha	6/32
5. Mr. Hughes:	
Two coatings of gutta-percha, with a viscid substance, said to be prepared from coal, interposed between the two coatings	6/32

6. Mr. Radcliffe :

Thickness of coating.

A preparation of gutta-percha with carbon, for the purpose of increasing its durability.................. A   6/32

Do. .... B   6/32

7. Mr. Godefroy :

A compound of gutta-percha and pounded cocoanut-shell   3/32

As standards with which to compare the lines above enumerated, two wires coated with ordinary gutta-percha, one 3/32 and the other 6/32 of an inch, were similarly experimented upon at the same time.

In the following table the wires coated with different materials are arranged according to the thickness of their coatings. In all cases the copper wire was one sixteenth of an inch in diameter.

Column A gives the discharges according to my own experiments. A battery of 512 elements and the galvanometer were employed. I made a great number of experiments to measure the discharges from these lines, employing batteries of different powers and galvanometers of various kinds ; the results were in satisfactory accordance, especially so far as the order of the inductive capacities was concerned.

Column B.—Another series of measures of discharges. These results are taken from Mr. Bartholomew's tables, temperature 52° F. A battery of 128 elements and the galvanometer were employed.

Column C.—Loss of insulation for dynamic electricity at 52° F., extracted from Mr. Rowland's table.

Column D.—The number of minutes and seconds in which the line lost half its charge of static electricity. Taken from Mr. Bartholomew's table ; temperature 52° F.

The extra thickness of Silver's india-rubber covered wire was due, in each case, to an external covering of tarred tape.

By comparing the results given in the following table, several interesting conclusions may be deduced.

	Thickness of coating, unit = 1/32 of an inch.	A.	B.	C.	D.
		°	°	°	m. s.
Silver's C.............	6·25	4·5	10·2	0·	79 0
Improved gutta-percha C.	6	4·5	10·7	0·	11 15
Wray's	6	4·5	11·4	0·	50 0
Gutta-percha alternate..	6	6	14·4	·43	6 40
Hughes's	6	6	15·1	10·3	0 40
Standard gutta-percha ..	6	6	15·5	6·16	0 55
Radcliffe's B...........	6	6·5	16·6	14·16	0 30
Radcliffe's A...........	6	7	17·0	10·16	0 29
Silver's B	3·86	6	11·2	0·	23 2
Improved gutta-percha B.	3	7	15·8	4·6	4 57
Improved gutta-percha A.	3	6·5	16·1	5·	3 85
Hall and Wells's B	3	6·5	16·2	3·	3 42
Hall and Wells's A	3	8·25	22·4	9·	0 51
Standard gutta-percha ..	3	9	20·	15·	0 18
Godefroy's	3	9·5	22·4	18·16	0 21

1. That india-rubber surpasses all other materials in the smallness of the amount of its inductive discharge and the perfectness of its insulation. A coating of india-rubber is fully equal to a gutta-percha coating of double its thickness.

2. That two artificial compositions, formed by the addition of other highly insulating materials to india-rubber or gutta-percha, viz. Wray's compound and the improved gutta-percha, closely resemble india-rubber in both these respects.

3. That the mixture of imperfectly conducting materials with gutta-percha, as carbon in Mr. Radcliffe's composition, or pounded cocoanut-shell as in Mr. Godefroy's, has the disadvantage of greatly reducing the insulation and increasing the induction.

4. That the interposition of cotton thread between the wire and an india-rubber coating, as in Hall and Wells's preparation, also considerably increases the induction and diminishes the insulation. The induction is augmented because the cotton thread, which is a bad insulator, increases the surface of the conductor; and the insulation is impaired, not only because the insulating coating is diminished by the thickness of the cotton, but probably also in consequence of the greater inductive action. The interposition of cotton

between two layers of gutta-percha is equally disadvantageous, as is proved by the experiments on Mr. Hearder's short line of 450 yards.

5. That the interposition of a viscid insulator between two coatings of gutta-percha neither decreases the induction nor improves the insulation of the line. If Mr. Hughes's process should be found to possess any advantage, it will be in the tendency of the viscid fluid to fill up air-holes or flaws in the gutta-percha coatings.

6. Generally speaking the more perfect the insulating property of the material is, the less is its inductive capacity. There are, however, several apparent exceptions to this rule among the experiments quoted; but there are so many causes to affect the experiments on insulation, that we are not warranted to infer, from these exceptions, the entire independence of the two properties.

10. *Influence of Temperature on the Amount of Discharge.*

Table II. [App. 6 to Report of the Joint Committee] shows the influence of temperatures varying from 32° F. to 92° F. on 29 coated wires of different descriptions. I will confine myself to a few remarks on the results there given.

The imperfectness of insulation increases with the temperature, but there is a great difference in the rate of increase, according to the insulating substance which is made the subject of experiment. Silver's india-rubber maintains an insulation almost perfect up to 92° F., whilst gutta-percha is very considerably affected. The degree in which other substances are affected will be seen in the Table.

Temperature does not, in general, appear to affect the charge in an immediate manner. In most of the lines in which the insulation is but slightly influenced by temperature, the amount of charge remains apparently the same; but when the augmentation of heat occasions a loss of insulation, the charge also increases with the temperature. Wray's compound is the most remarkable exception to this rule; between 32° F. and 92° F. its insulation is very little affected, but its induction varies about 3°/₀. On the contrary,

Silver's india-rubber, which retains its high insulation with little or no change, preserves nearly the same amount of discharge from 32° F. to 165° F.

From Mr. Bartholomew's Tables it is obvious that the amount of the discharge is generally less than that of the charge. This difference increases as the temperature becomes higher; and if the temperature remains the same, the difference is greater in proportion to the loss of insulation. It cannot be doubted that this difference between the charge and discharge arises solely from imperfect insulation, whether owing to the quality of the material or the increase of temperature. The *charge* is effected instantaneously, and the current arising from imperfect insulation is added to it; while the *discharge* is made after a short interval, during which some portion of the charge has had time to escape In the best insulating lines, as Silver's C, Wray's, and the improved gutta-percha C, the difference is unapparent at any temperature.

11. *Influence of Pressure on the Inductive Discharge and Insulation.*

It does not appear evident that pressure exerts any influence on the amount of the inductive discharge. Comparing together the results [recorded in Table VII., App. 6 to Report of the Joint Committee] of the experiments made with the wires not subjected to pressure with those made with the same wires when the water in which they were immersed was subjected to a pressure of three tons on the square inch, there appear to be greater discrepancies between the experiments made under the same circumstances of pressure at different times than there do between the comparative experiments made under pressure and without pressure.

The influence of pressure on the static insulation is, however, very decided. Pressure greatly improves the insulation, but its effect is more obvious as the material is a worse insulator. There appears to be less difference in india-rubber, when subjected and not subjected to pressure, than upon any other of the materials experimented upon.

12. *Influence of interposed Resistances on the Time of charging or discharging.*

When the charge is communicated to the wire through a very considerable resistance intervening between it and the battery, or when the discharge is effected through a great resistance to the earth, the time of charging or discharging the wire is very greatly augmented.

A wire from one of the poles of a battery of 512 elements was attached to a binding-screw on an insulated mahogany frame, the other pole communicating with the earth; to another binding-screw, about half an inch apart from the first, was attached the end of an insulated wire two miles in length. The charge was allowed to accumulate in the wire during different intervals of time, and the discharge current was conducted to the earth through the long wire galvanometer (B).

After $\frac{1}{4}$ minute the deflection was		12°
$\frac{1}{2}$ minute	,,	19°
1 minute	,,	32°
2 minutes	,,	42°
4 minutes	,,	55°

The experiment was not tried beyond. When the mahogany frame was removed from the circuit the deflection was immediate and beyond 90°.

In like manner, on interposing a very imperfect conductor between the charged wire and the earth the discharge was retarded so as to last upwards of two minutes. The effect was particularly striking, and presented some instructive particulars, when the discharge was effected through one of Mr. Gassiot's small vacuum-tubes.

From these experiments it might be inferred that the interposition of a galvanometer wire, especially when it is of great length, would sensibly retard the time of charging or discharging. It no doubt does retard it; but when the current is so instantaneous as only to produce a momentary deflection of the needle the same quantity affects the needle in the same manner whether the time of charging or discharging be greater or less. A direct experimental proof of this is given in § 6. The following experiment is also in accordance with this explanation.

One mile of wire 2/3 was charged by a battery of 144 elements, and then discharged to the earth through the long wire galvanometer (B); the needle deflected 67°. Four, eight, and sixteen miles of similar wire were successively interposed between the galvanometer and the earth, but in all cases the discharge occasioned precisely the same deflection of 67°.

Great as was the resistance of sixteen miles of wire added to that of the galvanometer wire (stated by the constructor to be equal to five miles of ordinary telegraphic wire), it was very small in comparison with the resistances first mentioned. When the time of charging or discharging is less than the time of deflection of the needle, that deflection is a measure of the quantity of dynamic electricity in the instantaneous currents; but when the time is greater, as in the first quoted examples, the quantity is to be measured by the intensity indicated by the permanent deviation of the needle multiplied by the time.

13. *Discharges from one end of a Wire when the other communicates with the Earth.*

The current produced by discharging to the earth the static electricity of a charged insulated wire has been considered in the preceding sections, and it has been shown to be governed by simple and well-ascertained laws. But in the actual transmission of telegraphic signals this condition of the insulation of the wire at its remote end never obtains. When a telegraph is in operation the wire is always connected with the earth at its opposite extremity. The state of tension is in this case never arrived at, because the current continues to flow, and at the moment of discharging the wire only one portion of the electricity at that moment existing in it is discharged through the new path, the other portion passing from the opposite end of the wire to the earth, with which it is permanently connected. It is of great practical importance that the proportion of the discharge from a wire connected with the earth, as compared with the discharge under the same circumstances from the same wire disconnected from the earth, should be determined. For this purpose I instituted a number of experiments with the wires with which the previous experiments were made; but I met with the most anomalous results, from causes to which I shall hereafter refer. I therefore found it necessary, in order to obtain reliable results, to make the comparison on an actual submarine or subterranean telegraphic line. With this view, W. Andrews, Esq., the Engineer of the Submarine Telegraph Company, kindly placed at my disposal four subterranean wires extending from London to Dover, a distance of 89 miles. They were insulated by gutta-percha, and enclosed in the same protective covering. The instruments employed in the ensuing experiments were the following :—The torsion galvanometer; a Peltier's electroscope; the discharging-key, fig. 1 (Plate XVII.). The wires were marked respectively 2, 4, 5, and 10 ; and a battery of 81 Daniell's elements, which had been some time in action, was employed.

The electroscope applied to a short wire attached to the copper terminal of the battery, while the zinc terminal was connected with the earth, gave a deviation of 11°. When either of the wires, No. 2, 4, or 5, insulated at its opposite end, was attached to the battery, and the electroscope touched at the point of junction, the deviation remained the same, viz. 11°. The wires retained no charge when removed from the battery. The battery, however, supplied the electricity of tension faster than it was dissipated from the wires. The battery was very imperfectly insulated ; the elements were contained in porcelain jars standing in wooden boxes with covers.

To ascertain the leakages of the respective wires for dynamic electricity, the torsion galvanometer (E) was interposed in succession between Nos. 5, 4, and 2, and the battery, the opposite ends of the wires remaining insulated. The following were the deflections and corresponding degrees of torsion :—

	Deflections.	Degrees of torsion.
No. 4	81°	570
No. 2	51½°	80
No. 5	47°	61

No. 4 leaked, therefore, above three times more than No. 5, the force of the current being as the square root of the torsion.

I then proceeded to ascertain the amount of discharge from each wire when insulated at its opposite end. The same torsion galvanometer was employed to measure the instantaneous deflections, and the charging and discharging was effected by the discharging key, fig. 1 (Plate XVII.). The following were the results :—

	Instantaneous Deflections*
No. 4	72, 85, 80, 87
No. 2	130, 130, 130
No. 5	141, 140, 141

It is obvious here that the wire which was the worst insulated gave the smallest amount of discharge, owing, no doubt, to the escape of a portion of the charge.

It appears also, from the above experiments, that when the insulation is imperfect, the amount of the discharge is less constant.

The ends of the wires at Dover were then connected with the earth, and similar experiments were made, with the following results :—

	Instantaneous Deflections.
No. 4................	18°, 20°, 31°, 20°.
No. 2................	17°, 18°, 20°.
No. 5................	14°, 12°, 18°, 12°, 15°.

The amount of discharge is much less than when the opposite ends of the wires are insulated, in the case of No. 2 it is about seven times less. There is also greater variability when the insulation is tolerably good, and imperfect insula-

---

* When a torsion galvanometer is employed, the force of the instantaneous current is not as the chord of the angle of deflection, as in the case of an ordinary galvanometer; but, as I have ascertained by experiment, it is directly as the angle of torsion with a small constant subtracted.

tion, instead of reducing the discharge, appears to augment it. The discharge from a perfectly insulated wire connected with the earth is by no means so accurate a measure of the inductive action as the discharge from a charged wire disconnected from the earth.

Wishing to make similar experiments with a double length of wire, Nos. 2 and 5 were connected together at Dover, without communicating with the earth there; the two ends of the conductor were therefore at London. The battery and key were first applied to the end of 5, and afterwards to the end of 2; in the first case the end of 2 remained insulated, and in the second case the end of 5. The following measures were obtained :—

	Instantaneous Deflections.		
The end of 2 insulated; the battery and key applied to the end of 5	242,	200,	220.
The end of 5 insulated; do. to the end of 2 .	175,	172,	166.

The first gives an amount of discharge nearly double that of a single wire, as might be expected; the second shows a result considerably less. The cause of this discrepancy, from the complicated conditions of the experiments, it is difficult to explain : the wires were not equal in insulation, and they might exert some inductive action on each other from being close to each other in the same sheaf. Corresponding experiments were then made with the ends previously insulated communicating with the earth.

The end of 2 connected with the earth; the battery and key applied to the end of 5	97°,	84°,	110°,	112°.
The end of 5 connected with the earth; battery and key to the end of 2	70°,	69°,	71°.	

The diminution of the discharge from the earth connexion, though considerable, is not so great as when half the length of wire is employed. The same proportion is kept between the measures obtained when the battery is applied to the ends 5 and 2 as in the preceding experiments.

14. *True and apparent Discharges.*

I will now revert to the anomalous results which occur when it is attempted to obtain a discharge from a wire in connexion with the earth, under the same circumstances as when the discharges are made from charged wires previously insulated at their two ends, as detailed in sections 1–9.

In general, strong deviations of the galvanometer are produced, very much stronger than those resulting from the charge of the same wire when insulated at its ends; but these deflections are capriciously variable, occurring sometimes in the deviation of the true discharge-current, and at other times in the opposite direction. These effects have been observed by Mr. Webb, and Mr. Fleeming Jenkin, and they have also attracted the attention of Professor W. Thomson. All these gentlemen attribute them to a conflict between the true discharge and a discharge of an inductive effect arising from the mutual action of the coils of the wire upon each other, the experiments from which they have drawn their conclusions having been made on large coils of insulated wire.

To ascertain the real causes of these perplexing phenomena, I instituted a series of experiments, which determined the following points:—1st, the length of wire remaining the same, the deviations of the galvanometer increased with the electro-motive force of the battery up to a certain limit. When the electro-motive force was small, *i. e.,* when one or two elements were employed, the deviations were generally constant in one direction; but when it was increased, the variations of direction became very capricious. 2nd, I compared the true discharge from wires of 1, 8, and 16 miles in length, insulated at their ends, with the apparent discharge from the same wires in connexion with the earth. The following Table shows the results:—

No. of Elements of the Battery.	One Mile.	Eight Miles.	Sixteen Miles.
	°	°	°
True Discharge.			
32		5	10
64		10	20
128	2·5	20	41
256	5	41	87
Apparent Discharge.			
1	90	70	25

While the true discharge increases with the length of the wire, the apparent discharge decreases; the latter therefore cannot depend on an inductive action. It has been proved in section 13 that the discharge from a wire with its end in connexion with the earth is much less than that of the same wire disconnected from it; but the above table shows that the apparent discharge from one mile of wire, when one cell of the battery is employed, is incomparably greater than the real discharge from the same wire when 32 or even 256 cells are brought into action.

The real origin of these currents will become evident, if instead of momentarily depressing the discharging-key, which is all that is necessary in experiments on the real discharge, the key be kept permanently depressed; a series of changes will then be observed which will leave no doubt of the cause of the phenomena. In the following experiments a battery of 512 elements, the galvanometer (A), and the discharging-key (fig. 1) were employed; the copper pole of the battery was connected with the earth, the zinc pole to 1; one mile of wire 2/3 in the water-tank was connected with 2, the other end remaining insulated; the binding-screw b of the galvanometer was connected with 3, and the binding-screw a with another earth-wire different from that of the battery. The real discharge-current was ascertained to be 7°·2 towards the left. The insulated end of the mile of wire was then connected with the earth, 512 elements still continuing to be employed. Pressing the discharging-key and keeping it depressed for some time, a very strong deflection of the needle of the galvanometer towards the left occurred, striking the pin at 90°; the needle then fell continuously, but occasionally with slight interruptions, towards 0°, but very much more slowly than the needle left to itself would fall; it took 28 seconds, and then slowly deflected, with occasional jerks, towards the right, and acquired a permanent deviation of 36°. The key may be relieved so as to disconnect the wires 1 and 2 from each other without completing the connexion with the battery; if this be done before the inversion has taken place, that is before the needle has receded to 0°, on renewing the contact the needle will again move towards the left, showing that the effect has a certain permanence in the wire 2; but if this contact be interrupted, and then restored after the deflection has taken place towards the right, the deflection always remains on the right, provided the battery contact be not renewed. Similar experiments were tried with 256, 128, 64, 32, 16, 8, 4, and 1 cells of the battery. The greater the number of cells the greater is the first deflection of the needle towards the left, and of longer continuance. With one cell the first deflection towards the left was 80°, it fell in a second or two to 0°, and afterwards attained a permanent deflection of 15° towards the right. If the wire 2 has been charged with a considerable number of cells, and a strong permanent deviation towards the right been attained, charging it with fewer cells, and then discharging it through the galvanometer, will occasion no deflection towards the left, the first effect from a few cells not being sufficient to overcome the second effect from a larger number.

The following experiment places it beyond doubt that the inductive influence of the coils of the cable in the tank on each other had nothing to do with the effects in question. I substituted for the coiled mile of insulated wire in the tank four feet only of uncoiled wire; the first deflection of the needle was 90° towards the left, which was followed by a permanent deflection of 65° towards the right; a similar experiment was made with a single cell with like results, the deflections only being less.

The cause of these effects is the polarization of the plates or masses of metal connecting the ends of the wires with the earth. The first effect is produced by the polarization of the terminal which connects the telegraphic wire with the earth, the second by the polarization of the terminal which connects the galvanometer with the earth. When the wire is sufficiently long for the discharge-current from it to affect the galvanometer, the first effect is a complex one, the discharge-current being added to the current of polarization. The electro-motive forces of the currents arising from polarization can never exceed that of one or two cells of a voltaic battery; and as the strength of these currents diminishes with the length of the wire, they are not observable in long telegraphic lines unless very delicate indicating instruments be employed.

15. *The Magnetic Rheometer.*

The method, already described, of measuring inductive charges and discharges by the momentary deflections of the needle of a galvanometer requires the employment of batteries of considerable electro-motive force, or wires of great length, in order that the charge may be accumulated, or galvanometers of extreme multiplying power. It is desirable to have a means by which short lengths of wire may be conveniently experimented upon, and I have employed with advantage for this purpose Marianini's rheo-electrometer, or magnetic rheometer. A description of this instrument and the details of numerous experiments made with it to measure the momentary currents resulting from the discharge of a Leyden jar will be found in " Memorie di fisica sperimentale scritte dal Professore Marianini dopo il 1836." The instrument I employed was thus constructed :—Two yards of insulated copper wire, 1/120th of an inch in diameter, were coiled round a cylindrical bar of soft iron, three inches in length and one-eighth of an inch in diameter. This bar was fixed horizontally over a graduated circle at a right angle to the line joining its zero points; at the centre of the bar there was an aperture to allow the free passage of a silk fibre, from which was suspended an extremely delicate magnetic needle, weighing less than half a grain. At the top of the glass cage the upper end of the fibre was attached to a pin, which, by means of an adjusting-screw, might be raised or depressed without causing it to rotate, in order to regulate the distance from the needle to the electro-magnetic bar. The ends of the coil of wire were connected to binding-screws placed outside the frame, and the graduated circle with the bar was movable by means of a lever b (Plate XVII. fig. 3), placed beneath the frame, in order that the zero point might be accurately brought to the magnetic meridian without moving the instrument bodily. Four small pins, c, c, were fixed on the circle in the vicinity of the bar, in order to prevent the contact of the needle with the bar, which it would be otherwise sometimes difficult to disengage from each other.

The mode of using this instrument is as follows :—The zero line of the scale is brought, by means of the lever, into the magnetic meridian; if then the bar is entirely deprived of magnetism, the needle will place itself in that direction. On a discharge being made through the coils, the bar becomes magnetic in proportion to the amount of the discharge; the two poles of the needle are attracted towards the two ends of the electro-magnet, causing an instantaneous deflection, and subsequently a permanent deviation, owing to the residual magnetism, to a less amount. To prepare the instrument for a fresh experiment the bar must be deprived of its magnetism, which is easily effected in the following manner:— One of the poles of a very weak single cell of a voltaic battery is to be attached to one of the binding-screws of the instrument, while the other pole is to be brought into successive momentary contacts with the other binding-screw, the current being in the opposite direction to that which produced the magnetism of the bar; each contact causes the approach of the needle towards zero, and when it has arrived at that point the magnetism of the bar is completely destroyed, but should the zero be overpassed the same operation must be repeated after inverting the current.

With this instrument the following experiments, among others, were made :—

A Leyden jar, having 380 square inches of tin-foil surface, was charged by a Daniell's battery of 512 elements. On discharging the jar, the rheo-electrometer showed 10° instantaneous and 4° permanent deflection.

One hundred yards of the standard wire, coated with guttapercha and immersed in water, were charged by a battery of 60 elements; when the discharge was made through the rheo-electrometer, a deflection of several degrees was shown. In this case the discharge-current could not be detected by any of the galvanometers employed in the course of these investigations, except the suspension galvanometer (F) with 30,500 coils employed in Mr. Bartholomew's experiments.

The permanent deflection occasioned by the discharge from one mile of the standard insulated wire when charged by 144 cells of the battery was 8°·25.

16. *Means of accumulating the effects of Charges and Discharges.*

The deflection of the needle of a galvanometer increases with the number of discharges in a given time so long as the time of a single discharge is less than that of a single contact; when a greater rapidity is attained, the deflection· remains constant, for then, although each charge or discharge occupies a less time, there is a greater number in the same time, and these two circumstances compensate each other A succession of instantaneous currents, therefore, produces a permanent deviation of the needle of a galvanometer as a continuous current does, provided that the above condition be fulfilled, and that they follow each other so rapidly tha the needle has not time to fall in the intervals between them.

It is therefore possible to accumulate the effects of a succession of charges or discharges, and to produce a very considerable deviation of the needle of a galvanometer, when a single charge or discharge would scarcely produce any appreciable motion. For this purpose I have constructed an instrument which I call the *accumulating discharger*, which will also be found useful for a great variety of other experiments.

Fig. 4 (Plate XVII.), A, B, is a mahogany board; C and D, two brass pillars which support an insulating plate of ebonite, in which are fixed six brass binding-screws (1, 2, 3, 4, 5, 6). A vertical axis carries a small fly-wheel E, a pulley F, and at its upper end a circular disk, at the circumference of which are attached two ebonite links, G, H, freely moving round their common axis; the other ends of these links are attached respectively to two brass arms, which have the pins of the binding-screws 2 and 5 for their axes, and each carries a cross piece terminating with springs, which, during the motion of the instrument, press alternately against the binding-screws 1 and 3, and 4 and 6. The distances of the springs and pins are so adjusted that each contact lasts during one-fourth of a revolution of the eccentric. A catgut band connects the pulley F with the multiplying wheel K, which is provided with a handle for the purpose of setting it in motion. When this wheel is turned, the eccentric is caused to revolve rapidly, and, in consequence of the arrangements above described, the binding-screws 2 and 5 are brought alternately, in rapid succession, in metallic contact, the former with the binding-screws 1 and 3, the latter with 4 and 6.

With this instrument, as constructed, it is easy to make 68 double contacts in a second; by employing extra multiplying wheels, a far greater number may be obtained; and, if needful, mechanism with a maintaining power may be employed to regularize the motion.

By connecting, in various manners, the poles of the battery, the wires to be experimented upon, the earth, and the galvanometers, or other measuring instruments, with the binding-screws of this apparatus, it may be disposed for obtaining a variety of different results, some of which I will briefly enumerate.

1. To accumulate the *charges* of a wire.

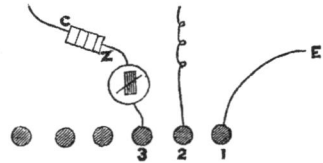

Connect 1 with the earth, 2 with the insulated wire to be charged, 3 with the galvanometer, which is then to be connected with one of the poles of the battery, the other proceeding to the earth. The instrument being put in motion, when the arm connects 2 and 3, the wire becomes charged, and the needle of the galvanometer is acted upon by the electricity in motion; when it connects 2 and 1, the electricity in the wire is discharged to the earth without acting on the galvanometer. The wire is thus alternately charged and discharged, but the accumulated *charges* only act on the galvanometer.

If the earth wire be removed from 1, and the time of charging is less than the duration of a single contact, as is generally the case, no effect is produced on the galvanometer, except from the first contact, because the wire afterwards remains permanently charged.

2. To accumulate the *discharges* of a wire.

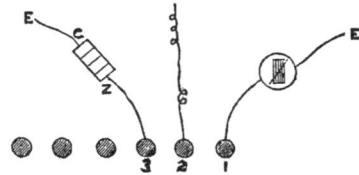

Connect the insulated wire with 2; interpose the galvanometer between the earth and 1, and the battery between the earth and 3. The contacts with 3 charge the wire, and the alternate contacts with 1 discharge it. The galvanometer being placed between the earth and 1, is affected only by the accumulated discharges.

With the short wire galvanometer A and 512 cells, the following results were obtained :—

Wire covered with Wray's compound, 170 yards .... 3°

Wire with Hearder's covering, 440 yards.............:...... 20°

With the long wire galvanometer (D) and 512 cells, the accumulated discharges from two inches of gutta-percha covered wire, coated externally with tin-foil, were rendered evident.

The following results were obtained with the nine experimental wires of different diameters and gutta-percha coatings of different thicknesses. The measures were made with the torsion galvanometer (E), and two cells of the battery only were employed. Three readings were taken for each wire.

	2	4	8
3	4·7	7·0	9·0
	4·7	7·0	9·0
	4·6	7·0	9·0
6	3·2	4·9	7·0
	3·1	4·9	7·0
	3·1	4·9	7·0
12	2·0	3·0	4·5
	2·0	3·0	4·5
	2·0	3·0	4·5

With the same measuring instrument no indication would have been obtained with a single discharge.

Experiments were made to ascertain if, by this means, the discharge-currents could be rendered evident from one or two miles of wire, without any insulating covering, suspended freely in the air by means of the ordinary insulating posts of a telegraphic line. For these experiments the long wire galvanometer (D) was employed. With 512 cells of the battery, and two miles of wire, a slight tremor only of the needle was observed on a single discharge; but the intermitting discharger occasioned a permanent deflection of 75°; this, however, was by no means constant; repetitions at different times during the same afternoon gave 65°, 70°, 70 , and even

40°. There is not the same steadiness in these results as when a wire with an insulating covering, and immersed in water, is experimented upon. Similar experiments, both with two miles and one mile of wire, were made, employing different battery powers; but with the same variable results. All the measures in the following table were obtained within four hours in the same day :—

NUMBER OF CELLS OF THE BATTERY.

	512	128	64	32
2 miles	40 75 65 70 70	13 32 40	15 10	
1 mile	22 60 65 55 65 70	28	11 9 10 7	4

The accumulated discharges from one mile of gutta-percha covered wire immersed in water, when 64 cells of the battery were employed, gave 65° with the same galvanometer.

During these experiments trials were made of the insulation of the aërial wire. The insulation for dynamic electricity measured by the same galvanometer employed for the discharges was very variable; the escape would sometimes be *nil*, and after the lapse of some time would attain to 72° when 512 cells of the battery were in action* This amount of loss would vitiate any result obtained by repeatedly *charging* the wire, for the rapidly reiterated contacts of the battery with the wire produce an effect similar to that of a permanent contact, and the needle of the galvanometer would be affected by the current arising from the escape of electricity

---

* It must be observed that the galvanometer employed in these experiments was one of extreme sensibility, and that this amount of escape would be insensible with the instruments ordinarily employed in telegraphic communication.

along the wire; it would be difficult therefore to distinguish, in this case, what portion of the effect should be attributed to the current arising from imperfect insulation, and what to the current produced by the charges. But with repeated *discharges* the case is different, the action on the galvano-meter is then entirely due to the discharges of the electri-city retained in the wire after contact with the battery has ceased.

The state of insulation for static electricity of the aërial wire was also ascertained during these experiments. The tension of the battery of 512 cells was shown by a Peltier's electrometer to be 58°. When two miles of the wire were connected with the battery the same tension was maintained. On removing the battery from the wire and electrometer, the index of the latter fell immediately to 42°, and after the lapse of two minutes to 40°; this experiment was repeated with nearly the same result.

3. To accumulate the charges and discharges of the wire separately at the same time.

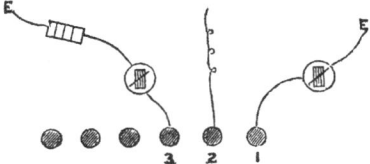

Connect the insulated wire with 2; interpose a galvano-meter between the battery and 3, and another similar one between the earth and 1.

4. Effect of the alternate charging and discharging of a wire upon the needle of a galvanometer.

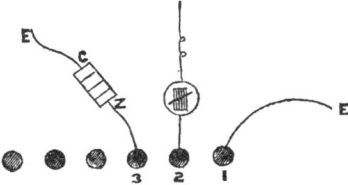

Connect the earth with 1; interpose the galvanometer

between 2 and the insulated wire; and connect 3 with one pole of the battery, the other proceeding to the earth.

If the charges and discharges be of equal intensity, we might expect that the rapid succession of the currents in opposite directions occasioned thereby would produce no change in the position of the needle of a galvanometer; and such is the case when either current separately deviates the needle only a few degrees. But a very anomalous effect takes place when the current is stronger. The needle then assumes indifferently a permanent position on either side the zero point; thus with a battery of 512 cells, a wire one mile in length, and the short wire galvanometer (A) this deflection was 75°. It will soon be perceived that the direction of the deflection is determined by the initial impulse. The cause of this singular effect is not difficult to explain. When the needle of the galvanometer is parallel to the coil, currents in either direction act with equal energy, though in opposite directions, on the needle; but when the needle is in any other position, the currents in opposite directions act with unequal energy upon it, and the effect is the same as if it were influenced by two series of alternating currents of different intensities.

I had formerly observed a similar effect in experimenting with Becquerel's differential galvanometer; this instrument is provided with two intermingled coils, each of which transmits a current in the opposite direction. The differential arrangement of an ordinary galvanometer, which I have described in my memoir on "New Instruments and Processes for determining the Constants of a Voltaic Circuit"*, is, however, free from this defect; the neutralization is in that arrangement effected in the coil itself, and all action on the needle is prevented; whereas in the other two instances the needle is influenced, either successively or simultaneously, by two actions in opposite directions.

In the preceding cases, where the wire is charged by one of the poles of the battery only, no more than three of the

---

* Philosophical Transactions, 1843, Part II. [vid. supr. p. 97].

binding-screws are employed; were the use of the instrument limited to these experiments the other three binding-screws, with the parts in connexion with them, might be dispensed with. But some very useful and interesting information may be obtained by charging the wire simultaneously by the two poles of the battery; and then it will be requisite to bring more of the binding-screws into use. I will briefly indicate a few of these applications.

5. Currents produced by discharge from one or both ends of the wire after its separation from the battery.

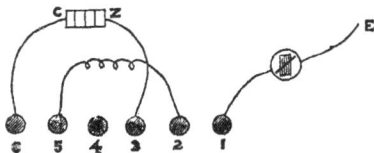

Connect the battery poles with 3 and 6, the insulated wire to 2 and 5, and interpose the galvanometer between the earth and either 1 or 4; or galvanometers may be interposed between both.

This experiment must be made with a very long wire. The contact of the battery poles with the two ends of the wire give rise to discharge-currents in the same direction with respect to the wire; the severance of the wire from the two poles of the battery, and its subsequent contact with the earth wires, give rise to return discharge-currents.

6. To ascertain what portion of its charge one charged wire communicates to another.

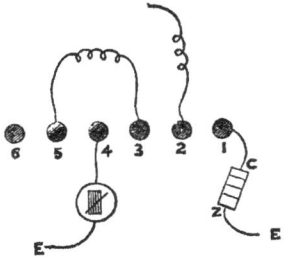

One of the poles of the battery being in communication

with the earth, the other is to be connected with 1; the wire to be first charged is to be connected with 2, the two ends of the wire to be secondarily charged with 3 and 5, and the galvanometer is to be interposed between the earth and 4. A wire coated with any insulating material may in this manner have its charge divided with an equal wire, or with one the inductive conditions of which are different. By employing copper wires of the same lengths and diameters, the relative specific inductive capacities of different insulating materials may be determined.

7. To obtain an intermitting current from the accumulated discharges of a Leyden jar or condenser.

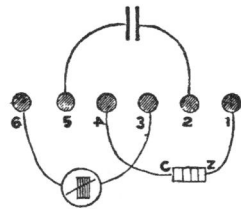

Connect one pole of the battery with 1 and the other with 4; the Leyden jar must be insulated, and a wire must connect its inner coating with 2, whilst another wire must unite its outer coating with 5; the galvanometer must be interposed in a wire uniting 3 and 6.

8. The accumulated discharges from a Leyden jar may be converted into an intermitting current, even when one of the poles of the battery communicates with the earth. The following is the arrangement for effecting this purpose :—

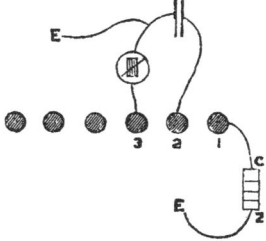

Interpose the battery between the earth and 1; let a wire

proceed from 2 to the inner coating of the jar, while the outer coating communicates with the earth and 3; the galvanometer must be interposed between 3 and the earth.

The " accumulating discharger " was completed and used in April 1860.   Mr. Latimer Clark's " inductometer " described in the ' Engineer ' of September 28th, 1850, which resembles this instrument in many respects, and was suggested by it, was not made by the Messrs. Elliots until the beginning of September 1860.

Mr. Siemens has also constructed a self-acting instrument for obtaining a rapid succession of charges and discharges, which he exhibited at the meeting of the British Association at Oxford in July 1860.

Though I was unaware of the fact at the time, an instrument for accumulating discharges for a special purpose (the same as that described in paragraph 7), was invented in 1849 by M. Guillemin*; and this gentleman has recently applied the same principle to measure the accumulated discharges from short lengths of coated telegraphic wires, covered externally with tin-foil †.   To M. Guillemin, therefore, the priority of this useful method must be conceded.

* " Courant dans une pile isolée et sans communication entre les deux poles," Comptes Rendus de l'Académie des Sciences, Nov. 12, 1849.

† Comptes Rendus, Oct. 8, 1860.

Description of the Telegraph Thermometer.

[From the ' Report of the British Association,' 1843, pp. 128, 129.]

THE Telegraph Thermometer, which is intended to be carried up by the balloon, weighs, with its case, about four pounds. It is thus constructed :—The movement of a small clock causes a vertical rack to ascend and descend regularly in six minutes, three minutes being occupied in the ascent and three in the descent. The rack carries a fine platina wire, which moves within the tube of a thermometer; the extent of motion of this wire corresponds with 28° of the thermometric scale, but it is capable of adjustment so that it may pass over any 28° of the range. Two very fine copper wires, covered with silk, and of sufficient length to reach from the ground to the balloon when at its greatest elevation, are connected with the instrument in the following manner :— The extremity of one wire is connected with the mercury in the bulb of the thermometer, and that of the other wire with the frame of the clock, which is in metallic continuity with the platina wire. On the ground the lower extremities are united together; in the wire whose opposite end is connected with the mercury in the thermometer, a sensible galvanometer is interposed, and in the course of the other wire a single, very small voltaic element is introduced. The galvanometer having been properly adjusted to its zero point, it will remain so during the time that the platina wire is not in contact with the mercury in the tube; but the needle will deviate as soon as the contact takes place, and will remain deflected until contact is again broken during the ascent of the rack. During each half-second of time, corresponding with the beats of the clock, the wire moves through the 360th part of its range, and a different point of the range consequently corresponds with a different beat or half-second of each alternate three minutes. If, therefore, an observer below be furnished with a chronometer timed to coincide with the clock in the balloon above, and note at what instant the needle of the galvanometer is deflected, he may infer from that observation the temperature indicated by the thermometer in the balloon; for according to the different expansion of the mercury in the thermometer the contact is broken at a different half-second. Should the rates of the two time-pieces not exactly correspond at the conclusion of a series of observations, the results will not be vitiated, as a correction may be easily made.

It is intended to add to this apparatus a wet-bulb thermometer; this will involve only the addition of another platina wire to the rack, and of another insulated wire, reaching from the balloon to the earth, with its interposed galvanometer.

For other meteorological instruments, the indications of which are to be transmitted to a distance, I occasionally employ the agency of electro-magnetism to ring a bell, to mark with a type or pencil, &c.; but for the purpose in question such methods cannot be so conveniently employed as the deflection of the needle of a galvanometer, on account of the necessity of having the long conducting wire extremely fine in order to avoid adding too much to the weight of the balloon. If the electro-motive force of the rheomotor were increased, which it would be necessary to do were stronger currents required, sparks would occur at the surface of contact of the mercury, which would produce injurious effects.

On a New Telegraphic Thermometer, and on the Application of the Principle of its construction to other Meteorological Indicators.

[From the 'Report of the British Association,' 1867, pp. 11–13.]

THE telegraphic thermometer which I constructed in 1843, and which is described in the Report of the Thirteenth Meeting of the British Association [*vid. supr.* p. 206], depended on the simultaneous action of two isochronous chronometer or clock movements—one at the remote station regulating the motion of a plunger in the bore of a thermometer, and the other at the near or observing station, marking, by the motion of the needle of a galvanometer, the moment at which the contact of the plunger with the mercury of the distant thermometer completed or broke the circuit. The clock movements required to be periodically wound up, and therefore the affected instrument could not be left to itself for an indefinite time.

There are, however, many situations in which it might be desirable to have meteorologic indications when the instruments would not be accessible for very long periods. I have therefore devised a new class of telegraphic meteorometers which shall be independent of clockwork, and may remain in any situation of difficult access as long as the instrument endures. This principle is applicable to all instruments which indicate by means of a revolving hand, and I have already devised its application to a Breguet's metallic thermometer, an aneroid barometer, and an hygrometer, depending on the absorption of moisture by a thin membrane. It is also applicable to a bar-magnet in a fixed position, and to a variety of other indicators.

The apparatus consists of two distinct instruments, connected only by telegraphic wires : the first I will call the questioner (A) ; the second, the responder (B).

The questioner (A) is a rectangular box presenting externally a circular dial face, round which are engraved the degrees both of the Fahrenheit and Centigrade thermometric scales—the former ranging from 20° below zero F. to 220° above that point, and the latter from 0° to 110° C. It shows besides three binding-screws for the purpose of connecting the telegraphic wires, and a handle which causes the rotation of the armature of a magnetomotor in the interior. This magnetomotor is similar in its construction to that employed in my alphabetic magnetic telegraph ; a soft iron armature rotating before the four poles of the magnet occasions, when the circuit is completed, alternate currents of equal intensity. The box also contains a small electromagnet which acts by means of mechanism similar to that employed in the indicator of the aforesaid telegraph, and causes the revolution of the index of the dial.

The responder (B) is a cylindrical brass box which presents on its upper surface a similar dial with its thermometric scales and index ; at its base three binding-screws, corresponding to those of the questioner, are fixed for connecting the telegraphic wires, and it is furnished with a brass cover that it may be hermetically sealed when lowered in the sea or buried in the ground. Its interior contains three essentially distinct parts :—1. The metallic thermometer, which consists of a spiral ribbon of two dissimilar metals, with its hand capable of ranging through the extent of the circular thermometric scale of the dial. 2. A small electromagnet, acting by means of a propelment on a disk, making as many stops in one rotation as there are half-degrees on the scale. 3. An axis, to which is fixed a delicate spiral spring which causes a pin to bear lightly against the hand of the thermometer, however it may vary in position.

The two instruments are connected by means of two telegraphic wires. The first proceeds from an earth-plate at the near station, passes through the coil of the electromotor in A, joins the coil of the small electromagnet in B, and then

P

proceeds to another earth-plate at the distant station. The second wire is permanently connected with the first between the earth-plate and the coil of the magnetomotor, and includes that of the electromagnet in B, and its opposite end is brought close to the remote end of the first wire. The mechanism is so disposed that when the first wire is disconnected from its earth terminal it is brought into circuit with the second wire.

By this arrangement, when the dial of A is brought to 0° and the handle turned, at the first moment the circuit is completed through the first wire, containing the coil of the electromagnet in B, and the return earth. A disk is thereby caused to revolve in an opposite direction to the graduation of the scale, until a pin, originally starting from 0°, comes into contact with the pin pressing against the thermometer hand, and thereby completes the circuit of the second wire and breaks the connexion with the earth-plate. At first only the electromagnet in B is acted upon, but when the currents are diverted into the new channel, both the electromagnets act simultaneously. In consequence of the action of the electromagnet in A the hand of its dial passes over a space corresponding with that between 0° and that indicated by the thermometer, and the hand of the dial ultimately accords with that of the distant thermometer. When the hand of the dial on A comes to rest, the disk in B arrives at 0°, and a catch permits the spiral spring to unwind itself, and its pin flies to and presses against the thermometer hand.

It must be observed that instruments thus constructed are not capable of marking every possible gradation; but they may be made to indicate divisions of the scale of any required minuteness. It is advisable to limit the extent of the scale when more minute divisions are deemed necessary.

The only circumstance that can affect the accuracy of the indications of the instrument is this. The pin pressing against the thermometric index displaces it a little, and causes it to assume a position about a degree in advance; but as this pressure is a constant one, the inconvenience is remedied by a slight corresponding shifting of the scale.

On the Augmentation of the Power of a Magnet by the reaction thereon of Currents induced by the Magnet itself.

[From the 'Proceedings of the Royal Society,' vol. xv. pp. 369–372.]

THE magneto electric machines which have been hitherto described are actuated either by a permanent magnet or by an electro-magnet deriving its power from a rheomotor placed in the circuit of its coil.   In the present note I intend to show that an electro-magnet, if it possess at the commencement the slightest polarity, may become a powerful magnet by the gradually augmenting currents which itself originates.

The following is a description of the form and dimensions of the electro magnet I have employed.   The construction, it will be seen, is the same as that of the electro-magnetic part of Mr. Wilde's machine.

The core of the electro-magnet is formed of a plate of soft iron 15 inches in length and $\frac{1}{2}$ an inch in breadth, bent at the middle of its length into a horseshoe form.   Round it is coiled, in the direction of its breadth, 640 feet of insulated copper wire $\frac{1}{12}$ of an inch in diameter.   The armature, which is according to Siemens's ingenious construction, consists of a rotating cylinder of soft iron $8\frac{1}{2}$ inches in length, grooved at two opposite sides so as to allow the wire to be coiled upon it longitudinally ; the length of the wire thus coiled is 80 feet, and its diameter is the same as that of the electro-magnet coil.

When this electro-magnet is excited by any rheomotor the current from which is in a constant direction, during the rotation of the armature currents are generated in its coil during each semirevolution which are alternately in opposite directions ; these alternate currents may be transmitted un-

P 2

changed to another part of the circuit, or by means of a rheotrope be converted to the same direction.

If now, while the circuit of the armature remains completed, the rheomotor be removed from the electro-magnet, on causing the armature to revolve, however rapidly, it will be found by the interposition of a galvanometer, or any other test, that but very slight effects take place. Though these effects become stronger in proportion to the residual magnetism left in the electro-magnet from the previous action of a current, they never attain any considerable amount.

But if the wires of the two circuits be so joined as to form a single circuit, in which the currents generated by the armature, after being changed to the same direction, act so as to increase the existing polarity of the electro-magnet, very different results will be obtained. The force required to move the machine will be far greater, showing a great increase of magnetic power in the horseshoe; and the existence of an energetic current in the wire is shown by its action on a galvanometer, by its heating 4 inches of platinum wire ·0067 in diameter, by its making a powerful electro-magnet, by its decomposing water, and by other tests.

The explanation of these facts is as follows :—The electro-magnet always retains a slight residual magnetism, and is therefore in the condition of a weak permanent magnet; the motion of the armature occasions feeble currents in alternate directions in the coils thereof, which, after being reduced to the same direction, pass into the coil of the electro-magnet in such manner as to increase the magnetism of the iron core; the magnet having thus received an accession of strength, produces in its turn more energetic currents in the coil of the armature; and these alternate actions continue until a maximum is attained, depending on the rapidity of the motion and the capacity of the electro-magnet.

If the two coils be connected in such manner that the rectified current from the coil of the armature passes into the coil of the electro-magnet in the direction which would impart a contrary magnetism to the iron core, no current is produced, and consequently there is no augmentation of magnetism.

It is easy to prove that the residual magnetism of the electro-magnet is the determining cause of these powerful effects. For this purpose it is sufficient to pass a current from a voltaic battery, a magneto-electric machine, or any other rheomotor, into the coil of the electro-magnet in either direction, and it will invariably be found that the direction of the current, however powerful it may eventually become, is in accordance with the polarity of the magnetism impressed on the iron core.

If, instead of the currents in the coil of the rotating armature being reduced to the same uniform direction, they retain their alternations, no effects, or at most very small differential ones, are produced, as no accumulation of magnetism then takes place.

I will now call attention to the fact that stronger effects are produced at the first moment of completing the combined circuit than afterwards. The machine having been put in motion, at the first moment of completing the circuit 4 inches of platina wire were made red-hot, but immediately afterwards the glow disappeared, and only about one inch of the wire could be permanently kept at a red heat. This diminution of effect was accompanied by a great increase of the resistance of the machine. The cause of the momentary strong effect was, that the machine from its acquired momentum continued its motion for a few seconds, though it required a stronger force than could be applied to maintain that motion. Each time the circuit is broken and recompleted the same effect recurs.

On bringing the primary coil of an inductorium (Ruhmkorff's coil) into the circuit formed by connecting the coils of the electro-magnet and rotating armature, no spark occurs in the secondary coil. On account of the great resistance of the circuit, which now also includes the primary coil of the inductorium, the current is not in sufficient quantity to produce any noticeable inductive effect.

[Reducing, however, the resistance of the wire of the

electro-magnet, by uniting the four coils into a single one, a spark of considerable length may be obtained in the secondary wire of the inductorium, without the interposition of the cross wire.—*Added by the Author after publication in Proc. Roy. Soc.*]

A very remarkable increase of all the effects, accompanied by a diminution in the resistance of the machine, is observed when a cross wire is placed so as to divert a great portion of the current from the electro-magnet. The four inches of platinum wire, instead of flashing into redness and then disappearing, remains permanently ignited. The inductorium, which before gave no spark, now gave one a quarter of an inch in length; water was more abundantly decomposed; and all the other effects were similarly increased.

I account for this augmentation of the effects in the following way :—

Though so much of the current is diverted from the electromagnet by the cross wire, the magnetic effect still continues to accumulate, though not to so high a degree; but the current generated by the armature, passing through the short circuit formed by the armature-branch and cross wire, experiences a far less resistance than if it had passed through the armature and electro-magnet branches; and though the electromotive force is less, the resistance having been rendered less in a much greater proportion, the resultant effect is greater.

I must observe that a certain amount of resistance in the cross wire is necessary to produce the maximum effect. If the resistance be too small, the electro-magnet does not acquire sufficient magnetism; and if it be too great, though the magnetism becomes stronger, the increase of resistance more than counterbalances its effect.

But the effects already described are far inferior to those obtained by causing them to take place in the cross wire itself. With the same application of force, 7 inches of platinum wire were made red hot, and sparks were elicited in the inductorium $2\frac{1}{2}$ inches in length.

The force of two men was employed in these, as well as in

the other experiments. When the interrupter of the primary coil was fixed, the machine was much easier to move than when it acted. For when the interrupter acted, at each moment of interruption the cross wire being, as it were, removed, the whole of the current passed through the electro-magnet, and consequently a greater amount of magnetic energy was excited, while in the intervals during which the cross wire was complete the current passed mainly through the primary coil.

The effects are much less influenced by a resistance in the electro-magnet branch than in either of the other branches.

To reduce the length of the spark in the inductorium (the primary coil of which was placed in the cross wire) to $\frac{3}{4}$ of an inch, it required the resistance of $5\frac{1}{4}$ inches of the fine platinum wire in the cross wire, 5 inches in the armature-branch, or* 4 feet in the electro-magnet branch.

When there was no extra resistance in either of the branches, the length of the cross wire being only about a few feet, the intensity of the current in the electro-magnet branch, compared with that in the cross wire, was as 1 : 60; and when the resistance of the primary coil of the inducto-rium was interposed in the cross wire, the relative intensities were as 1 : 42.

In conclusion I will mention that there is an evident ana-logy between the augmentation of the power of a weak magnet by means of an inductive action produced by itself, and that accumulation of power shown in the static electric machines of Holtz and others which have recently excited considerable attention, in which a very small quantity of electricity directly excited is, by a series of inductive actions, augmented so as to equal, and even exceed, the effects of the most powerful machines of the ordinary construction.

---

* ["and" in Proc. Roy. Soc.; altered to "or" by the author after publication.]

On a cause of Error in Electroscopic Experiments.

[From the 'Proceedings of the Royal Society,' vol. xviii. pp. 330–333.]

To arrive at accurate conclusions from the indications of an electroscope or electrometer, it is necessary to be aware of all the sources of error which may occasion these indications to be misinterpreted.

In the course of some experiments on electrical conduction and induction which I have recently resumed, I was frequently delayed by what at first appeared to be very puzzling results. Occasionally I found that I could not discharge the electrometer with my finger, or only to a certain degree, and that it was necessary, before commencing another experiment, to place myself in communication with a gaspipe which entered the room. How I became charged I could not at that time explain; the following chain of observations and experiments, however, soon led me to the true solution.

I was sitting at a table not far from the fireplace with the electrometer (one of Peltier's construction) before me, and was engaged in experimenting with disks of various substances. To ensure that the one I had in hand, which was of tortoiseshell, should be perfectly dry, I rose and held it for a minute before the fire; returning and placing it on the plate of the electrometer, I was surprised to find that it had apparently acquired a strong charge, deflecting the index of the electrometer beyond 90°. I found that the same thing took place with every disk I thus presented to the fire, whether of metal or any other substance. My first impression was that the disk had been rendered electrical by heat, though it would have been extraordinary that, if so, such a result had not been observed before; but on placing it in contact with a vessel of boiling water, or heating it by a gas-lamp, no such effect was produced. I next conjectured that the phenomenon might arise from a difference in the electrical state of the air in the room and that at the top of the chimney; and to put this to the proof, I adjourned to the adjacent room where there was no fire, and bringing my disk to the fireplace I obtained precisely the same result. That this conjecture, however, was not tenable was soon evident, because I was able to produce the same deviation of the needle of the electrometer by bringing my disk near any part of the wall of the room. This seemed to indicate that different parts of the room were in different electrical states; but this again was disproved by finding that when the position of the electrometer and the place where the disk was supposed to be charged were interchanged, the charge of the electrometer was still always negative. The last resource was to assume that my body had become charged by walking across the carpeted room, though the effect was produced even by the most careful treading. This ultimately proved to be the case; for resuming my seat at the table and scraping my foot on the rug, I was able at will to move the index to its greatest extent.

Before I proceed further I may state that a gold-leaf electrometer shows the phenomena as readily.

When I first observed these effects the weather was frosty; but they present themselves, as I have subsequently found, almost equally well in all states of the weather, provided the room be perfectly dry.

I will now proceed to state the conditions which are necessary for the complete success of the experiments, and the absence of which has prevented them from being hitherto observed in the striking manner in which they have appeared to me.

The most essential condition appears to be that the boot or shoe of the experimenter must have a thin sole and be perfectly dry; a surface polished by wear seems to augment the effect. By rubbing the sole of the boot against the carpet or rug, the electricities are separated, the carpet assumes the positive state and the sole the negative state; the former being a tolerable insulator, prevents the positive electricity from running away to the earth, while the sole of the foot, being a much better conductor, readily allows the charge of negative electricity to pass into the body.

So effective is the excitation, that if three persons hold each other by the hands, and the first rubs the carpet with his foot while the third touches the plate of the electrometer with his finger, a strong charge is communicated to the instrument.

Even approaching the electrometer by the hand or body, it becomes charged by induction at some distance.

A stronger effect is produced on the index of the instrument if, after rubbing the foot against the carpet, it be immediately raised from it. When the two are in contact, the electricities are in some degree coerced or dissimulated; but when they are separated, the whole of the negative electricity becomes free and expands itself in the body. A single stamp on the carpet followed by an immediate removal of the foot causes the index of the electrometer to advance several degrees, and by a reiteration of such stamps the index advances 30° or 40°

The opposite electrical states of the carpet and the sole of the boot were thus shown: after rubbing, I removed the boot from the carpet, and placed on the latter a proof-plate (i. e. a small disk of metal with an insulating handle), and then transferred it to the plate of the electrometer; strong positive electricity was manifested. Performing the same operation with the sole of the boot a very small charge was carried, by reason of its ready escape into the body.

The negative charge assumed by sole-leather when rubbed with animal hair was thus rendered evident. I placed on the plate of the electrometer a disk of sole-leather and brushed it lightly with a thick camel's-hair pencil; a negative charge was communicated to the electrometer, which charge was principally one of conduction, on account of the very imperfect insulating power of the leather.

Various materials, as India-rubber, gutta percha, &c.,

were substituted for the sole of the boot; metal plates were also tried; all communicated negative electricity to the body. Woollen stockings are a great impediment to the transmission of electricity from the boot; when these experiments were made I wore cotton ones.

When I substituted for the electrometer a long wire galvanometer, such as is usually employed in physiological experiments, the needle was made to advance several degrees.

At the Meeting of the British Association at Dublin in 1857, Professor Loomis, of New York, attracted great attention by his account of some remarkable electrical phenomena observed in certain houses in that city. It appears that in unusually cold and dry winters, in rooms provided with thick carpets and heated by stoves or hot-air apparatus to 70°, electrical phenomena of great intensity are sometimes produced. A lady walking along a carpeted floor drew a spark one quarter of an inch in length between two metal balls, one attached to a gas-pipe, the other touched by her hand; she also fired ether, ignited a gaslight, charged a Leyden jar, and repelled and attracted pith-balls similarly or dissimilarly electrified. Some of these statements were received with great incredulity at the time both here and abroad, but they have since been abundantly confirmed by the Professor himself and by others. (See Silliman's 'American Journal of Science, July 1858.)

My experiments show that these phenomena are exceptional only in degree. The striking effects observed by Professor Loomis were feeble unless the thermometer was below the freezing-point, and most energetic when near zero, the thermometer in the room standing at 70°. Those observed by myself succeed in almost any weather, when all the necessary conditions are fulfilled. Some of these conditions must frequently be present, and experimentalists cannot be too much on their guard against the occurrence of these abnormal effects. I think I have done a service to them, especially to those engaged in the delicate investigations of animal electricity, by drawing their attention to the subject.

Experimental Verification of Bernouilli's Theory of Wind Instruments.

[From the 'Report of the British Association,' 1835, p. 558.]

M R. WHEATSTONE exhibited an experimental proof, which he had devised, of the following result of Bernouilli's theory of wind instruments :—*that in the fundamental sound of a tube, open at both ends, the portions of air on opposite sides of the centre of the tube move in contrary directions to each other.*

It consisted of a leaden tube, about an inch in diameter and thirteen inches long, bent nearly into a circle, so that its two ends were near, and opposite to, each other. Between these ends was held a vibrating part of a square plate of glass, put into vibration either with a violin-bow or a hammer, so as to produce its lowest sound, corresponding with Chladni's first figure. By this arrangement, the plate, advancing in its vibration towards one end of the tube, and receding at the same instant from the other, the effects neutralize each other, and no resonance or augmentation of the original sound takes place. In the middle of the tube was a joint, which allowed each half to move independently round the axis of the tube; by this means the two ends were capable of being brought to the opposite sides of portions of the plate vibrating at the same moment on contrary sides of the neutral plane; in this case, the impulses were made at the same instant *towards* both ends of the tube, and the augmentation of sound was very considerable. It is obvious that these effects would be reversed were Bernouilli's theory wrong.

Remarks on Purkinje's Experiments.

[From the 'Report of the British Association,' 1835, pp. 551-553.]

AFTER the reading of Sir David Brewster's paper, Mr. Wheatstone said, that having been the first person to introduce Purkinje's beautiful experiment into this country, and having repeated it a great number of times under a variety of forms, he would take the opportunity of stating a few particulars respecting it, which appeared not to be generally known.—The experiment succeeds best in a dark room, when, one eye being excluded from the light, the flame of a candle is placed by the side of the unshaded eye, but so as not to occupy any of the central part of the field of view. So long as the flame of the candle remains stationary, nothing further occurs than a diminution of the sensibility of the retina to light; but after the flame has been moved upwards and downwards, through a small space, for a length of time, varying with the susceptibility of the individual on whom the experiment is tried, the phenomenon presents itself. The blood-vessels of the retina, with all their ramifications, exactly as represented in the engravings of Sœmmerring, are distinctly seen, apparently projected on a plane before the eye, and greatly magnified. The image continues only while the flame is in motion; directly, or soon after, the flame becomes stationary, it dissolves into fragments and disappears.

Mr. Wheatstone dissented from the ingenious explanation of this appearance offered by Sir David Brewster, and also from that opinion, stated to be the generally received one; and begged to repeat the solution he had published, and which he had not since been induced to relinquish. Mr. W. observed, that there was no difficulty in accounting for the image; it evidently was a shadow resulting from the obstruction of light by the blood-vessels spread over the retina; the real difficulty was to explain why this shadow is not always visible. To account for this, Mr. W. adduced several facts, which tended to prove *that an object, either more or less luminous than the ground on which it is placed, when continuously presented to the same point of the retina, becomes invisible; and the rapidity of its disappearance is greater as the difference of luminous intensity between the object and the ground is less; but by continually shifting the place of the image of the object on the retina, or by making it act intermittently on the same point, the object may be rendered permanently visible.* To apply this explanation to the phenomenon in question, Mr. W. observed, that whenever the flame of the candle changes its place, the shadows of the vessels fall on different parts of the retina; which is evident from the motion of the figure while the eye remains still, which is always in a contrary direction to that of the flame. Hence the shadow, being thus made to change its place on the retina, remains, according to the law above stated, permanently visible; but instantly the flame is at rest, the shadow also becomes stationary, and consequently disappears.

Mr. Wheatstone then exhibited an instrument for showing an original variation of this experiment: it consisted of a circular plate of metal, about two inches in diameter, blackened at its outer side, and perforated at its centre with an aperture about as large as an ordinary gun-hole; to the inner face was fixed a similar plate of ground glass. On placing the aperture between the eye and the flame of a candle, and keeping the plate in motion, so as to displace continually the image of the aperture on the retina, the blood-vessels will be seen distributed as before, but will now appear brighter, and the spaces between the ramifications will be seen filled with innumerable minute vessels, anastomosing with each other in every direction, which were invisible in the former experiment. In the very centre of the field of vision there is a small circular space, in which no traces of these vessels appear. Mr. W. remarked, that the absence of these minute obstructions to light will probably account for the greater distinctness with which small objects are there seen, and also for the difference of colour observed by anatomists in that spot of the retina.

On the Prismatic Decomposition of Electrical Light.

[From the 'Report of the British Association,' 1835, pp. 11, 12.]

THE following is a brief notice of the principal results stated in this communication: 1. The spectrum of the electro-magnetic spark taken from mercury consists of seven definite rays only, separated by dark intervals from each other; these visible rays are two orange lines close together, a bright green line, two bluish-green lines near each other, a very bright purple line, and lastly, a violet line. The observations were made with a telescope furnished with a measuring apparatus; and to ensure the appearance of the spark invariably in the same place, an appropriate modification of the electro-magnet was employed. 2. The spark taken in the same manner from zinc, cadmium, tin, bismuth, and lead, in the melted state, gives similar results; but the number, position, and colours of the lines vary in each case; the appearances are so different, that, by this mode of examination, the metals may be readily distinguished from each other. A table accompanied the paper, showing the position and colour of the lines in the various metals used. The spectra of zinc and cadmium are characterized by the presence of a red line in each, which occurs in neither of the other metals. 3. When the spark of a voltaic pile is taken from the same metals still in the melted state, precisely the same appearances are presented. 4. The voltaic spark from mercury was taken successively, in the ordinary vacuum of the air-pump, in the Torricellian vacuum, in carbonic acid gas, &c., and the same results were obtained as when the experiment was performed in the air or in oxygen gas. The light, therefore, does not arise from the combustion of the metal. Professor Wheatstone also examined, by the prism, the light which accompanies the ordinary combustion of the metals in oxygen gas and by other means, and found the appearances totally dissimilar to the above. 5. Fraunhofer having found that the ordinary electric spark examined by a prism presented a spectrum crossed by numerous bright lines, Professor Wheatstone examined the phenomena in different metals, and found that these bright lines differ in number and position in every different metal employed. When the spark is taken between balls of dissimilar metals, the lines appertaining to both are simultaneously seen. 6. The peculiar phenomena observed in the voltaic spark taken between different metallic wires connected with a powerful battery were then described, and the paper concluded with a review of the various theories which have been advanced to explain the origin of electric light. Professor Wheatstone infers from his researches, that electric light results from the volatilization and ignition (not combustion) of the ponderable matter of the conductor itself; a conclusion closely resembling that arrived at by Fusinieri from his experiments on the transport of ponderable matter in electric discharges.

Contributions to the Physiology of Vision.—Part the First. On some remarkable, and hitherto unobserved, Phenomena of Binocular Vision.

[From the ‘Philosophical Transactions,’ 1838.]

[Plates XVIII. & XIX.]

§ 1.

WHEN an object is viewed at so great a distance that the optic axes of both eyes are sensibly parallel when directed towards it, the perspective projections of it, seen by each eye separately, are similar, and the appearance to the two eyes is precisely the same as when the object is seen by one eye only. There is, in such case, no difference between the visual appearance of an object in relief and its perspective projection on a plane surface; and hence pictorial representations of distant objects, when those circumstances which would prevent or disturb the illusion are carefully excluded, may be rendered such perfect resemblances of the objects they are intended to represent as to be mistaken for them; the Diorama is an instance of this. But this similarity no longer exists when the object is placed so near the eyes that to view it the optic axes must converge: under these conditions a different perspective projection of it is seen by each eye, and these perspectives are more dissimilar as the convergence of the optic axes becomes greater. This fact may be easily verified by placing any figure of three dimensions, an outline cube for instance, at a moderate distance before the eyes, and while the head is kept perfectly steady, viewing it with each eye successively while the other is closed. Plate XIX. fig. 13 represents the two perspective projections of a cube; *b* is that seen by the right eye, and *a* that presented to the left eye, the figure being supposed to be placed about seven inches immediately before the spectator.

The appearances, which are by this simple experiment rendered so obvious, may be easily inferred from the established laws of perspective; for the same object in relief is, when viewed by a different eye, seen from two points of sight at a distance from each other equal to the line joining the two eyes. Yet they seem to have escaped the attention of every philosopher and artist who has treated of the subjects of vision and perspective. I can ascribe this inattention to a phenomenon leading to the important and curious consequences, which will form the subject of the present communication, only to this circumstance—that the results being contrary to a principle which was very generally maintained by optical writers, viz. that objects can be seen single only when their images fall on corresponding points of the two retinæ, an hypothesis which will be hereafter discussed, if the consideration ever arose in their minds, it was hastily discarded under the conviction that if the pictures presented to the two eyes are under certain circumstances dissimilar, their differences must be so small that they need not be taken into account.

It will now be obvious why it is impossible for the artist to give a faithful representation of any near solid object, that is, to produce a painting which shall not be distinguished in the mind from the object itself. When the painting and the object are seen with both eyes, in the case of the painting two *similar* pictures are projected on the retinæ, in the case of the solid object the pictures are *dissimilar*; there is therefore an essential difference between the impressions on the organs of sensation in the two cases, and consequently between the perceptions formed in the mind; the painting therefore cannot be confounded with the solid object.

After looking over the works of many authors who might be expected to have made some remarks relating to this subject, I have been able to find but one, which is in the ' Trattato della Pittura ' of Leonardo da Vinci *. This great artist

---

* See also a Treatise of Painting, p. 178 (London, 1721); and Dr. Smith's ' Complete System of Optics,' vol. ii. r. 244, where the passage is quoted.

and ingenious philosopher observes, " that a painting, though conducted with the greatest art and finished to the last perfection, both with regard to its contours, its lights, its shadows, and its colours, can never show a relievo equal to that of the natural objects, unless these be viewed at a distance and with a single eye.   For," says he, " if an object C (Plate XVIII. fig. 1) be viewed by a single eye at A, all objects in the space behind it, included as it were in a shadow E C F cast by a candle at A, are invisible to the eye at A ; but when the other eye at B is opened, part of these objects become visible to it, those only being hid from both eyes that are included, as it were, in the double shadow C D, cast by two lights at A and B, and terminated in D, the angular space E D G beyond D being always visible to both eyes.   And the hidden space C D is so much the shorter as the object C is smaller and nearer to the eyes.   Thus the object C seen with both eyes becomes, as is were, transparent, according to the usual definition of a transparent thing—namely, that which hides nothing beyond it.   But this cannot happen when an object, whose breadth is bigger than that of the pupil, is viewed by a single eye.   The truth of this observation is therefore evident, because a painted figure intercepts all the space behind its apparent place, so as to preclude the eyes from the sight of every part of the imaginary ground behind it."

Had Leonardo da Vinci taken, instead of a sphere, a less simple figure for the purpose of his illustration, a cube for instance, he would not only have observed that the object obscured from each eye a different part of the more distant field of view, but the fact would also perhaps have forced itself upon his attention that the object itself presented a different appearance to each eye.   He failed to do this, and no subsequent writer within my knowledge has supplied the omission ; the projection of two obviously dissimilar pictures on the two retinæ when a single object is viewed, while the optic axes converge, must therefore be regarded as a new fact in the theory of vision.

§ 2.

It being thus established that the mind perceives an object of three dimensions by means of the two dissimilar pictures projected by it on the two retinæ, the following question occurs: What would be the visual effect of simultaneously presenting to each eye, instead of the object itself, its projection on a plane surface as it appears to that eye? To pursue this inquiry it is necessary that means should be contrived to make the two pictures, which must necessarily occupy different places, fall on similar parts of both retinæ. Under the ordinary circumstances of vision the object is seen at the concourse of the optic axes, and its images consequently are projected on similar parts of the two retinæ; but it is also evident that two exactly similar objects may be made to fall on similar parts of the two retinæ, if they are placed one in the direction of each optic axes, at equal distances before or beyond their intersection.

Fig. 2 represents the usual situation of an object at the intersection of the optic axes. In fig. 3 the similar objects are placed in the direction of the optic axes before their intersection, and in fig 4 beyond it. In all these three cases the mind perceives but a single object, and refers it to the place where the optic axes meet. It will be observed that when the eyes converge beyond the objects, as in fig. 3, the right-hand object is seen by the right eye, and the left-hand object by the left eye; while when the axes converge nearer than the objects, the right-hand object is seen by the left eye, and conversely. As both of these modes of vision are forced and unnatural, eyes unaccustomed to such experiments require some artificial assistance. If the eyes are to converge beyond the objects, this may be afforded by a pair of tubes (fig. 5) capable of being inclined towards each other at various angles, so as to correspond with the different convergences of the optic axes. If the eyes are to converge at a nearer distance than that at which the objects are placed, a box (fig. 6) may be conveniently employed: the objects $a\ a'$ are placed distant from each other, on a stand capable of

being moved nearer the eyes if required, and the optic axes
being directed towards them will cross at c, the aperture $b\ b'$
allowing the visual rays from the right-hand object to reach
the left eye, and those from the left-hand object to fall on
the right eye; the coincidence of the images may be facilitated by placing the point of a needle at the point of intersection of the optic axes c, and fixing the eyes upon it.   In
both these instruments (figs. 5 and 6) the lateral images
are hidden from view, and much less difficulty occurs in
making the images unite than when the naked eyes are employed.

Now if, instead of placing two exactly similar objects to
be viewed by the eyes in either of the modes above described,
the two perspective projections of the same solid object be so
disposed, the mind will still perceive the object to be single,
but instead of a representation on a plane surface, as each
drawing appears to be when separately viewed by that eye
which is directed towards it, the observer will perceive a
figure of three dimensions, the exact counterpart of the object
from which the drawings were made.   To make this matter
clear I will mention one or two of the most simple cases.

If two vertical lines near each other, but at different distances from the spectator, be regarded first with one eye and
then with the other, the distance between them when referred
to the same plane will appear different; if the left-hand line
be nearer to the eyes, the distance as seen by the left eye
will be less than the distance as seen by the right eye: fig. 7
will render this evident; $a\ a'$ are vertical sections of the two
original lines, and $b\ b'$ the plane to which their projections
are referred.   Now if the two lines be drawn on two pieces
of card, at the respective distances at which they appear to
each eye, and these cards be afterwards viewed by either of
the means above directed, the observer will no longer see two
lines on a plane surface, as each card separately shows; but
two lines will appear, one nearer to him than the other, precisely as the original vertical lines themselves.   Again, if a
straight wire be held before the eyes in such a position that
one of its ends shall be nearer to the observer than the other

is, each eye separately referring it to a plane perpendicular to the common axis, will see a line differently inclined; and then if lines having the same apparent inclinations be drawn on two pieces of card, and be presented to the eyes as before directed, the real position of the original line will be correctly perceived by the mind.

In the same manner the most complex figures of three dimensions may be accurately represented to the mind, by presenting their two perspective projections to the two retinæ. But I shall defer these more perfect experiments until I describe an instrument which will enable any person to observe all the phenomena in question with the greatest ease and certainty.

In the instruments above described the optic axes converge to some point in a plane before or beyond that in which the objects to be seen are situated. The adaptation of the eye, which enables us to see distinctly at different distances, and which habitually accompanies every different degree of convergence of the optic axes, does not immediately adjust itself to the new and unusual condition; and to persons not accustomed to experiments of this kind, the pictures will either not readily unite, or will appear dim and confused. Besides this, no object can be viewed according to either mode when the drawings exceed in breadth the distance of the two points of the optic axes in which their centres are placed.

These inconveniences are removed by the instrument I am about to describe; the two pictures (or rather their reflected images) are placed in it at the true concourse of the optic axes, the focal adaptation of the eye preserves its usual adjustment, the appearance of lateral images is entirely avoided, and a large field of view for each eye is obtained. The frequent reference I shall have occasion to make to this instrument, will render it convenient to give it a specific name; I therefore propose that it be called a Stereoscope, to indicate its property of representing solid figures.

§ 3.

The stereoscope is represented by figs. 8 and 9, the former

being a front view, and the latter a plan of the instrument.
A A' are two plane mirrors, about four inches square, inserted
in frames, and so adjusted that their backs form an angle of
90° with each other; these mirrors are fixed by their common
edge against an upright B, or, which was less easy to represent
in the drawing, against the middle line of a vertical board,
cut away in such manner as to allow the eyes to be placed
before the two mirrors. C C' are two sliding boards, to which
are attached the upright boards D D', which may thus be
removed to different distances from the mirrors. In most of
the experiments hereafter to be detailed, it is necessary that
each upright board shall be at the same distance from the
mirror which is opposite to it. To facilitate this double adjustment, I employ a right- and a left-handed wooden screw,
$r\ l$; the two ends of this compound screw pass through the
nuts $e\ e'$, which are fixed to the lower parts of the upright
boards D D', so that by turning the screw pin p one way the
two boards will approach, and by turning it the other they
will recede from each other, one always preserving the same
distance as the other from the middle line f. E E' are pannels, to which the pictures are fixed in such manner that
their corresponding horizontal lines shall be on the same
level: these pannels are capable of sliding backwards and
forwards in grooves on the upright boards D D'. The apparatus having been described, it now remains to explain the
manner of using it. The observer must place his eyes as
near as possible to the mirrors, the right eye before the right-
hand mirror, and the left eye before the left-hand mirror,
and he must move the sliding pannels E E' to or from him
until the two reflected images coincide at the intersection of
the optic axes, and form an image of the same apparent magnitude as each of the component pictures. The pictures will
indeed coincide when the sliding pannels are in a variety of
different positions, and consequently when viewed under different inclinations of the optic axes; but there is only one
position in which the binocular image will be immediately
seen single, of its proper magnitude, and without fatigue to
the eyes, because in this position only the ordinary relations

between the magnitude of the pictures on the retina, the inclination of the optic axes, and the adaptation of the eye to
distinct vision at different distances are preserved.   The
alteration in the apparent magnitude of the binocular images,
when these usual relations are disturbed, will be discussed in
another paper of this series, with a variety of remarkable
phenomena depending thereon.   In all the experiments detailed in the present memoir I shall suppose these relations
to remain undisturbed, and the optic axes to converge about
six or eight inches before the eyes.

If the pictures are all drawn to be seen with the same inclination of the optic axes, the apparatus may be simplified
by omitting the screw r l and fixing the upright boards D D'
at the proper distances.   The sliding pannels may also be
dispensed with, and the drawings themselves be made to slide
in the grooves.

§ 4.

A few pairs of outline figures, calculated to give rise to the
perception of objects of three dimensions when placed in the
stereoscope in the manner described, are represented from
figs. 10 to 20.   They are one half the linear size of the
figures actually employed.   As the drawings are reversed by
reflection in the mirrors, I will suppose these figures to be
the reflected images to which the eyes are directed in the
apparatus—those marked b being seen by the right eye, and
those marked a by the left eye.   The drawings, it has been
already explained, are two different projections of the same
object seen from two points of sight, the distance between
which is equal to the interval between the eyes of the observer; this interval is generally about $2\frac{1}{2}$ inches.

a and b, fig. 10, will, when viewed in the stereoscope,
present to the mind a line in the vertical plane, with its
lower end inclined towards the observer.   If the two
component lines be caused to turn round their centres
equally in opposite directions, the resultant line will,
while it appears to assume every degree of inclination
to the referent plane, still seem to remain in the same
vertical plane.

Fig. 11. A series of points all in the same horizontal plane, but each towards the right hand successively nearer the observer.

Fig. 12. A curved line intersecting the referent plane, and having its convexity towards the observer.

Fig. 13. A cube.

Fig. 14. A cone, having its axis perpendicular to the referent plane and its vertex towards the observer.

Fig. 15. The frustum of a square pyramid; its axis perpendicular to the referent plane, and its base furthest from the eye.

Fig. 16. Two circles at different distances from the eyes, their centres in the same perpendicular, forming the outline of the frustum of a cone.

The other figures require no observation.

For the purposes of illustration I have employed only outline figures, for had either shading or colouring been introduced it might be supposed that the effect was wholly or in part due to these circumstances, whereas by leaving them out of consideration no room is left to doubt that the entire effect of relief is owing to the simultaneous perception of the two monocular projections, one on each retina. But if it be required to obtain the most faithful resemblances of real objects, shadowing and colouring may properly be employed to heighten the effects. Careful attention would enable an artist to draw and paint the two component pictures, so as to present to the mind of the observer, in the resultant perception, perfect identity with the object represented. Flowers, crystals, busts, vases, instruments of various kinds, &c. might thus be represented so as not to be distinguished by sight from the real objects themselves.

It is worthy of remark, that the process by which we thus become acquainted with the real forms of solid objects is precisely that which is employed in descriptive geometry, an important science we owe to the genius of Monge, but which is little studied or known in this country. In this science, the position of a point, a right line, or a curve, and con-

sequently of any figure whatever, is completely determined
by assigning its projections on two fixed planes, the situations of which are known, and which are not parallel to each
other.   In the problems of descriptive geometry the two referent planes are generally assumed to be at right angles to
each other, but in binocular vision the inclination of these
planes is less according as the angle made at the concourse
of the optic axes is less; thus the same solid object is represented to the mind by different pairs of monocular pictures,
according as they are placed at a different distance before
the eyes, and the perception of these differences (though we
seem to be unconscious of them) may assist in suggesting to
the mind the distance of the object.   The more inclined to
each other the referent planes are, with the greater accuracy
are the various points of the projections referred to their
proper places; and it appears to be a useful provision that
the real forms of those objects which are nearest to us are
thus more determinately apprehended than those which are
most distant.

§ 5.

A very singular effect is produced when the drawing originally intended to be seen by the right eye is placed at the
left-hand side of the stereoscope, and that designed to be seen
by the left eye is placed on its right-hand side.   A figure of
three dimensions, as bold in relief as before, is perceived, but
it has a different form from that which is seen when the
drawings are in their proper places.   There is a certain relation between the proper figure and this, which I shall call its
converse figure.   Those points which are nearest the observer
in the proper figure are the most remote from him in the
converse figure, and *vice versâ,* so that the figure is, as it
were, inverted; but it is not an exact inversion, for the near
parts of the converse figure appear smaller, and the remote
parts larger than the same parts before the inversion.   Hence
the drawings which, properly placed, occasion a cube to be
perceived, when changed in the manner described, represent
the frustum of a square pyramid with its base remote from
the eye; the cause of this is easy to understand.

This conversion of relief may be shown by all the pairs of drawings from fig. 10 to 19. In the case of simple figures like these the converse figure is as readily apprehended as the original one, because it is generally a figure of as frequent occurrence; but in the case of a more complicated figure, an architectural design, for instance, the mind, unaccustomed to perceive its converse, because it never occurs in nature, can find no meaning in it.

§ 6.

The same image is depicted on the retina by an object of three dimensions as by its projection on a plane surface, provided the point of sight remain in both cases the same. There should be, therefore, no difference in the binocular appearance of two drawings, one presented to each eye, and of two real objects so presented to the two eyes that their projections on the retina shall be the same as those arising from the drawings. The following experiments will prove the justness of this inference.

I procured several pairs of skeleton figures, *i. e.* outline figures of three dimensions, formed either of iron wire or of ebony beading about one tenth of an inch in thickness. The pair I most frequently employed consisted of two cubes whose sides were three inches in length. When I placed these skeleton figures on stands before the two mirrors of the stereoscope, the following effects were produced, according as their relative positions were changed :—1st. When they were so placed that the pictures with their reflected images projected on the two retinæ were precisely the same as those which would have been projected by the cube placed at the concourse of the optic axes, a cube in relief appeared before the eyes. 2ndly. When they were so placed that their reflected images projected exactly similar pictures on the two retinæ, all effect of relief was destroyed, and the compound appearance was that of an outline representation on a plane surface. 3rdly. When the cubes were so placed that the reflected image of one projected on the left retina the same picture as in the first case was projected on the right retina, and conversely, the converse figure in relief appeared.

§ 7.

If a symmetrical object (that is, one whose right and left sides are exactly similar to each other but inverted) be placed so that any point in the plane which divides it into these two halves is equally distant from the two eyes, its two monocular projections are, it is easy to see, inverted fac-similes of each other. Thus fig. 15, *a* and *b* are symmetrical monocular projections of the frustum of a four-sided pyramid, and figs. 13, 14, and 16 are corresponding projections of other symmetrical objects. This being kept in view, I will describe an experiment which, had it been casually observed previous to the knowledge of the principles developed in this paper, would have appeared an inexplicable optical illusion.

M and M' (fig. 21) are two mirrors, inclined so that their *faces* form an angle of 90° with each other. Between them in the bisecting plane is placed a plane outline figure, such as fig. 15 *a*, made of card, all parts but the lines being cut away, or of wire. A reflected image of this outline, placed at A, will appear behind each mirror at B and B', and one of these images will be the inversion of the other. If the eyes be made to converge at C, it is obvious that these two reflected images will fall on corresponding parts of the two retinæ, and a figure of three dimensions will be perceived; if the outline placed in the bisecting plane be reversed, the converse skeleton form will appear : in both these experiments we have the singular phenomenon of the conversion of a single plane outline into a figure of three dimensions. To render the binocular object more distinct, concave lenses may be applied to the eyes; and to prevent the two lateral images from being seen, screens may be placed at D and D'.

§ 8.

An effect of binocular perspective may be remarked in a plate of metal, the surface of which has been made smooth by turning it in a lathe. When a single candle is brought

near such a plate, a line of light appears standing out from
it, one half being above, and the other half below the surface; the position and inclination of this line changes with
the situation of the light and of the observer, but it always
passes through the centre of the plate.   On closing the left
eye the relief disappears, and the luminous line coincides
with one of the diameters of the plate; on closing the right
eye the line appears equally in the plane of the surface, but
coincides with another diameter; on opening both eyes it
instantly starts into relief*.   The case here is exactly analogous to the vision of two inclined lines (fig. 10) when each
is presented to a different eye in the stereoscope.   It is
curious that an effect like this, which must have been seen
thousands of times, should never have attracted sufficient
attention to have been made the subject of philosophic observation.   It was one of the earliest facts which drew my
attention to the subject I am now treating.

Dr. Smith † was very much puzzled by an effect of binocular perspective which he observed, but was unable to explain.   He opened a pair of compasses, and while he held
the joint in his hand, and the points outwards and equidistant
from his eyes, and somewhat higher than the joint, he looked
at a more distant point; the compasses appeared double.
He then compressed the legs until the two inner points coincided; having done this the two inner legs also entirely
coincided, and bisected the angle formed by the outward
ones, appearing longer and thicker than they did, and reaching from the hand to the remotest object in view.   The explanation offered by Dr. Smith accounts only for the coincidence of the points of the compasses, not for that of the
entire leg.   The effect in question is best seen by employing
a pair of straight wires about a foot in length.   A similar
observation, made with two flat rulers, and afterwards with

---

* The luminous line seen by a single eye arises from the reflection of
the light from each of the concentric circles produced in the operation of
turning; when the plate is not large the arrangement of these successive
reflections does not differ from a straight line.

† System of Optics, vol. ii. p. 388 and r. 526.

silk threads, induced Dr. Wells to propose a new theory of
visible direction in order to explain it, so inexplicable did it
seem to him by any of the received theories.

§ 9.

The preceding experiments render it evident that there is
an essential difference in the appearance of objects when seen
with two eyes and when only one eye is employed, and that
the most vivid belief of the solidity of an object of three dimensions arises from two different perspective projections of
it being simultaneously presented to the mind. How happens
it then, it may be asked, that persons who see with only one
eye form correct notions of solid objects, and never mistake
them for pictures ? and how happens it, also, that a person
having the perfect use of both eyes perceives no difference
in objects around him when he shuts one of them ? To explain these apparent difficulties, it must be kept in mind
that although the simultaneous vision of two dissimilar pictures suggests the relief of objects in the most vivid manner,
yet there are other signs which suggest the same ideas to the
mind, which, though more ambiguous than the former, become less liable to lead the judgment astray in proportion to
the extent of our previous experience. The vividness of relief
arising from the projection of two dissimilar pictures, one on
each retina, becomes less and less as the object is seen at a
greater distance before the eyes, and entirely ceases when it
is so distant that the optic axes are parallel while regarding
it. We see with both eyes all objects beyond this distance
precisely as we see near objects with a single eye; for the
pictures on the two retinæ are then exactly similar, and the
mind appreciates no difference whether two identical pictures
fall on the corresponding parts of the two retinæ, or whether
one eye is impressed with only one of these pictures. A
person deprived of the sight of one eye sees, therefore, all external objects, near and remote, as a person with both eyes
sees remote objects only; but that vivid effect arising from
the binocular vision of near objects is not perceived by the
former; to supply this deficiency he has recourse uncon-

sciously to other means of acquiring more accurate information. The motion of the head is the principal means he employs. That the required knowledge may be thus obtained will be evident from the following considerations. The mind associates with the idea of a solid object every different projection of it which experience has hitherto afforded; a single projection may be ambiguous, from its being. also cne of the projections of a picture or of a different solid object; but when different projections of the same object are successively presented, they cannot all belong to another object, and the form to which they belong is completely characterized. While the object remains fixed, at every movement of the head it is viewed from a different point of sight, and the picture on the retina consequently continually changes.

Every one must be aware how greatly the perspective effect of a picture is enhanced by looking at it with only. one eye, especially when a tube is employed to exclude the vision of adjacent objects, whose presence might disturb the illusion. Seen under such circumstances from the proper point of sight, the picture projects the same lines, shades, and colours on the retina as the more distant scene which it represents would do were it substituted for it. The appearance which would make us certain that it is a picture is excluded from the sight, and the imagination has room to be active. Several of the older writers erroneously attributed this apparent superiority of monocular vision to the concentration of the visual power in a single eye*.

There is a well-known and very striking illusion of perspective which deserves a passing remark, because the reason of the effect does not appear to be generally understood. When a perspective of a building is projected on a horizontal plane, so that the point of sight is in a line greatly inclined towards the plane, the building appears to a single eye placed

---

* " We see more exquisitely with one eye shut than with both, because the vital spirits thus unite themselves the more, and become the stronger; for we may find by looking in a glass whilst we shut one eye, that the pupil of the other dilates."—LORD BACON'S *Works, Sylva Sylvarum,* art. *Vision.*

at the point of sight to be in bold relief, and the illusion is almost as perfect as in the binocular experiments described in §§ 2, 3, 4. This effect wholly arises from the unusual projection, which suggests to the mind more readily the object itself than the drawing of it; for we are accustomed to see real objects in almost every point of view, but perspective representations being generally made in a vertical plane with the point of sight in a line perpendicular to the plane of projection, we are less familiar with the appearance of other projections. Any other unusual projection will produce the same effect.

§ 10.

If we look with a single eye at the drawing of a solid geometrical figure, it may be imagined to be the representation of either of two dissimilar solid figures, the figure intended to be represented, or its converse figure (§ 5). If the former is a very usual, and the latter a very unusual figure, the imagination will fix itself on the original without wandering to the converse figure; but if both are of ordinary occurrence, which is generally the case with regard to simple forms, a singular phenomenon takes place; it is perceived at one time distinctly as one of these figures, at another time as the other, and while one figure continues it is not in the power of the will to change it immediately.

The same phenomenon takes place, though less decidedly, when the drawing is seen with both eyes. Many of my readers will call to mind the puzzling effect of some of the diagrams annexed to the problems of the eleventh book of Euclid; which, when they were attentively looked at, changed in an arbitrary manner from one solid figure to another, and would obstinately continue to present the converse figures when the real figures alone were wanted. This perplexing illusion must be of common occurrence, but I have only found one recorded observation relating to the subject. It is by Professor Necker of Geneva, and I shall quote it in his own words from the 'Philosophical Magazine,' Third Series, vol. i. p. 337.

"The object I have now to call your attention to is an observation which has often occurred to me while examining figures and engraved plates of crystalline forms; I mean a sudden and involuntary change in the apparent position of a crystal or solid represented in an engraved figure. What I mean will be more easily understood from the figure annexed (fig. 22). The rhomboid A X is drawn so that the solid angle A should be seen the nearest to the spectator, and the solid angle X the farthest from him, and that the face A C D B should be the foremost, while the face X D C is behind. But in looking repeatedly at the same figure, you will perceive that at times the apparent position of the rhomboid is so changed that the solid angle X will appear the nearest, and the solid angle A the farthest; and that the face A C D B will recede behind the face X D C, which will come forward, which effect gives to the whole solid a quite contrary apparent inclination."

Professor Necker attributes this alteration of appearance, not to a mental operation, but to an involuntary change in the adjustment of the eye for obtaining distinct vision. He supposed that whenever the point of distinct vision on the retina is directed on the angle A, for instance, this angle, seen more distinctly than the others, is naturally supposed to be nearer and foremost, while the other angles, seen indistinctly, are supposed to be farther and behind, and that the reverse takes place when the point of distinct vision is brought to bear on the angle X.

That this is not the true explanation, is evident from three circumstances: in the first place, the two points A and X being both at the same distance from the eyes, the same alteration of adjustment which would make one of them indistinct would make the other so; secondly, the figure will undergo the same changes whether the focal distance of the eye be adjusted to a point before or beyond the plane in which the figure is drawn; and thirdly, the change of figure frequently occurs while the eye continues to look at the same angle. The effect seems entirely to depend on our mental contemplation of the figure intended to be repre-

R

sented, or of its converse. By following the lines with the eye with a clear idea of the solid figure we are describing, it may be fixed for any length of time; but it requires practice to do this or to change the figure at will. As I have before observed, these effects are far more obvious when the figures are regarded with one eye only.

No illusion of this kind can take place when an object of three dimensions is seen with both eyes while the optic axes make a sensible angle with each other, because the appearance of the two dissimilar images, one to each eye, prevents the possibility of mistake. But if we regard an object at such a distance that its two projections are sensibly identical, and if this projection be capable of a double interpretation, the illusion may occur. Thus a placard on a pole carried in the streets, with one of its sides inclined towards the observer, will, when he is distant from it, frequently appear inclined in a contrary direction. Many analogous instances might be adduced, but this will suffice to call others to mind; it must, however, be observed, that when shadows or other means capable of determining the judgment are present, these fallacies do not arise.

§ 11.

The same indetermination of judgment which causes a drawing to be perceived by the mind at different times as two different figures, frequently gives rise to a false perception when objects in relief are regarded with a single eye. The apparent conversion of a cameo into an intaglio, and of an intaglio into a cameo, is a well-known instance of this fallacy in vision; but the fact does not appear to me to have been correctly explained, nor the conditions under which it occurs to have been properly stated.

This curious illusion, which has been the subject of much attention, was first observed at one of the early meetings of the Royal Society*. Several of the members looking through a compound microscope of a new construction at a guinea, some of them imagined the image to be depressed, while

* Birch's History, vol. ii. p. 348.

others thought it to be embossed, as it really was. Professor Gmelin, of Wurtemburg, published a paper on the same subject in the Philosophical Transactions for 1745 ; his experiments were made with telescopes and compound microscopes which inverted the images ; and he observed that the conversion of relief appeared in some cases and not in others, at some times and not at others, and to some eyes also and not to others. He endeavoured to ascertain some of the conditions of the two appearances ; " but why these things should so happen," says he, " I do not pretend to determine."

Sir David Brewster accounts for the fallacy in the following manner* :—" A hollow seal being illuminated by a window or a candle, its shaded side is of course on the same side with the light. If we now invert the seal with one or more lenses, so that it may look in the opposite direction, it will appear to the eye with the shaded side furthest from the window. But as we know that the window is still on our left hand, and as every body with its shaded side furthest from the light must necessarily be convex or protuberant, we immediately believe that the hollow seal is now a cameo or bas-relief. The proof which the eye thus receives of the seal being raised, overcomes the evidence of its being hollow, derived from our actual knowledge and from the sense of touch. In this experiment the deception takes place from our knowing the real direction of the light which falls on the seal ; for if the place of the window, with respect to the seal, had been inverted as well as the seal itself, the illusion could not have taken place. The illusion, therefore, under our consideration is the result of an operation of our own minds, whereby we judge of the forms of bodies by the knowledge we have acquired of light and shadow. Hence the illusion depends on the accuracy and extent of our knowledge on this subject ; and while some persons are under its influence, others are entirely insensible to it."

These considerations do not fully explain the phenomenon, for they suppose that the image must be inverted, and that

* Natural Magic, p. 100.

the light must fall in a particular direction; but the conversion of relief will still take place when the object is viewed through an open tube without any lenses to invert it, and also when it is equally illuminated in all parts. The true explanation I believe to be the following. If we suppose a cameo and an intaglio of the same object, the elevations of the one corresponding exactly to the depressions of the other, it is easy to show that the projection of either on the retina is sensibly the same. When the cameo or the intaglio is seen with both eyes, it is impossible to mistake an elevation for a depression, for reasons which have been already amply explained; but when either is seen with one eye only, the most certain guide of our judgment, viz. the presentation of a different picture to each eye, is wanting; the imagination therefore supplies the deficiency, and we conceive the object to be raised or depressed according to the dictates of this faculty. No doubt in such cases our judgment is in a great degree influenced by accessory circumstances, and the intaglio or the relief may sometimes present itself according to our previous knowledge of the direction in which the shadows ought to appear; but the principal cause of the phenomenon is to be found in the indetermination of the judgment arising from our more perfect means of judging being absent.

Observers with the microscope must be particularly on their guard against illusions of this kind. Raspail observes* that the hollow pyramidal arrangement of the crystals of muriate of soda appears, when seen through a microscope, like a striated pyramid in relief. He recommends two modes of correcting the illusion. The first is to bring successively to the focus of the instrument the different parts of the crystal; if the pyramid be in relief, the point will arrive at the focus sooner than the base will; if the pyramid be hollow, the contrary will take place. The second mode is to project a strong light on the pyramid in the field of view of the microscope, and to observe which sides of the crystal are illuminated, taking, however, the inversion of the image into consideration if a compound microscope be employed.

* Nouveau Système de Chimie Organique, 2me édit. t. i. p. 333.

The inversion of relief is very striking when a skeleton cube is looked at with one eye, and the following singular results may in this case be observed. So long as the mind perceives the cube, however the figure be turned about, its various appearances will be but different representations of the same object, and the same primitive form will be suggested to the mind by all of them : but it is not so if the converse figure fixes the attention; the series of successive projections cannot then be referred to any figure to which they are all common, and the skeleton figure will appear to be continually undergoing a change of shape.

§ 12.

I have given ample proof that objects whose pictures do not fall on corresponding points of the two retinæ may still appear single. I will now adduce an experiment which proves that similar pictures falling on corresponding points of the two retinæ may appear double and in different places.

Present, in the stereoscope, to the right eye a vertical line, and to the left eye a line inclined some degrees from the perpendicular (fig. 23) ; the observer will then perceive, as formerly explained, a line, the extremities of which appear at different distances before the eyes. Draw on the left-hand figure a faint vertical line exactly corresponding in position and length to that presented to the right eye, and let the two lines of this left-hand figure intersect each other at their centres. Looking now at these two drawings in the stereoscope, the two strong lines, each seen by a different eye, will coincide, and the resultant perspective line will appear to occupy the same place as before ; but the faint line which now falls on a line of the left retina, which corresponds with the line of the right retina on which one of the coinciding strong lines, viz. the vertical one, falls, appears in a different place. The place this faint line apparently occupies is the intersection of that plane of visual direction of the left eye in which it is situated with the plane of visual direction of the right eye, which contains the strong vertical line.

This experiment affords another proof that there is no

necessary physiological connection between the corresponding points of the two retinæ,—a doctrine which has been
maintained by so many authors.

§ 13. *Binocular Vision of Images of different Magnitudes.*

We will now inquire what effect results from presenting
similar images, differing only in magnitude, to analogous
parts of the two retinæ. For this purpose two squares or
circles, differing obviously but not extravagantly in size, may
be drawn on two separate pieces of paper, and placed in the
stereoscope so that the reflected image of each shall be
equally distant from the eye by which it is regarded. It will
then be seen that, notwithstanding this difference, they coalesce and occasion a single resultant perception. The limit
of the difference of size within which the single appearance
subsists may be ascertained by employing two images of
equal magnitude, and causing one of them to recede from
the eye while the other remains at a constant distance; this
is effected merely by pulling out the sliding board C (fig. 8)
while the other C′ remains fixed, the screw having previously
been removed.

Though the single appearance of two images of different
sizes is by this experiment demonstrated, the observer is unable to perceive what difference exists between the apparent
magnitude of the binocular image and that of the two monocular images. To determine this point the stereoscope must
be dispensed with, and the experiment so arranged that all
three shall be simultaneously seen, which may be done in
the following manner :—The two drawings being placed side
by side on a plane before the eyes, the optic axes must be
made to converge to a nearer point, as at fig. 4, or to a more
distant one, as at fig. 3, until the three images are seen at
the same time, the binocular image in the middle and the
monocular images at each side. It will thus be seen that the
binocular image is apparently intermediate in size between
the two monocular ones.

If the pictures be too unequal in magnitude, the binocular
coincidence does not take place. It appears that if the in-

equality of the pictures be greater than the difference which exists between the two projections of the same object when seen in the most oblique position of the eyes (*i. e.* both turned to the extreme right or to the extreme left), ordinarily employed, they do not coalesce. Were it not for the binocular coincidence of two images of different magnitude, objects would appear single only when the optic axes converge immediately forwards ; for it is only when the converging visual lines form equal angles with the visual base (the line joining the centres of the two eyes), as at fig. 2, that the two pictures can be of equal magnitude ; but when they form different angles with it, as at fig. 24, the distance from the object to each eye is different, and consequently the picture projected on each retina has a different magnitude. If a piece of money be held in the position *a* (Pl. XVIII. fig. 24), while the optic axes converge to a nearer point *c*, it will appear double, and that seen by the left eye will be evidently smaller than the other.

§ 14. *Phenomena which are observed when pictures, which are neither similar nor the binocular complements of each other, are simultaneously presented to corresponding parts of the two retinæ.*

If we regard one picture with the right eye alone for a considerable length of time it will be constantly perceived ; if we look at the other with the left eye alone its effect will be equally permanent ; it might therefore be expected that if each of these pictures were presented to its corresponding eye at the same time the two would appear permanently superposed on each other. This, however, contrary to expectation, is not the case.

If *a* and *b* (fig. 25) are each presented at the same time to a different eye, the common border will remain constant, while the letter within it will change alternately from that which would be perceived by the right eye alone to that which would be perceived by the left eye alone. At the moment of change the letter which has just been seen breaks into fragments, while fragments of the letter which is about to appear mingle with them, and are immediately after re-placed by the entire letter. It does not appear to be in the power of the will to determine the appearance of either of the letters, but the duration of the appearance seems to de-pend on causes which are under our control : thus if the two pictures be equally illuminated, the alternations appear in general of equal duration; but if one picture be more illu-minated than the other, that which is less so will be perceived during a shorter time. I have generally made this experi-ment with the apparatus, fig. 6. When complex pictures are employed in the stereoscope, various parts of them alternate differently.

There are some facts intimately connected with the subject of the present article which have already been frequently observed. I allude to the experiments, first made by Du Tour, in which two different colours are presented to corresponding parts of the two retinæ. If a blue disk be presented to the right eye and a yellow disk to the corresponding part of the left eye, instead of a green disk which would appear if these two colours had mingled before their arrival at a single eye, the mind will perceive the two colours distinctly, one or the other alternately predominating either partially or wholly over the disk. In the same manner the mind perceives no trace of violet when red is presented to one eye and blue to the other, nor any vestige of orange when red and yellow are separately presented in a similar manner. These experi-ments may be conveniently repeated by placing the coloured disks in the stereoscope, but they have been most usually made by looking at a white object through differently coloured glasses, one applied to each eye.

In some authors we find it stated, contrary to fact, that if similar objects of different colour be presented one to each eye, the appearance will be that compounded of the two colours. Dr. Reid* and Janin are among the writers who have fallen into this error.

* Enquiry, Sect. xiii.

§ 15.

No question relating to vision has been so much debated as the cause of the single appearance of objects seen by both eyes. I shall in the present section give a slight review of the various theories which have been advanced by philosophers to account for this phenomenon, in order that the remarks I have to make in the succeeding section may be properly understood.

The law of visible direction for monocular vision has been variously stated by different optical writers. Some have maintained, with Drs. Reid and Porterfield, that every external point is seen in the direction of a line passing from its picture on the retina through the centre of the eye; while others have supposed, with Dr. Smith, that the visible direction of an object coincides with the visual ray, or the principal ray of the pencil which flows from it to the eye. D'Alembert, furnished with imperfect data respecting the refractive densities of the humours of the eye, calculated that the apparent magnitudes of objects would differ widely on the two suppositions, and concluded that the visible point of an object was not seen in either of these directions, but sensibly in the direction of a line joining the point itself and its image on the retina; but he acknowledged that he could assign no reason for this law. Sir David Brewster, provided with more accurate data, has shown that these three lines so nearly coincide with each other, that "at an inclination of 30°, a line perpendicular to the point of impression on the retina passes through the common centre, and does not deviate from the real line of visible direction more than half a degree, a quantity too small to interfere with the purposes of vision." We may therefore assume in all our future reasonings the truth of the following definition given by this eminent philosopher :—"As the interior eyeball is as nearly as possible a perfect sphere, lines perpendicular to the surface of the retina must all pass through one single point, namely the centre of its spherical surface. This one point may be called the centre of visible direction, because every point of a visible object will be seen in the direction of a line drawn from this centre to the visible point."

It is obvious that the result of any attempt to explain the single appearance of objects to both eyes, or, in other words, the law of visible position for binocular vision, ought to contain nothing inconsistent with the law of visible direction for monocular vision.

It was the opinion of Aguilonius that all objects seen at the same glance with both eyes appear to be in the plane of the horopter. The horopter he defines to be a line drawn through the point of intersection of the optic axes, and parallel to the line joining the centres of the two eyes; the plane of the horopter to be a plane passing through this line at right angles to that of the optic axes. All objects which are in this plane must, according to him, appear single because the lines of direction in which any point of an object is seen coincide only in this plane and nowhere else; and as these lines can meet each other only in one point, it follows from the hypothesis, that all objects not in the plane of the horopter must appear double, because their lines of direction intersect each other, either before or after they pass through it. This opinion was also maintained by Dechales and Porterfield. That it is erroneous, I have given, I think, sufficient proof, in showing that, when the optic axes converge to any point, objects before or beyond the plane of the horopter are, under certain circumstances, equally seen single as those in that plane.

Dr. Wells's "new theory of visible direction" was a modification of the preceding hypothesis. This acute writer held with Aguilonius, that objects are seen single only when they are in the plane of the horopter, and consequently that they appear double when they are either before or beyond it; but he attempted to make this single appearance of objects only in the plane of the horopter to depend on other principles, from which he deduced, contrary to Aguilonius, that the objects which are doubled do not appear in the plane of the horopter, but in other places which are determined by these

principles. Dr. Wells was led to his new theory by a fact
which he accidentally observed, and which he could not
reconcile with any existing theory of visible direction : this
fact had, though he was unaware of it, been previously
noticed by Dr. Smith; it is already mentioned in § 8, and
is the only other instance of binocular vision of relief which
I have found recorded previous to my own investigations. So little does Dr. Wells's theory appear to have been
understood, that no subsequent writer has attempted either
to confirm or disprove his opinions. It would be useless
here to discuss the principles of this theory, which was
framed to account for an anomalous individual fact, since it
is inconsistent with the general rules on which that fact has
been now shown to depend. Notwithstanding these erroneous views, the " essay upon single vision with two eyes "
contains many valuable experiments and remarks, the truth
of which is independent of the theory they were intended
to illustrate.

The theory which has obtained greatest currency is that
which assumes that an object is seen single because its pictures
fall on corresponding points of the two retinæ—that is, on
points which are similarly situated with respect to the two
centres both in distance and position. This theory supposes
that the pictures projected on the retinæ are exactly similar
to each other, corresponding points of the two pictures falling on corresponding points of the two retinæ. Authors
who agree with regard to this property, differ widely in explaining why objects are seen in the same place, or single,
according to this law. Dr. Smith makes it to depend entirely
on custom, and explains why the eyes are habitually directed
towards an object so that its pictures fall on corresponding
parts in the following manner :—" When we view an object
steadily, we have acquired a habit of directing the optic axes
to the point in view; because its pictures falling upon the
middle points of the retinas, are then distincter than if they
fell upon any other places ; and since the pictures of the
whole object are equal to one another, and are both inverted
with respect to the optic axes, it follows that the pictures of

any collateral point are painted upon corresponding points of the retinas."

Dr. Reid, after a long dissertation on the subject, concludes, " that by an original property of human eyes, objects painted upon the centres of the two retinæ, or upon points similarly situated with regard to the centres, appear in the same visible place; that the most plausible attempts to account for this property of the eyes have been unsuccessful; and therefore, that it must be either a primary law of our constitution, or the consequence of some more general law which is not yet discovered."

Other writers who have admitted this principle have regarded it as arising from anatomical structure and dependent on connexion of nervous fibres; among these stand the names of Galen, Dr. Briggs, Sir Isaac Newton, Rohault, Dr. Hartley, Dr. Wollaston, and Professor Müller.

Many of the supporters of the theory of corresponding points have thought, or rather have admitted, *without thinking*, that it was not inconsistent with the law of Aguilonius; but very little reflection will show that both cannot be maintained together; for corresponding lines of visible direction, that is, lines terminating in corresponding points of the two retinæ, cannot all at the same time meet in the plane of the horopter unless the optic axes be parallel, and the plane be at an infinite distance before the eyes. Some of the modern German writers* have inquired what is the curve in which objects appear single while the optic axes are directed to a given point, on the hypothesis that objects are seen single only when they fall on corresponding points of the two retinæ. An elegant proposition has resulted from their investigations, which I shall need no apology for introducing in this place, since it has not yet been mentioned in any English work.

R and L (fig. 26) are the two eyes; CA, C'A the optic axes converging to the point A; and CABC' is a circle drawn through the point of convergence A and the centres of visible

* *Tortual,* Die Sinne des Menschen: Münster, 1827. *Bartels,* Beiträge zur Physiologie des Gesichtssinnes: Berlin, 1834.

direction CC'.   If any point be taken in the circumference of this circle, and lines be drawn from it through the centres of the two eyes CC', these lines will fall on corresponding points of the two retinæ DD'; for the angles ACB, AC'B being equal, the angles DCE, DC'E are also equal; therefore any point placed in the circumference of the circle CABC' will, according to the hypothesis, appear single while the optic axes are directed to A, or to any other point in it.

I will mention two other properties of this binocular circle: 1st. The arc subtended by two points on its circumference contains double the number of degrees of the arc subtended by the pictures of these points on either retina, so that objects which occupy 180° of the supposed circle of single vision are painted on a portion of the retina extended over 90° only; for the angle DCE or DC'E being at the centre, and the angle BCA or BC'A at the circumference of a circle, this consequence follows. 2ndly. To whatever point of the circumference of the circle the optic axes be made to converge, they will form the same angle with each other; for the angles CAC' CBC are equal.

In the eye itself, the centre of visible direction, or the point at which the principal rays cross each other, is, according to Dr. Young and other eminent optical writers, at the same time the centre of the spherical surface of the retina, and that of the lesser spherical surface of the cornea; in the diagram (fig. 26), to simplify the consideration of the problem, R and L represent only the circle of curvature of the bottom of the retina, but the reasoning is equally true in both cases.

The same reasons, founded on the experiments in this memoir, which disprove the theory of Aguilonius, induce me to reject the law of corresponding points as an accurate expression of the phenomena of single vision.   According to the former, objects can appear single only in the plane of the horopter; according to the latter, only when they are in the circle of single vision; both positions are inconsistent with the binocular vision of objects in relief, the points of which they consist appearing single though they are at different distances before the eyes.   I have already proved that the assumption made by all the maintainers of the theory of corresponding points, namely that the two pictures projected by any object on the retinæ are exactly similar, is quite contrary to fact in every case except that in which the optic axes are parallel.

Gassendus, Porta, Tacquet, and Gall maintained that we see with only one eye at a time though both remain open, one, according to them, being relaxed and inattentive to objects while the other is upon the stretch. It is a sufficient refutation of this hypothesis that we see an object double when one of the optic axes is displaced either by squinting or by pressure on the eyeball with the finger; if we saw with only one eye, one object only should under such circumstances be seen. Again, in many cases which I have already explained, the simultaneous affection of the two retinæ excites a different idea in the mind to that consequent on either of the single impressions, the latter giving rise to the idea of a representation on a plane surface, the former to that of an object in relief; these things could not occur did we see with only one eye at a time.

Du Tour* held that though we might occasionally see at the same time with both eyes, yet the mind cannot be affected simultaneously by two corresponding points of the two images. He was led to this opinion by the curious facts alluded to in § 14. It would be difficult to disprove this conjecture by experiment; but all that the experiments adduced in its favour, and others relating to the disappearance of objects to one eye, really proves is, that the mind is inattentive to impressions made on one retina when it cannot combine the impressions on the two retinæ together so as to occasion a perception resembling that of some external object; but they afford no ground whatever for supposing that the mind cannot under any circumstances attend to impressions made simultaneously on points of the two retinæ, when they harmonize with each other in suggesting to the mind the same idea.

A perfectly original theory has been recently advanced by

* Act. Par. 1743. M. p. 334.

M. Lehot*, who has endeavoured to prove that instead of pictures on the retinæ, images of three dimensions are formed in the vitreous humour which we perceive by means of nervous filaments extending thence from the retina. This theory would account for the single appearance to both eyes of objects in relief, but it would be quite insufficient to explain why we perceive an object of three dimensions when two pictures of it are presented to the eyes; according to it, also, no difference should be perceived in the relief of objects when seen by one or both eyes, which is contrary to what really happens. The proofs, besides, that we perceive external objects by means of pictures on the retinæ are so numerous and convincing, that a contrary conjecture cannot be entertained for a moment. On this account it will suffice merely to mention two other theories. Vallée†, without denying the existence of pictures on the retina, has advocated that we perceive the relief of objects by means of anterior foci on the hyaloid membrane; and Raspail ‡ has developed at considerable length the strange hypothesis, that images are not painted on the retina, but are immediately perceived at the focus of the lenticular system of which the eye is formed.

§ 16.

It now remains to examine *why* two dissimilar pictures projected on the two retinæ give rise to the perception of an object in relief. I will not attempt at present to give the complete solution of this question, which is far from being so easy as at a first glance it may appear to be, and is indeed one of great complexity. I shall in this place merely consider the most obvious explanation which might be offered, and show its insufficiency to explain the whole of the phenomena.

It may be supposed that we see but one point of an object distinctly at the same instant, the one namely to which the optic axes are directed, while all other points are seen so in-

---

* Nouvelle Théorie de la Vision, Par. 1823.

† Traité de la Science du Dessein, Par. 1821, p. 270.

‡ Nouveau Système de Chimie Organique, t. ii. p. 329.

distinctly, that the mind does not recognize them to be either single or double, and that the figure is appreciated by successively directing the point of convergence of the optic axes successively to a sufficient number of its points to enable us to judge accurately of its form.

That there is a degree of indistinctness in those parts of the field of view to which the eyes are not immediately directed, and which increases with the distance from that point, cannot be doubted, and it is also true that the objects thus obscurely seen are frequently doubled. It may be said, this indistinctness and duplicity is not attended to, because the eyes shifting continually from point to point, every part of the object is successively rendered distinct; and the perception of the object is not the consequence of a single glance, during which only a small part of it is seen distinctly, but is formed from a comparison of all the pictures successively seen while the eyes are changing from one point of the object to another.

All this is in some degree true; but were it entirely so, no appearance of relief should present itself when the eyes remain intently fixed on one point of a binocular image in the stereoscope. But on performing the experiment carefully, it will be found, provided the pictures do not extend too far beyond the centres of distinct vision, that the image is still seen single and in relief when this condition is fulfilled. Were the theory of corresponding points true, the appearance should be that of the superposition of the two drawings, to which, however, it has not the slightest similitude. The following experiment is equally decisive against this theory.

Draw two lines inclined towards each other, as in Plate XIX. fig. 10, on a sheet of paper, and having caused them to coincide by converging the optic axes to a point nearer than the paper, look intently on the upper end of the resultant line, without allowing the eyes to wander from it for a moment. The entire line will appear single and in its proper relief, and a pin or a piece of straight wire may without the least difficulty be made to coincide exactly in position with it; or, if while the optic axes continue to be directed to the upper and nearer end, the point of a pin be made to coincide with the lower and further end or with any intermediate point of the resultant line, the coincidence will remain exactly the same when the optic axes are moved and meet there. The eyes sometimes become fatigued, which causes the line to appear double at those parts to which the optic axes are not fixed, but in such case all appearance of relief vanishes. The same experiment may be tried with more complex figures, but the pictures should not extend too far beyond the centres of the retinæ.

Another and a beautiful proof that the appearance of relief in binocular vision is an effect independent of the motions of the eyes, may be obtained by impressing on the retinæ ocular spectra of the component figures. For this purpose the drawings should be formed of broad coloured lines on a ground of the complementary colour, for instance red lines on a green ground, and be viewed either in the stereoscope or in the apparatus, fig. 6, as the ordinary figures are, taking care, however, to fix the eyes only to a single point of the compound figure; the drawings must be strongly illuminated, and after a sufficient time has elapsed to impress the spectra on the retinæ, the eyes must be carefully covered to exclude all external light. A spectrum of the object in relief will then appear before the closed eyes. It is well known that a spectrum impressed on a single eye and seen in the dark, frequently alternately appears and disappears : these alternations do not correspond in the spectra impressed on the two retinæ, and hence a curious effect arises ; sometimes the right-eye spectrum will be seen alone, sometimes that of the left eye, and at those moments when the two appear together, the binocular spectrum will present itself in bold relief. As in this case the pictures cannot shift their places on the retinæ in whatever manner the eyes be moved about, the optic axes can during the experiment only correspond with a single point of each.

When an object, or a part of an object, thus appears in relief while the optic axes are directed to a single binocular

s

point, it is easy to see that each point of the figure that
appears single is seen at the intersection of the two lines of
visible direction in which it is seen by each eye separately,
whether these lines of visible direction terminate at corresponding points of the two retinæ or not.

But if we were to infer the converse of this, viz. that every
point of an object in relief is seen by a single glance at the
intersection of the lines of visible direction in which it is
seen by each eye singly, we should be in error.   On this
supposition, objects before or beyond the intersection of the
optic axes should never appear double, and we have abundant evidence that they do.   The determination of the
points which shall appear single seems to depend in no
small degree on previous knowledge of the form we are
regarding.   No doubt, some law or rule of vision may be
discovered which shall include all the circumstances under
which single vision by means of non-corresponding points
occurs and is limited.   I have made numerous experiments
for the purpose of attaining this end, and have ascertained
some of the conditions on which single and double vision
depend, the consideration of which, however, must at present
be deferred.

Sufficient, however, has been shown to prove that the laws
of binocular visible position hitherto laid down are too restricted to be true.   The law of Aguilonius assumes that
objects in the plane of the horopter are alone seen single;
and the law of corresponding points carried to its necessary
consequences, though these consequences were unforeseen
by its first advocates, many of whom thought that it was
consistent with the law of Aguilonius, leads to the conclusion that no object appears single unless it is seen in a circle
passing through the centres of visible direction in each eye
and the point of convergence of the optic axes.   Both of
these are inconsistent with the single vision of objects whose
points lie out of the plane in one case and the circle in the
other; and that objects do appear single under circumstances
that cannot be explained by these laws, has, I think, been

placed beyond doubt by the experiments I have brought forward. Should it be hereafter proved, that all points in the plane or in the circle above mentioned are seen single, and from the great indistinctness of lateral images it will be difficult to give this proof, the law must be qualified by the admission that points out of them do not always appear double.

Contributions to the Physiology of Vision.—Part the Second. On some remarkable, and hitherto unobserved, Phenomena of Binocular Vision.

[From the 'Philosophical Transactions,' 1852.]

[Plate XX.]

§ 17.

In § 3 of the first part of my "Contributions to the Physiology of Vision," published in the Philosophical Transactions for 1838, speaking of the stereoscope, I stated, "The pictures will indeed coincide when the sliding pannels are in a variety of different positions, and consequently when viewed under different inclinations of the optic axes; but there is only one position in which the binocular image will be immediately seen single, of its proper magnitude, and without fatigue to the eyes, because in this position only the ordinary relations between the magnitude of the pictures on the retina, the inclination of the optic axes, and the adaptation of the eye to distinct vision at different distances are preserved. The alteration in the apparent magnitude of the binocular images, when these usual relations are disturbed, will be discussed in another paper of this series, with a variety of remarkable phenomena depending thereon."

In 1833, five years before the publication of the memoir just mentioned, these yet unpublished investigations were announced, in the third edition of Herbert Mayo's 'Outlines of Human Physiology,' in the following words :—"Mr. Wheatstone has shown, in a paper he is about to publish, that if by artificial means the usual relations which subsist between the degree of inclination of the optic axes and the visual angle which the object subtends on the retina be disturbed, some extraordinary illusions may be produced. Thus, the magnitude of the image remaining constant on the retina its apparent size may be made to vary with every alteration of the angular inclination of the optic axes."

I shall resume the consideration of the phenomena of binocular vision with this subject, because the facts I have ascertained regarding it are necessary to be understood before entering on the new experiments relating to stereoscopic appearances which I intend to bring forward on the present occasion.

Under the ordinary conditions of vision, when an object is placed at a certain distance before the eyes, several concurring circumstances remain constant, and they always vary in the same order when the distance of the object is changed. Thus, as we approach the object, or as it is brought nearer to us, the magnitude of the picture on the retina increases, the inclination of the optic axes, required to cause the pictures to fall on corresponding places of the retinæ, becomes greater; the divergence of the rays of light proceeding from each point of the object, and which determines the adaptation of the eyes to distinct vision of that point, increases; and the dissimilarity of the two pictures projected on the retinæ also becomes greater. It is important to ascertain in what manner our perception of the magnitude and distance of objects depends on these various circumstances, and to inquire which are the most, and which the least influential in the judgments we form. To advance this inquiry beyond the point to which it has hitherto been brought, it is not sufficient to content ourselves with drawing conclusions from observations on the circumstances under which vision naturally occurs, as preceding writers on this subject mostly have done, but it is necessary to have more extended recourse to the methods so successfully employed in experimental philosophy, and to endeavour, wherever it be possible, not only to analyse the elements of vision, but also to recombine them in unusual manners, so that they may be associated under circumstances that never naturally occur.

The instrument I shall proceed to describe enables these abnormal combinations to be made in a very simple and effectual manner. Its principal object is to cause the binocular pictures to coincide, with any inclination of the optic axes, while their magnitudes on the retinæ remain the same; or inversely, while the optic axes remain at the same angle, to cause the size of the pictures on the retinæ to vary in any manner.

Two plane mirrors inclined 90° to each other are placed together and fixed vertically upon a horizontal board. Two wooden arms move round a common centre situated on this board in the vertical plane which bisects the angle of the mirrors, and about $1\frac{1}{2}$ inch beyond their line of junction. Upon each of these arms is placed an upright pannel, at right angles thereto, for the purpose of receiving its appropriate picture, and each pannel is made to slide to and from the opposite mirror. The eyes being placed before the mirrors, the right eye to the right mirror and the left eye to the left mirror, and the pannels being adjusted to the same distances, however the arms be moved round their centre, the distance of the reflected image of each picture from the eye will remain exactly the same, and consequently its retinal magnitude will be unchanged. But as the two reflected images do not occupy the same place when the pictures are in different positions, to cause the former to coincide the optic axes must converge differently. When the arms are in the same straight line, the images coincide while the optic axes are parallel; and as they form a less angle with each other, the optic axes converge more to occasion the coincidence. When the arms remain in the same positions, while the pannels slide towards or from the mirrors, the convergence of the optic axes remains the same, but the magnitude of the pictures on the retinæ increases as the distance decreases. By the arrangement described, and which is represented by figs. 1 and 2, Plate XX., the reflected pictures are always perpendicular to the optic axes, and the corresponding points of the pictures, when they are exactly similar, fall upon corresponding points of the retinæ. The instrument has an adjustment for otherwise inclining them if it be required.

Let us now attend to the effects produced. The pictures being fixed at the same distance from the mirrors, there is a certain adjustment of the arms at which the binocular image will appear of its natural size, that is, the size we judge the picture itself to be when we look at it directly; in this case the magnitude of the pictures on the retinæ and the inclination of the optic axes preserve their usual relation to each other. If now the arms be moved back, so as to cause a less convergence of the axes, the image will appear to increase in magnitude until the arms are in a straight line and the optic axes are parallel; and, on the other hand, if the arms be moved forwards, so as to form a less angle, the optic axes will converge more, and the image will appear gradually smaller. In this manner, while the retinal magnitude remains the same, the perceived magnitude of the binocular object varies through a very considerable range.

The instrument being again adjusted so that the image shall be seen of its natural size, on sliding the pictures nearer the mirrors its perceived magnitude will be augmented, and on sliding them from the mirrors it will appear diminished in size. During these variations of magnitude the inclination of the optic axes remains the same.

The perceived magnitude of an object, therefore, diminishes as the inclination of the axes becomes greater, while the distance remains the same; and it increases, when the inclination of the axes remains the same, while the distance diminishes. When both these conditions vary inversely, as they do in ordinary vision when the distance of an object changes, the perceived magnitude remains the same*.

Before I proceed further it will be proper to explain the meaning of some of the terms I employ. I call the magnitude of the object itself, the real or objective magnitude; the

---

* Several cases of the alteration of the perceived magnitude of objects are mentioned by Dr. R. Smith (Complete System of Optics, 1738, vol. ii. p. 388, and rem. 526 and 532); and Dr. R. Darwin (Philosophical Transactions, vol. lxxvi. p. 313) observed that when an ocular spectrum was impressed on both eyes it appeared magnified when they were directed to a wall at a considerable distance. The facts noticed by these authors are satisfactorily explained by the above considerations.

magnitude of the picture on the retina, the retinal magnitude; and the magnitude we estimate the object to be from its retinal magnitude and the inclination of the optic axes conjointly, I name the perceived magnitude. I do not use the term apparent magnitude, because, according to its ordinary acceptation, it sometimes means what I call retinal, and at other times what I name perceived magnitude.

We have seen in what manner our perception of magnitude is modified by the new associations which this instrument enables us to form; let us now examine how our perception of distance is affected by them. If we continue to observe the binocular picture whilst it apparently increases or decreases, in consequence of the inclination of the optic axes varying while the magnitude of the impressions on the retina remains the same, it does not appear either to approach or to recede; and yet if we attentively regard it in any fixed position, it is perceived to be at a different distance. On the other hand, if we continue to regard the binocular picture, enlarging and diminishing in consequence of the change of retinal magnitude while the convergence of the axes remains the same, we perceive it to approach or recede in the most evident manner; but on fixing the attention to it, when it is stationary, at any instant, it appears to be at the same distance at one time as it is at another.

Convergence of the optic axes therefore suggests fixed distance to the mind; variation of retinal magnitude suggests change of distance. We may, as I have above shown, perceive an object approach or recede without appearing to change its distance, and an object to be at a different distance without appearing to approach or recede; these paradoxical effects render it difficult, until the phenomena are well apprehended, to know, or to express, what we actually do perceive.

It is the prevalent opinion that the sensation which accompanies the inclination of the optic axes immediately suggests distance, and that the perceived magnitude of an object is a judgment arising from our consciousness of its distance and of the magnitude of its picture on the retina. From the experiments I have brought forward, it rather appears to me that what the sensation which is connected with the convergence of the axes immediately suggests is a correction of the retinal magnitude to make it agree with the real magnitude of the object; and that distance, instead of being a simple perception, is a judgment arising from a comparison of the retinal and perceived magnitudes. However this may be, unless other signs accompany this sensation the notion of distance we thence derive is uncertain and obscure, whereas the perception of the change of magnitude it occasions is obvious and unmistakable.

To see, in their full extent, the variations of magnitude exhibited by the instrument I have described, it is necessary to attend to the following observations.

As the inclination of the optic axes corresponding to a different distance is habitually, under ordinary circumstances, accompanied with the particular adaptation of the eyes required for distinct vision at that distance, it is difficult to disassociate these two conditions so as to see with equal distinctness the binocular picture when the optic axes are parallel and when they converge greatly, although the pictures remain, in both cases, at the same distance from the eyes. The adaptation is therefore not entirely dependent on the divergence of the rays of light which proceed from the object regarded, but also, in some degree, on the inclination of the optic axes. I have acquired by practice considerable power of adjustment, or rather disadjustment, of the eyes, and can, without having recourse to artificial means, see the binocular picture distinctly when its perceived magnitude is widely different. Those to whom such an effort is painful may employ short-sighted spectacles to see the binocular picture when the eyes converge within the limit of distinct vision for the distance at which the pictures are placed; and long-sighted spectacles when the eyes converge beyond that limit, or become parallel.

There is a means of avoiding to a very considerable extent the influence of the adjustment of the eyes, and thereby enabling the pictures to be seen distinctly within the entire range of the inclination of the optic axes. This is by looking at the reflected images in the mirrors through two very minute apertures, not larger than fine pin-holes, placed near each eye, and illuminating the pictures by a very strong light; sunshine in the middle of the day answers the purpose very well. By this expedient the divergence of the rays of light is greatly diminished, and the adaptation of the eyes does not materially influence the result.

§ 18.

Leaving this subject, I will now revert to the stereoscope and its effects.

Since 1838 numerous modifications of the stereoscope have occurred to me, and several ingenious arrangements have also been proposed by Sir David Brewster and Professor Dove; but there is no form of the instrument which has so many advantages for investigating the phenomena of binocular vision as the original reflecting stereoscope. Pictures of any size may be placed in it, and it admits of every kind of adjustment.

I have constructed a very portable reflecting stereoscope which is represented at fig. 3 (Plate XX.). The sides fold over the mirrors, and the mirrors then fold into a box, which is not larger than six inches in any of its dimensions. To avoid the second feeble reflection from the anterior surface of the silvered glass, which has a bad effect when the attention is attracted to it, I have sometimes employed reflecting prisms. The reflecting surfaces of the prisms should be silvered in order to obviate the unequal brightness of the field of view on each side of the limit of total reflexion; and as it would be too costly to employ very large prisms, they should have an adjustment to accommodate their distance to the width between the eyes of the observer.

I have, for many years past, employed also another means to occasion, without any straining of the eyes, the coincidence of the pictures, so that the image in relief shall appear of the same magnitude and at the same distance as the object which they represent would do if it were itself directly regarded. In this apparatus, prisms being employed to deflect the rays of light proceeding from the pictures, so as to make them appear to occupy the same place, I have called it the refracting stereoscope.

It is represented by fig. 4. It consists of a base 6 inches long and 4 inches broad, upon which stands an upright partition, 5 inches high, dividing it equally; this partition is capable of extension by means of a slide to double the length, and carries at its upper extremity a board placed parallel to the base and of the same dimensions. In this upper board there are two apertures an inch square, one on each side of the partition, the centres of which are $2\frac{1}{2}$ inches from each other; in these apertures are fixed a pair of glass prisms having their faces inclined 15°, and their refractive angles turned towards each other. The stereoscope pictures are to be placed on the base, and their centres ought not to exceed the distance of $2\frac{1}{2}$ inches.

A pair of plate-glass prisms, their faces making with each other an angle of 12°, will bring two pictures, the corresponding points of which are $2\frac{1}{2}$ inches apart, to coincide at a distance of 12 inches, and a pair with an angle of 15° will occasion coincidence at 8 inches.

The refracting stereoscope has the advantage of portability, but it is limited to pictures of small dimensions. It is well suited for Daguerreotypes, which are usually of small size, and, on account of the nature of their reflecting surface, must be viewed in a particular direction with respect to the light which falls upon them; whereas in the reflecting stereoscope it is somewhat difficult to render the two Daguerreotypes equally visible. For drawings and Talbotypes it, however, offers no advantages, though it is equally well suited for them when their dimensions are small.

Stereoscopic drawings afford a means of illustrating works with figures of three dimensions, instead of with mere plane representations. Works on crystallography, solid geometry, spherical trigonometry, architecture, machinery, &c. might be thus rendered more instructive, from the perfect counterpart of the solid figure seen from a single point of view being represented instead of merely one of its plane projections. For this purpose the corresponding binocular figures must be engraved in parallel vertical columns, and their coalescence may be effected by viewing them through a pair of prisms, similar to those employed in the refracting stereoscope, placed in a frame at the proper distance from each other. If the engravings should be less than $2\frac{1}{2}$ inches apart, the prisms may be dispensed with by persons who have command over the adaptation of their eyes, particularly if they be short-sighted.

§ 19.

At the date of the publication of my experiments on binocular vision, the brilliant photographic discoveries of Talbot, Niepce, and Daguerre had not been announced to the world. To illustrate the phenomena of the stereoscope I could therefore, at that time, only employ drawings made by the hands of an artist. Mere outline figures, or even shaded perspective drawings of simple objects, do not present much difficulty; but it is evidently impossible for the most accurate and accomplished artist to delineate, by the sole aid of his eye, the two projections necessary to form the stereoscopic relief of objects as they exist in nature with their delicate differences of outline, light, and shade. What the hand of the artist was unable to accomplish, the chemical action of light, directed by the camera, has enabled us effect.

It was at the beginning of 1839, about six months after the appearance of my memoir in the Philosophical Transactions, that the photographic art became known, and soon after, at my request, Mr. Talbot, the inventor, and Mr. Collen (one of the first cultivators of the art) obligingly prepared for me stereoscopic Talbotypes of full-sized statues, buildings, and even portraits of living persons. M. Quetelet, to whom I communicated this application and sent specimens, made mention of it in the Bulletins of the Brussels Academy of October 1841. To M. Fizeau and M. Claudet I was indebted for the first Daguerreotypes executed for the stereoscope. The beautiful stereoscopic representations of statuary, architecture, machinery, natural history specimens, portraits of living persons, single and in groups, &c., which have recently been produced by M. Soleil and M. Claudet, are now too well known to the public to need more than a slight reference to them.

With respect to the means of preparing the binocular photographs (and in this general term I include both Talbotypes and Daguerreotypes), little requires to be said beyond a few directions as to the proper positions in which it is necessary to place the camera in order to obtain the two required projections.

We will suppose that the binocular pictures are required to be seen in the stereoscope at a distance of 8 inches before the eyes, in which case the convergence of the optic axes is about 18°. To obtain the proper projections for this distance, the camera must be placed, with its lens accurately directed towards the object, successively in two points of the circumference of a circle of which the object is the centre, and the points at which the camera is so placed must have the angular distance of 18° from each other, exactly that of the optic axes in the stereoscope. The distance of the camera from the object may be taken arbitrarily; for, so long as the same angle is employed, whatever that distance may be, the pictures will exhibit in the stereoscope the same relief, and be seen at the same distance of 8 inches, only the magnitude of the picture will appear different. Miniature stereoscopic representations of buildings and full-sized statues are therefore obtained merely by taking the two projections of the object from a considerable distance, but at the same angle as if the object were only 8 inches distant, that is, at an angle of 18°.

To produce the best effect, it is necessary that the pictures be so placed in the stereoscope that each eye shall see its respective picture at the proper point of sight: if this condition be not attended to, the binocular perspective will be incorrect.

For obtaining binocular photographic portraits, it has been found advantageous to employ, simultaneously, two cameras fixed at the proper angular positions.

I subjoin a Table of the inclinations of the optic axes which correspond to different distances; it also shows the angular positions of the camera required to obtain binocular pictures which shall appear at a given distance in the stereoscope in their true relief.

Inclination of the optic axes	2°	4°	6°	8°	10°	12°	14°	16°	18°	20°	22°	24°	26°	28°	30°
Distance in inches	71·5	35·7	23·8	17·8	13·2	11·8	10·1	8·8	7·8	7·0	6·4	5·8	5·4	5·0	4·6

The distance is equal to $\frac{a}{2}$ cotang $\frac{\theta}{2}$; a denoting the distance between the two eyes, and θ the inclination of the optic axes.

§ 20.

As the inclination of the optic axes diminishes by the removal of an object to which they are directed to a greater distance, not only does the magnitude of the pictures projected by it on the retinæ proportionately diminish, but the dissimilarity of the pictures becomes less. The difference of distance between any two points of each of the pictures will diminish until the projections become sensibly similar. Under the usual circumstances attending the vision of a solid object placed at a given distance, a particular inclination of the axes is invariably accompanied by a specific pair of dissimilar projections; and if the distance be changed, a different inclination of the axes is accompanied by another pair of projections; but, by means of the stereoscope, we have it within our power to associate these circumstances abnormally, and to cause any degree of inclination of the axes to coexist with any dissimilarity of the two pictures. To ascertain experimentally what takes place under these circumstances, M. Claudet prepared for me a number of Daguerreotypes of the same bust, taken at a variety of different angles, so that I was enabled to place in the stereoscope two pictures taken at any angular distance from 2° to 18°, the former corresponding with a distance of about 6 feet, and the latter with a distance of about 8 inches. The effect of a pair of near projections seen with a distant convergence of the optic

axes is to give an undue elongation to lines joining two unequally distant points, so that all the features of a bust
appear to be exaggerated in depth. The effect, on the contrary, of a pair of distant projections seen with a *near*
convergence of the axes is to give an undue shortening to
the same lines, so that the appearance of a bas-relief is obtained from the two projections of the bust. The apparent
dimensions in breadth and height remain in both cases the
same.

§ 21.

To reproduce the conditions of the binocular vision of a
solid object as completely as possible by means of its two
plane projections, it is necessary, as I have before stated,
that the projections shall be such as correspond exactly with
the inclination of the optic axes under which they are viewed.
I have already shown in § 20 what takes place when this
condition is not strictly observed, and I may add that the
mind is not unpleasantly affected by a considerable incon-
gruity in this respect; on the contrary, the effect in many
cases seems heightened by viewing the solid appearance, intended for a determinate degree of inclination of the axes,
under an angle several degrees less; the reality is as it were
exaggerated. When the optic axes are parallel, in strictness
there should be no difference between the pictures presented
to each eye, and in this case there would be no binocular
relief; but I find that an excellent effect is produced when
the axes are nearly parallel by pictures taken at an inclination of 7° or 8°, and even a difference of 16° or 17° has no
decidedly bad effect.

This circumstance enables us to combine the ideal ampli-
fication arising from viewing pictures placed near the eyes
under a small inclination, or even parallelism, of the optic
axes mentioned in § 17, with the perception of solidity arising
from the dissimilarity of the projections; for this purpose,
the pictures, in the refracting stereoscope, or their reflected
images in the reflecting instrument, must be viewed through
lenses the focal distance of which is equal to the distance
between them and the pictures; the perceived magnitude of

the binocular image will increase with the nearness of the
pictures, and depends almost entirely on the disassociation
of the retinal magnitude from its usually accompanying inclination of the optic axes, the actual magnifying-power of
the lenses having a very small influence.

The sole use of the lenses is to render the rays of light
parallel, which it is necessary they should be for distinct
vision when the optic axes are parallel.   When the reflecting
stereoscope is employed, this means of magnifying the effect
is not of much utility, as pictures of any size may be adapted
to that instrument.   But in the case of the refracting stereoscope it may be advantageously made use of.   By combining
lenses with the refracting stereoscope, described in § 18,
Daguerreotypes somewhat wider than the width between the
eyes may be employed.   Sir David Brewster has used, to
effect the same purpose, semi-lenses with their edges directed
towards each other, which serve at the same time to render
the rays less convergent and slightly to displace the pictures
towards each other.   Two corresponding Daguerreotypes,
each not exceeding in breadth the width between the eyes,
being placed close to each other, and viewed with lenses of
short focal distance, will, even without the aid of the prisms,
give an apparently highly magnified binocular image in bold
relief.

There is a peculiarity in such images worthy of remark;
although the optic axes are parallel, or nearly so, the image
does not appear to be referred to the distance we should,
from this circumstance, suppose it to be, but it is perceived
to be much nearer, and indeed more so, as the pictures are
nearer the eyes, though the inclination of the optic axes
remains the same, and should therefore suggest the same
distance; it seems as if the dissimilarity of the projections,
corresponding as they do to a nearer distance than that
which would be suggested by the former circumstance alone,
alters in some degree the perception of distance.

I recommend, as a convenient arrangement of a refracting
stereoscope for viewing Daguerreotypes of small dimensions,
the instrument represented, Pl. XX. fig. 4, shortened in its

length from 8 inches to 5, and lenses of 5 inches focal distance placed before and close to the prisms.

§ 22.

I now proceed to another subject—to the consideration of those phenomena which I have termed Conversions of Relief.

In § 5 of my first memoir I noticed the remarkable circumstance, that when the drawing intended to be seen by the right eye is presented to the left eye in the stereoscope, and *vice versâ*, a totally different solid figure is perceived to that seen before the transposition. I called this the converse figure, and showed that it differs from the normal figure in the circumstance that those points which appear the most distant in the latter, appear the nearest in the former.

The pictures being, in the first place, presented directly to their corresponding eyes, as in the refracting stereoscope, and exhibiting therefore the resultant image in its normal relief, the conversion of the relief may be effected in three different ways,—1st, by transposing the pictures from one eye to the other, as mentioned above; 2ndly, by reflecting the pictures, while they remain presented to the same eye, as in the reflecting stereoscope; and 3rdly, by inverting the position of the pictures without transposing them.

The following considerations will explain the cause of the conversion of relief in the preceding cases.

If two different objects, or parts of an object (fig. 5 *a*), have a greater lateral distance between them on the right-hand picture than that which they have on the left-hand picture, the optic axes must converge more to make the left-hand than to make the right-hand objects coincide, and the left-hand object will appear the nearest.

If the pictures be now transposed from one eye to the other (fig. 5 *a'*), the greatest distance will be between the corresponding points of the picture presented to the left eye; the optic axes must therefore converge less to make the left-hand objects coincide, and the right-hand object will appear the nearest.

If the pictures, remaining untransposed, be each separately

T

reflected (fig. 5 *b*), the relative distances of the corresponding objects remain the same to each eye, and the left-hand object will still appear nearest; but in consequence of the lateral inversion of the objects in each picture by reflection, that which was previously on the left will now be on the right, and therefore the object which before appeared nearest will now appear furthest.

When the pictures are turned upside down, still remaining untransposed (fig. 5 *c*), the objects are reversed with respect to the right and left, in the same manner as they are when reflected, and the lateral distances between the objects remaining the same to each eye, precisely the same conversion of relief is produced as in the preceding case, except that the resultant image is inverted. The diagram (fig. 5) represents all the possible changes of the two binocular pictures; those marked N show the normal relief, and those marked C the converse relief.

But it may be asked why, if the reflexion or inversion of the binocular pictures of an object gives rise to the mental idea of the converse relief, the same converse relief is not observed when the object itself is reflected in a mirror or inverted. The reason is this : that in the former cases the projections to each eye are separately reflected or inverted, still remaining presented to the same eye, whereas, by the reflexion or inversion of the object itself, not only are the projections reflected or inverted, but they are also transposed from one eye to the other ; and these circumstances occurring simultaneously reproduce the normal relief.

Fig. 6 will render this evident in the case of reflexion : A is the object, B its reflection in the mirror CD; RB and LB are the directions in which the right and left eyes view the reflected image respectively, and *l*A and *r*A the directions in which the eyes would view the corresponding face of the object directly.

In the case of an inverted object, it is obvious that that projection which was before seen by the right eye must be seen by the left eye, and the contrary.

It is possible to make this normal or converse relief appear while one of the pictures remains constantly presented to the same eye. This result may be thus obtained. Having taken a photograph of the object, which should be one the converse of which has a meaning, take two others at the same angular distance (say 18°), one on the right side, the other on the left side of the original. Of the three pictures thus taken, if the middle one be presented to the right eye, and the left picture to the left eye, a normal relief will be seen; but if the right picture be presented to the left eye, the other remaining unchanged, a converse relief will be seen. In like manner, if the middle picture be presented to the left eye, and the right picture to the right eye, a normal relief will appear; but if the left picture be presented to the right eye, the converse relief will present itself. It must be observed that the normal and converse reliefs, when the same picture remains presented to the same eye, belong to two different positions of the object.

§ 23.

Hitherto I have taken into consideration only those cases of the conversion of relief which are exhibited by binocular pictures in the stereoscope when they are transposed, reflected, or inverted. I shall now proceed to show how phenomena of the same kind may be elicited by regarding objects themselves, by means of an instrument adapted for the purpose. As this instrument conveys to the mind false perceptions of all external objects, I have called it the Pseudoscope. It is represented by fig. 7, and is thus constructed: two rectangular prisms of flint glass, the faces of which are 1·2 inch square, are placed in a frame with their hypothenuses parallel and 2·1 inches from each other; each prism has a motion on an axis corresponding with the angle nearest the eyes, that they may be adjusted so that their bases may have any inclination towards each other; and the frame itself is adjustable by a hinge at a, in order to bring the prisms nearer each other to suit the eyes of the observer.

The instrument being held to the eyes, and adjusted to an object so that it shall appear single, each eye will see a reflected image of that projection of the object which would be seen by the same eye without the pseudoscope. This is exactly the contrary of what occurs when the eyes regard the reflected image of an object in a looking-glass; the left eye then sees the reflected image of the right-hand projection, and the right eye the reflected image of the left projection, as shown by fig. 6.

Plane mirrors cannot be substituted for the reflecting prisms, for this reason : the refraction of the rays of light at the incident and emergent surfaces of the prisms enables the reflexion of an object to be seen when the object is even behind the prolongation of the reflecting surface, as shown at fig. 8, and thus the reflected binocular image may be seen in the same place as the object itself, whereas the images cannot be made by means of plane mirrors thus to coincide.

When the pseudoscope is so adjusted as to see a near object while the optic axes are parallel, to view a more distant object with the same adjustment, the axes must converge, and the more so as the object is more distant; all nearer objects than that seen when the axes are parallel will appear double, because the optic axes can never be simultaneously directed to them.   If this instrument be so adjusted that very distant objects are seen single when the eyes are parallel, *all* nearer objects will appear double, because the optic axes can never converge to make their binocular images coincide.   If the attention is required to be directed to an object at a particular distance, the best mode of viewing it with the pseudoscope is to adjust the instrument so that the object shall appear at the proper distance and of its natural size.   In this case the more distant objects will appear nearer and smaller, and the nearer objects will appear more distant and larger.

In ordinary vision, whenever the distance of an object varies, the magnitude of the picture on the retina, and the degree of convergence of the optic axes, always maintain a constant relation to each other, both increasing or decreasing together, and the perceived magnitude, suggesting to the mind the real magnitude of the object, in consequence thereof remains the same.   The instrument I described in § 17 shows what illusions arise when the usual relations of these elements of our perceptions are disturbed, by causing one to remain constant while the other varies.   The pseudoscope exhibits the still more curious illusions which result from combining these elements inversely; so that as an object becomes nearer, its larger picture on the retina is accompanied by a less convergence of the optic axes.   With the pseudoscope we have a glance, as it were, into another visible world, in which external objects and our internal perceptions have no longer their habitual relation with each other.

I will now proceed to describe some of the illusions produced by the aid of this instrument.   Those which may be strictly designated conversions of relief, in which the illusive appearance has the same relation to that of the real object as a cast to a mould, or a mould to a cast, are very readily perceived.   I must, however, remark, that it is necessary to illuminate the object equally, so as to allow no lights or shades to appear upon them; for their presence has a considerable influence on the judgment, and is one of the principal causes of the perception of the proper relief when a single eye is employed.

The inside of a tea-cup appears as a solid convex body; the effect is more striking if there are painted figures within the cup.

A china vase, ornamented with coloured flowers in relief, presents a very remarkable appearance; we apparently see a vertical section of the interior of the vase, with painted hollow impressions of the flowers.

A small terrestrial globe appears as a concave hemisphere; on turning it round on its axis, it was curious to see different portions of the spherical map appear and disappear in a manner that nothing in external nature can imitate.

A bust regarded in front becomes a deep hollow mask; the appearance when regarded in profile is equally striking.

A framed picture hanging against a wall appears as if imbedded in a cavity made in the wall.

A medal, or the impression of a seal, is perfectly converted into a representation of the die from which it has been struck; and, on the other hand, the mould or die of a medal, or an engraved seal, becomes a *fac-simile* of the medal or raised impression. It will also be observed that if the medal be placed on a flat surface, as a sheet of paper, it will appear sunk beneath the surface; and if it be placed in a hollow of the same size it will appear to stand above the surface as much as it actually is below it.

These appearances are not always immediately perceived, and some much more readily present themselves than others. Those converse forms which have a meaning, and resemble real forms we have been accustomed to see, are those which are the most easily apprehended. Viewed with the pseudo-scope, notwithstanding the inversion of the pictures on the retina, the natural appearance of the object continues to intrude itself, when sometimes suddenly, and at other times gradually, the converse occupies its place. The reason of this is, that the relief and distance of objects is not suggested to the mind solely by the binocular pictures and the convergence of the optic axes, but also by other signs, which are perceived by means of each eye singly; among which the most effective are the distributions of light and shade and the perspective forms which we have been accustomed to see accompany these appearances. One idea being therefore suggested to the mind by one set of signs, and another totally incompatible idea by another set, according as the mental attention is directed to the one and abstracted from the other, the normal form or its converse is perceived. This mental attention is involuntary; no immediate effort of the will can call up one idea while the other continues to present itself, though the transition may be facilitated by intentionally removing some of the signs which suggest the preponderating idea; thus the converse form being perceived, closing either eye will most frequently cause an instant reversion to the normal form; and always, if the monocular signs of relief are sufficiently suggestive.

I know of nothing more wonderful, among the phenomena of perception, than the spontaneous successive occurrence of these two very different ideas in the mind, while all external circumstances remain precisely the same. Thus a small

statuary group, an elegant and beautiful object, without any
apparent cause becomes converted into another totally dissimilar object uncouth in appearance, and which gives rise
to no agreeable emotions in the mind; yet in both cases all
the sensations that intervene between objective reality and
ideal conception continue unchanged.

The effects of the pseudoscope I have already mentioned
may be strictly called conversions of relief, because the illusive appearance is in each case the converse impression of
the relief of the real object.   If, however, the object consists
of parts detached from and behind each other, the preceding
term is inappropriate to denote the effects which result, but
the more general expression conversion or inversion of distance may be employed to designate them.   I proceed to call
attention to a few such effects.

Skeleton figures of geometrical solids, as cubes, pyramids,
&c., readily show their converse.

Two objects at different distances, being simultaneously
regarded, the most remote will appear the nearest and the
nearest the most remote.

An ivory foot-rule, held immediately before the eyes a
little inclined to the horizon with its remote end elevated,
appears inclined in the opposite way, its nearer end elevated,
and as if the observer were looking at its lower surface.   Its
form also undergoes a change.   Since the nearest end, the
retinal magnitude of which is the largest, appears farthest
from the eyes, and the farthest end, the retinal magnitude
of which is least, appears near the eyes, the rule will no
longer be perceived to be rectangular, but trapezoidal.   If
the rule be placed horizontally, and it be regarded with the
pseudoscope at an angle of 45°, it will appear with the form
just described standing vertically.

Any object placed before the wall of a room will appear
behind the wall, and as if an aperture of the proper dimensions had been made in the wall to allow it to be seen; if
the object be illuminated by a candle, its shadow will appear
as far before the object as in reality it is behind.

The appearance of a plant is very remarkable; as the

branches which are furthest from the eye are perceived to be the nearest, those parts which are actually obscured by the branches before them, appear broken away and allow the parts apparently behind them to be seen. A flowering shrub before a hedge appears to be transferred behind it; and a tree standing outside a window may be brought visibly within the room in which the observer is standing.

I have before observed that the transition from the normal to the converse perception is often gradual; I will give one instance of this as an illustration. The object was a page of medallions embossed on card-board, and the raised impressions were protected from injury by a thick piece of millboard having apertures in it made to correspond to each medallion. The page was placed horizontally, illuminated by a candle placed beyond it, and looked at through the pseudoscope at an angle of 45°; for the first moment the page appeared as it would have done without the instrument; soon after the medallions appeared level with the upper surface, and the shadows on the upper parts of the circular apertures were converted into deep depressions as if cut out with a tool; they next, from horizontal, became vertical, each standing erect on the horizontal plane, and immediately afterwards the reliefs were all changed into hollows; finally, the page itself stood vertical, but with that change of form which I indicated in the case of the rule, the upper edge appearing much shorter than the lower edge; the series of changes being now complete, the final form remained constant as long as the object was regarded.

In endeavouring to analyse the phenomena of converse perception, it must be borne in mind that the transposition of distances has reference only to distances from the retinæ, not to absolute horizontal distances in space. Thus, if a straight ruler be held in the vertical plane perpendicular to the optic base, and also inclined 45° to the horizon so that its upper end shall be the most distant, when the eyes are directed horizontally towards it, the rule will appear exactly in the converse position. If the rule be now removed lower down in the same vertical plane, its inclination remaining unchanged, so that to look upon it the plane of the optic axes must be inclined 45°, it will appear unaltered in position, because its two pictures are parallel on the retinæ, and the optic axes would require the same convergence to make the upper and lower ends coalesce. The rule being removed still lower down, instead of its position being apparently reversed, it will appear to have a greater inclination on the same side than the object itself has. In the first case the more distant end is actually furthest from the eyes; in the second, the near and remote ends are equally distant; and in the third the nearest end is most distant.

Attention to what I have just stated will explain many anomalous circumstances which occur when the eyes are differently directed towards the same object. It may also be necessary to remark that the conversion of distance takes place only within those limits in which the optic axes sensibly converge, or the pictures projected on the retinæ are sensibly dissimilar. Beyond this range there is no mutual transposition of the apparent distances of objects with the pseudoscope; a distant view therefore appears unchanged.

Some very paradoxical results are obtained when objects in motion are viewed through the pseudoscope. When an object approaches, the magnitude of its picture on the retinæ increases as in ordinary vision; but the inclination of the optic axes, instead of increasing, becomes less, as I have already explained. Now an enlargement of the picture on the retina invariably suggests approach, and a less convergence of the optic axes indicates that the object is at a greater distance; and we have thus two contradictory suggestions. Hence if two objects be placed side by side at a certain distance before the eyes, and one of them be moved forwards, so as to vary its distance from the other, its continually enlarging picture on the retina makes it appear to come towards the eyes, as it actually does, while at the same time it appears at every step at a greater distance beyond the fixed object; from one suggestion the object appears to approach, from the other to have receded. I again observe that retinal magnitude does not itself suggest distance, but from its changes we infer changes of distance.

I have hitherto only described the pseudoscope constructed with two reflecting prisms. This is the most convenient apparatus for effecting the conversion of distance and relief that has occurred to me; but other means may be employed, which I will briefly mention.

1st. Two plane mirrors are placed together so as to form a very obtuse angle towards the eyes of the observer; immediately before them the object is to be placed at such distance that a reflected image shall appear in each mirror. The eyes being placed before and a little above the object, must be caused to converge to a point between the object and the mirrors; the right-hand image of the left eye will then unite with the left-hand image of the right eye, and the converse relief will be perceived. The disadvantages of this method are that only particular objects can be examined, and it requires a painful adaptation of the eye to distinct vision.

2ndly. Place between the object and each eye a lens of small focal distance, and adjust the distances of the object and the lenses so that distinct inverted images of the object shall be seen by each eye; on directing the eyes to the place of the object, the two images will unite, and the converse relief be perceived. As the rays of light proceeding from the images have a greater divergence than those which would proceed from the point to which the optic axes are directed, long-sighted persons will see the binocular image more distinctly by wearing a pair of short-sighted spectacles. In this experiment the field of view is very small, on account of the distance at which it is necessary to place the lenses from the eyes; but I have been enabled in this manner to see beautifully the converse relief of a small ivory bust and of other small objects, which, however, should be inverted in order to see them direct.

3rdly. The inverted images of the lenses, instead of being received immediately by the eyes, as just described, may be thrown on a plate of ground glass, as in the case of the ordinary camera-obscura, and may be then caused to unite by the means employed in any form of the refracting stereo-scope.

§ 24.

The cases of the conversion of relief when the object is regarded with one eye only, some of which were known more than a century ago, were taken into consideration and endeavoured to be explained by me in § 11 of the first part of this memoir, and Sir David Brewster* has published some interesting and instructive observations on the same subject; I will therefore not revert to this matter here, but only to say that I have myself never observed the conversion of relief when looking with both eyes immediately on a solid object, and if it has been observed by others under such circumstances, I should be inclined to attribute the effect to an inequality in the impressions on the two eyes so that one only is attended to. But the plane shaded representation of a solid object, the relief of which is not very deep, may easily be made to appear at will, either as the solid which it is intended to represent or as its converse, even when both eyes are employed. This effect is strikingly observed in the glyptographic engravings of medals of low relief, and depends entirely on whether the light is so placed that it would cast the same shadows on the real object as are represented in the picture, or that it would cast shadows in the opposite direction. In the former case the picture appears with the relief it was intended to suggest; in the latter with the converse relief. I have observed similar effects with Daguerreotypes of medallions and cameos and with carefully shaded drawings of simple objects.

* Transactions of the Royal Society of Edinburgh, vol. xv. pp. 365 & 657.

On a singular Effect of the Juxtaposition of certain Colours under particular circumstances.

[From the 'Report of the British Association,' 1844, p. 10.]

HAVING had his attention drawn to the fact, that a carpet worked with a small pattern in green and red, when illuminated with gas-light, if viewed carelessly, produced an effect upon the eye as if all the parts of the pattern were in motion, he was led to have several patterns worked in various contrasted pairs of colours; and he found that in many of them the motion was perceptible, but in none so remarkably as those in red and green; it appeared also to be necessary that the illumination should be gas-light, as the effect did not appear to manifest itself in daylight, at least in diffused daylight. He accounted for it by the eye retaining its sensibility for various colours during various lengths of time.

On a means of determining the apparent Solar Time by the Diurnal Changes of the Plane of Polarization at the North Pole of the Sky.

[From the ' Report of the British Association,' 1848, pp. 10-12.]

A SHORT time after the important discovery by Malus of the polarization of light by reflexion, it was ascertained by Arago that the light reflected from different parts of the sky was polarized. The observation was made in clear weather with the aid of a thin film of mica and a prism of Iceland spar ; he saw that the two images projected on the sky were in general of dissimilar colours, which appeared to vary in intensity with the hour of the day and with the position, in relation to the sun, of the part of the sky from which the rays fell upon the film. The first attempt to assign a law to the phenomena of atmospheric polarization was made by Professor Quetelet of Brussels in 1825 in the following terms :—" If the observer consider himself as placed in the centre of a sphere of which the sun occupies one of the poles, the polarization is at its maximum at the different points of the equator of this sphere, and goes on diminishing in the ratio of the squares of the sines unto the poles where it is at zero." This law would be true did the reflected light proceeding from the part of the sky regarded arise solely from the direct light of the sun sent to that part; but other secondary reflexions occur which complicate the result and give rise to the neutral points since discovered by Arago, Babinet, and Brewster. But for the purpose of explaining the principle of the instrument now submitted to the examination of the Section, we need not take into consideration the intensity of the polarization of the part of the sky to which it is directed; the plane of polarization for the time being is the

only thing we need concern ourselves about, and a very
simple expression, stated first, I believe, by M. Babinet,
defines the position of this plane for any given point of the
sky; it is this:—"For a given point of the atmosphere the
plane of polarization of the portion of polarized light which
it sends to the eye coincides with the plane which passes
through this point, the eye of the observer and the sun."
The truth of this law may be easily demonstrated without
any refined apparatus in the following manner:—Let the
observer be provided with a Nicol's prism and a plate of Iceland spar cut perpendicularly to the axis, and stand with his
back towards the sun; keeping the diagonal of the prism
always in the same vertical plane, let him direct it successively to every point of the sky within that plane; the intensity of the polarization indicated by the brightness of the
coloured image will vary very considerably at these different
points, but the plane of polarization indicated by the upright
position of the black or white cross, as the case may be, will
remain unchanged. I leave out of consideration for the present the inversion of the plane of polarization observed occasionally near the horizon below the neutral point.

If we direct our analysing apparatus to the zenith during
the whole day, the change in the plane of polarization of
that point of the sky will correspond with the azimuths of
the sun. Let us now turn our attention to the north pole
of the sky: as the sun in its apparent daily course moves
equably in a circle round this pole, it is obvious that the
planes of polarization at the point in question change exactly
as the positions of the hour-circles do. The position of the
plane of polarization of the north pole of the sky will at any
period of the day therefore indicate the apparent or true
solar time. The point of intersection of the hour-circles, or
the north pole of the sky, corresponds on only two days of
the year with the maximum intensity of polarization; these
days are the equinoxes; on all other days the points of maximum polarization of the respective hour-circles describe a
circle round the point of intersection; but the angular distance thereof, which is greatest at the solstices, never exceed-

ing 23° 28', the polarization has always sufficient intensity to exhibit brilliant colours in films of selenite, &c.

These points being premised, I proceed to describe the new instrument, which I have called the Polar Clock or Dial. It is thus constructed.   At the extremity of a vertical pillar is fixed, within a brass ring, a glass disk, so inclined that its plane is perpendicular to the polar axis of the earth.   On the lower half of this disk is a graduated semicircle divided into twelve parts (each of which is again subdivided into five or ten parts), and against the divisions the hours of the day are marked, commencing and terminating with VI.   Within the fixed brass ring containing the glass dial-plate, the broad end of a conical tube is so fitted that it freely moves round its own axis; this broad end is closed by another glass disk, in the centre of which is a small star or other figure, formed of thin films of selenite, exhibiting when examined with polarized light strongly contrasting colours; and a hand is painted in such a position as to be a prolongation of one of the principal sections of the crystalline films.   At the smaller end of the conical tube a Nicol's prism is fixed so that either of its diagonals shall be 45° from the principal section of the selenite films.   The instrument being so fixed that the axis of the conical tube shall coincide with the polar axis of the earth, and the eye of the observer being placed to the Nicol's prism, it will be remarked that the selenite star will in general be richly coloured, but as the tube is turned on its axis the colours will vary in intensity, and in two positions will entirely disappear.   In one of these positions a small circular disk in the centre of the star will be a certain colour (red for instance), while in the other position it will exhibit the complementary colour.   This effect is obtained by placing the principal section of the small central disk $22\frac{1}{2}°$ from that of the other films of selenite which form the star.   The rule to ascertain the time by this instrument is as follows : the tube must be turned round by the hand of the observer until the coloured star entirely disappears while the disk in the centre remains red; the hand will then point accurately to the hour.   The accuracy with which the solar time may be indicated by this means will depend on the exactness with which the plane of polarization can be determined; one degree of change in the plane corresponds with four minutes of solar time.

The instrument may be furnished with a graduated quadrant for the purpose of adapting it to any latitude; but if it be intended to be fixed in any locality, it may be permanently adjusted to the proper polar elevation and the expense of the graduated quadrant be saved: a spirit-level will be useful to adjust it accurately. The instrument might be set to its proper azimuth by the sun's shadow at noon, or by means of a declination needle; but an observation with the instrument itself may be more readily employed for this purpose. Ascertain the true solar time by means of a good watch and a time equation table, set the hand of the polar clock to correspond thereto, and turn the vertical pillar on its axis until the colours of the selenite star entirely disappear. The instrument then will be properly adjusted.

The advantages a polar clock possesses over a sun-dial are : —1st. The polar clock being constantly directed to the same point of the sky, there is no locality in which it cannot be employed, whereas, in order that the indications of a sundial should be observed during the whole day, no obstacle must exist at any time between the dial and the places of the sun, and it therefore cannot be applied in any confined situation. The polar clock is consequently applicable in places where a sun-dial would be of no avail; on the north side of a mountain or of a lofty building for instance. 2ndly. It will continue to indicate the time after sunset and before sunrise; in fact, so long as any portion of the rays of the sun are reflected from the atmosphere. 3rdly. It will also indicate the time, but with less accuracy, when the sky is overcast, if the clouds do not exceed a certain density.

The plane of polarization of the north pole of the sky moves in the opposite direction to that of the hand of a watch; it is more convenient therefore to have the hours graduated on the lower semicircle, for the figures will then be read in their direct order, whereas they would be read backwards on an upper semicircle.   In the southern hemisphere the upper semicircle should be employed, for the plane of polarization of the south pole of the sky changes in the *same* direction as the hand of a watch.   If both the upper and lower semicircles be graduated, the same instrument will serve equally for both hemispheres.

Several other forms of the polar clock were then described. The following is a description of one among them, which, though much less accurate in its indications than the preceding, beautifully illustrates the principle.

On a plate of glass twenty-five films of selenite of equal thickness are arranged at equal distances radially in a semicircle; they are placed so that the line bisecting the principal sections of the films shall correspond with the radii respectively, and figures corresponding to the hours are painted above each film in regular order.   This plate of glass is fixed in a frame so that its plane is inclined to the horizon 38° 32′, the complement of the polar elevation; the light passing perpendicularly through this plate falls at the polarizing angle 56° 45′ on a reflector of black glass, which is inclined 18° 13′ to the horizon.   This apparatus being properly adjusted (that is, so that the glass dial-plate shall be perpendicular to the polar axis of the earth), the following will be the effects when presented towards an unclouded sky.   At all times of the day the radii will appear of various shades of two complementary colours, which we will assume to be red and green, and the hour is indicated by the figure placed opposite the radius which contains the most red; the half-hour is indicated by the equality of two adjacent tints.

Experiments on the Successive Polarization of Light, with the Description of a new Polarizing Apparatus.

[From the 'Proceedings of the Royal Society,' 1871, vol. xix. pp. 381–389.]

[Plate XXI.]

I.

THE term successive polarization was applied by Biot to denote the effects produced when a ray of polarized light is transmitted through a plate of rock-crystal cut perpendicularly to the axis, or through limited depths of certain liquids. In these cases the plane of polarization is found to be changed on emergence, and differently for each homogeneous ray; so that, when white light is employed, on turning the analyzer round continuously in one direction different colours successively appear, rising or falling in the scale according to the nature of the substance.

If, while the analyzer is turned from left to right, the tints ascend (*i. e.* follow the order R, O, Y, G, B, P, V), the substance is said to exhibit right-handed successive polarization; but if the tints descend, the successive polarization is said to be left-handed.

These phenomena were satisfactorily explained by Fresnel in the following way. The incident polarized ray, instead of resolving itself into two plane-polarized rays at right angles to each other, as in the ordinary cases of dipolarization, resolves itself in these instances into two circularly polarized rays, one right-handed, the other left-handed, which are transmitted with different velocities : each homogeneous ray, thus resolved into two opposite circularly polarized pencils, on emergence composes a ray polarized in a single plane, the deviation of which from the primitive plane of polarization depends on the difference of phase of the two circularly polarized rays on emergence.

The rotation of the planes of polarization is from left to right, or from right to left, according to whether the right-handed or left-handed circular rays are transmitted with the greater velocity.

II.

The term dipolarization, proposed by Dr. Whewell to express the bifurcation which a ray of polarized light suffers when it is transmitted through a crystallized plate, is a very appropriate one; but as there are different kinds of such separation, we may designate plane dipolarization the resolution into two plane-polarized rays at right angles to each other, and circular dipolarization the resolution into two circularly polarized rays, one right-handed, the other left-handed. In like manner, the term elliptic dipolarization may be employed to represent the phenomena shown by transmitting a polarized ray through a plate of rock-crystal obliquely to the axis.

The object of the present communication is to make known another means of producing successive polarization, both right-handed and left-handed, which, equally with the well-known modes, may be proved to arise from the interference of two opposite systems of circularly polarized rays.

III.

The polarizing-apparatus which I have employed for the experiments I am about to detail is represented by Pl. XXI.

A plate of black glass, G, is fixed at an angle of 3°* to the horizon. The film to be examined is to be placed on a diaphragm, D, so that the light reflected at the polarizing-angle from the glass plate shall pass through it at right angles and, after reflection at an angle of 18° from the surface of a

---

* [Thus in the original; but this does not agree with the figure (Plate XXI.), nor with what is said in the text as to the subsequent path of the light. In a separate impression of this Article found among the author's papers, the figure 3 has been altered with a pen, apparently into 17, but the correction cannot be deciphered with certainty.]

polished silver plate S, shall proceed vertically upwards. N is a Nicol's prism, or any other analyzer, placed in the path of the second reflection. The diaphragm is furnished with a ring, movable in its own plane, by which the crystallized plate to be examined may be placed in any azimuth. C is a small movable stand, by means of which the film to be examined may be placed in any azimuth and at any inclination; for the usual experiments this is removed.

If a lamina of quartz cut parallel to the axis, and sufficiently thin to show the colours of polarized light, be placed upon the diaphragm so that its principal section (*i. e.* the section containing the axis) shall be 45° to the *left* of the plane of reflection, on turning the analyzer from left to right, instead of the alternation of two complementary colours at each quadrant, which appear in the ordinary polarizing-apparatus, the phenomena of successive polarization, exactly similar to those exhibited in the ordinary apparatus by a plate of quartz cut perpendicular to the axis, will be exhibited; the colours follow in the order R, O, Y, G, B, P, V, or, in other words, ascend as in the case of a right-handed plate of quartz cut perpendicularly to the axis. If the lamina be now either inverted or turned in its own plane 90°, so that the principal section shall be 45° to the right of the plane of reflection, the succession of the colours will be reversed, while the analyzer moves in the same direction as before, presenting the same phenomena as a left-handed plate of quartz cut perpendicularly to the axis. Quartz is a positive doubly refracting crystal; and in it consequently the ordinary index of refraction is smaller than the extraordinary index. But if we take a lamina of a negative crystal, in which the extraordinary index is the least, as a film of Iceland spar split parallel to one of its natural cleavages, the phenomena are the reverse of those exhibited by quartz: when the principal section is on the *left* of the plane of reflection the colours descend, and when it is on the *right* of the same plane the colours ascend, the analyzer being turned from left to right.

It has been determined that the ordinary ray, both in positive and negative crystals, is polarized in the principal section, while the extraordinary ray is polarized in the section perpendicular thereto. It is also established that the index of refraction is inversely as the velocity of transmission. It follows from the above experimental results, therefore, that when the resolved ray whose plane of polarization is to the left of the plane of reflection is the quickest the successive polarization is right-handed, and when it is the slowest the successive polarization is left-handed—in the order R, O, Y, G, B, P, V, and in the second case in the reversed order.

The rule thus determined is equally applicable to laminæ of biaxal crystals.

As selenite (sulphate of lime) is an easily procurable crystal, and readily cleavable into thin laminæ capable of showing the colours of polarized light, it is most frequently employed in experiments on chromatic polarization. The laminæ into which this substance most readily splits contain in their planes the two optic axes; polarized light transmitted through such laminæ is resolved in two rectangular directions, which respectively bisect the angles formed by the two optic axes : the line which bisects the smallest angle is called the intermediate section ; and the line perpendicular thereto which bisects the supplementary angle is called the supplementary section. These definitions being premised, if a film of selenite is placed on the diaphragm with its intermediate section to the left of the plane of reflection, the successive polarization is direct or right-handed; if, on the contrary, it is placed to the right of that plane, the successive polarization is left-handed. The ray polarized in the intermediate section is therefore the most retarded ; and as that section is considered to be equivalent to a single optic axis, the crystal is positive.

In one kind of mica the optic axes are in a plane perpendicular to the laminæ. They are inclined $22\frac{1}{2}°$ on each side the perpendicular within the crystal, but, owing to the refraction, are seen respectively at an angle of $35°\cdot3$ therefrom. The principal section is that which contains the two optic axes. If the film is placed on the diaphragm with its principal section inclined 45° to the left of the plane of reflection, the successive polarization is right-handed. The ray, therefore, polarized in the section which contains the optic axes is the one transmitted with the greatest velocity.

Films of uniaxal crystals, whether positive or negative, and of biaxal crystals, all agree therefore in this respect :— that if the plane of polarization of the quickest ray is to the left of the plane of reflection, the successive polarization is right-handed when the analyzer moves from left to right; and if it is to the right of the plane of reflection, other circumstances remaining the same, the successive polarization is left-handed.

It must be taken into consideration that the principal section of the film is inverted in the reflected image; so that if the plane of polarization of the quickest ray in the film is to the left of the plane of reflection, it is to the right of that plane in the reflected image.

IV.

It may not be uninteresting to state a few obvious consequences of this successive polarization in doubly refracting laminæ, right-handed and left-handed according to the position of the plane of polarization of the quickest ray. They are very striking as experimental results, and will serve to impress the facts more vividly on the memory.

1. A film of uniform thickness being placed on the diaphragm with its principal section 45° on either side the plane of reflection, when the analyzer is at 0° or 90° the colour of the film remains unchanged, whether the film be turned in its own plane 90°, or be turned over so that the back shall become the front surface; but if the analyzer be fixed at 45°, 135°, 225°, or 315°, complementary colours will appear when the film is inverted from back to front, or rotated in its own plane either way 90°.

2. If a uniform film be cut across and the divided portions be again placed together, after inverting one of them, a compound film (fig. 4) is formed, which, when placed on the diaphragm, exhibits simultaneously both right-handed and

left-handed successive polarization. When the analyzer is at 0° or 90° the colour of the entire film is uniform; as it is turned round the tints of one portion ascend, while those of the other descend; and when the analyzer is at 45° or $n90° + 45°$, they exhibit complementary colours.

3. A film increasing in thickness from one edge to the other is well suited to exhibit at one glance the phenomena due to films of various thicknesses. It is well known that such a film placed between a polarizer and an analyzer will show, when the two planes are parallel or perpendicular to each other, and the principal section of the film is intermediate to these two planes, a series of parallel coloured bands, the order of the colours in each band from the thick towards the thin edge being that of their refrangibilities, or R, O, Y, G, B, P, V. The bands seen when the planes are perpendicular are intermediate in position to those seen when the planes are parallel; on turning round the analyzer these two systems of bands alternately appear at each quadrant, while in the intermediate positions they entirely disappear.

Now let us attend to the appearances of these bands when the wedge-form film is placed on the diaphragm of the instrument, fig. 1. As the analyzer is moved round, the bands advance toward or recede from the thin edge of the wedge without any changes occurring in the colours or intensity of the light, the same tint occupying the same place at every half revolution of the analyzer. If the bands advance toward the thin edge of the wedge, the successive polarization of each point is left-handed; and if they recede from it the succession of colours is right-handed; every circumstance, therefore, that with respect to a uniform film changes right-handed into left-handed successive polarization, in a wedge of the same substance transforms receding into advancing bands, and *vice versâ*. These phenomena are also beautifully shown by concave or convex films of selenite or rock-crystal, which exhibit concentric rings contracting or expanding in accordance with the conditions previously explained.

4. Few experiments in physical optics are so beautiful and

striking as the elegant pictures formed by cementing laminæ of selenite of different thicknesses (varying from $\frac{1}{2000}$ to $\frac{1}{50}$ of an inch) between two plates of glass. Invisible under ordinary circumstances, they exhibit, when examined in the usual polarizing-apparatus, the most brilliant colours, which are complementary to each other in the two rectangular positions of the analyzer. Regarded in the instrument, fig. 1, the appearances are still more beautiful; for, instead of a single transition, each colour in the picture is successively replaced by every other colour. In preparing such pictures it is necessary to pay attention to the direction of the principal section of each lamina when different pieces of the same thickness are to be combined together to form a surface having the same uniform tint; otherwise in the intermediate transitions the colours will be irregularly disposed.

5. A plate of rock-crystal cut perpendicular to the axis loses its successive polarization, and behaves exactly as an ordinary crystallized film through which rectilinear polarized light is transmitted.

6. A thick plate of unannealed glass undergoes a series of regular transformations, the principal phases of which are shown, fig. 5.

V.

The phenomena of successive or rotary polarization I have experimentally demonstrated admit of a very simple explanation.

The polarized light incident on the crystallized plate is resolved into two portions of equal intensity, polarized at right angles to each other, one in the principal section, the other perpendicular thereto. These resolved portions, when they fall on the silver plate, have their planes of polarization each at an azimuth of 45°, one to the right, the other to the left of the plane of reflection. These are again resolved in the plane of reflection and the plane perpendicular thereto, and are, in consequence of the unequal retardation, which in silver at an angle of 72° amounts to a quarter of an undulation, converted into circularly polarized beams, one right-handed, the other left-handed.

The various homogeneous rays being accelerated differently in their transmission through the two sections of the crystallized plate, this difference is preserved after reflection from the silver plate, and the oppositely circularly polarized beams are reflected with the same difference of phase as the two plane-polarized rays are when emerging from the crystallized lamina. The composition of two circular waves, one right-handed, the other left-handed, gives for resultant a plane wave the azimuth of which varies with the difference of phase of the two components.

When the plane of polarization does not lie equally between the two rectangular sections of the laminæ, these still remaining 45° from the plane of reflection of the silver plate, the beam is resolved into two unequal portions, the amplitudes of which are as sin a to cos a. Each therefore gives rise to a circular undulation of different amplitude. The resultant of two opposite circular undulations of different amplitudes is an ellipse of constant form, the axes of which vary in position according to the difference of phase. The same phenomena of successive polarization are therefore exhibited, in whatever azimuth the lamina is turned in its own plane; but the tints become fainter and fainter until ultimately, when the principal or perpendicular section is parallel to the plane of reflection of the polarizing plate, all colour disappears.

VI.

By means of the phenomena of successive polarization it is easy to determine which is the thicker of two films of the same crystalline substance. Place one of the films on the diaphragm (a) of the instrument (fig. 1 a) in the position to show, say right-handed polarization, then cross it with the other film : if the former be the thicker, the successive polarization will be still right-handed; if both be equal, there will be no polarization ; and if the crossed film be the thicker, the successive polarization will be left-handed. In this manner a series of films may be readily arranged in their proper order in the scale of tints.

VII.

In the experiments I have previously described the planes of reflection of the polarizing mirror and of the silver plate were coincident; some of the results obtained when the azimuth of the plane of reflection of the silver plate is changed are interesting.

I will confine my attention here to what takes place when the plane of reflection of the silver plate is 45° from that of the polarizing reflector.

When the principal sections of the film are parallel and perpendicular to the plane of reflection of the polarizing mirror, as the whole of the polarized light passes through one of the sections, no interference can take place, and no colour will be seen, whatever be the position of the analyzer.

When the principal sections of the film are parallel and perpendicular to the plane of reflection of the silver plate, they are 45° from the plane of reflection of the polarizing mirror.

The polarized ray is then resolved into two components polarized at right angles to each other: one component is polarized in the plane of reflection of the silver plate, the other perpendicular thereto; and one is retarded upon the other by a quarter of an undulation.

When the analyzer is at 0° or 90° no colours are seen, because there is no interference; but when it is placed at 45° or 135°, interference takes place, and the same colour is seen as if light circularly polarized had been passed through the film. The bisected and inverted film (fig. 4) shows simultaneously the two complementary colours.

But when the film is placed with one of its principal sections 22½° from the plane of reflection of the polarizing mirror, on turning round the analyzer the appearances of successive polarization are reproduced exactly as when the planes of reflection of the silver plate and of the polarizing mirror coincide. In this case the components of the light oppositely polarized in the two sections are unequal, being as cos $22\frac{1}{2}°$ to sin $22\frac{1}{2}°$; these components respectively fall $22\frac{1}{2}°$ from the plane of reflection of the silver plate and from the perpendicular plane, and are each resolved in the same proportion in these two planes. The weak component of the first, and the strong component of the second, are resolved into the normal plane, while the strong component of the first and the weak component of the second are resolved into the perpendicular plane.

VIII.

As bearing intimately on the subject of this paper, I will here quote a passage from a memoir presented by Fresnel to the French Academy of Sciences in 1817, and published, in abstract, in the 'Annales de Chimie,' t. xxviii. (1825) :—

" If a thin crystallized plate be placed between two parallelopipeds of glass crossed at right angles, in each of which the light previously polarized undergoes two total reflections at the incidence of $54\frac{1}{2}°$, first before its entrance into the plate (which we suppose perpendicular to the rays), and subsequently after its emergence, and if, besides, the plate be turned so that its axis makes an angle of 45° with the two planes of double reflection, this system will present the optical properties of plates of rock-crystal perpendicular to the axis, and of liquids which colour polarized light. When the principal section of the rhomboid with which the emergent light is analyzed is turned round, the two images will gradually change colour, instead of experiencing only simple variations in the vividness of their tints, as occurs in the ordinary case of thin crystallized plates ; besides, the nature of these colours depends only on the respective inclination of the primitive plane of polarization and the principal section of the rhomboid—that is to say, of the two extreme planes of polarization ; thus, when this angle remains constant, the system of the crystallized plate and the two parallelopipeds may be turned round the transmitted pencil without changing the colour of the images. It is this analogy between the optical properties of this little apparatus and those of plates of rock-crystal perpendicular to the axis which enabled M. Fresnel to foresee the peculiar characters of double refraction that rock-crystal exerts on rays parallel to the axis."

It does not appear that Fresnel, in any of his published memoirs, has given any further modifications of this experiment, the importance of which has been almost entirely overlooked in elementary treatises on light. He does not seem to have remarked that similar phenomena of successive polarization are exhibited when the light incident on the crystallized plate is plane-polarized, nor that the order of the succession of the colours depends on the position of the principal section with respect to the plane of polarization. These circumstances are indeed necessarily included in the beautiful theory established by this eminent philosopher; but I am not aware that they have hitherto been specifically deduced or experimentally shown.

IX.

The apparatus (fig. 1) affords also the means of obtaining large surfaces of uncoloured or coloured light in every state of polarization—rectilinear, elliptical, or circular.

It is for this purpose much more convenient than a Fresnel's rhomb, with which but a very small field of view can be obtained. It must, however, be borne in mind that the circular and elliptical undulations are inverted in the two methods: in the former case they undergo only a single, in the latter case a double reflection.

For the experiments which follow, the crystallized plate must be placed on the diaphragm E between the silver plate and the analyzer, instead of, as in the preceding experiments, between the polarizer and the silver plate.

By means of a moving ring within the graduated circle D the silver plate is caused to turn round the reflected ray, so that, while the plane of polarization of the ray remains always in the plane of reflection of the glass plate, it may assume every azimuthal position with respect to the plane of reflection of the silver plate. The film to be examined and the analyzer move consentaneously with the silver plate, while the polarizing mirror remains fixed.

In the normal position of the instrument the ray polarized by the mirror is reflected unaltered by the silver plate; but when the ring is turned to 45°, 135°, 225°, or 315°, the plane of polarization of the ray falls 45° on one side of the plane of reflection of the silver plate, and the ray is resolved into two others, polarized respectively in the plane of reflection and the perpendicular plane, one of which is retarded on the other by a quarter of an undulation, and consequently gives rise to a circular ray, which is right-handed or left-handed according to whether the ring is turned 45° and 225° or 135° and 315°. When the ring is turned so as to place the plane of polarization in any intermediate position between those producing rectilinear and circular light, elliptical light is obtained, on account of the unequal resolution of the ray into its two rectangular components.

Turning the ring of the graduated diaphragm from left to right when the crystallized film is between the silver plate and the analyzer, occasions the same succession of colours for the same angular rotation as rotating the analyzer from right to left when the instrument is in its normal position and the film is between the polarizer and the silver plate.

X.

To arrange the apparatus for the ordinary experiments of plane-polarized light without the intervention of the silver plate, all that is necessary is to remove the silver plate from the frame F, and to substitute for it a plate of black glass, which must be fixed at the proper polarizing-angle.

To convert it into a Norrenberg's polarizer, a silvered mirror must be laid horizontal at H, and the instrument straightened, as shown at fig. 3, so that a line perpendicular to the mirror shall correspond with the line of sight. The silver plate must be removed from the frame F, and a plate of transparent glass substituted for it, which must be so inclined that the light falling upon it shall be reflected at the polarizing-angle perpendicularly toward the horizontal mirror. The eye will receive the polarized ray reflected from the mirror; and the polarized ray will have passed, before it reaches the eye, twice through a crystallized plate placed between the mirror and the polarizer. The result is the same as if, in the ordinary apparatus, the polarized ray had passed through a plate of double the thickness.

Fig. 2 shows the addition to the apparatus when the coloured rings of crystals are to be examined by light circularly or eliptically polarized : a is the optical tube containing the lenses (which require no particular explanation), and b the condenser, over which the plate is to be placed.

Note relating to M. Foucault's new Mechanical Proof of the Rotation of the Earth.

[From the 'Proceedings of the Royal Society,' vol. vi. 1851, pp. 65–68.]

THE experiment which led M. Foucault to his ingenious and interesting researches relating to the rotation of the earth is stated by him thus :—" Having fixed on the arbor of a lathe and in the direction of the axis, a round and flexible steel rod, it was put in vibration by deflecting it from its position of equilibrium and leaving it to itself. A plane of oscillation is thus determined, which, from the persistence of the visual impressions, is clearly delineated in space; now it was remarked that on turning by the hand the arbor which serves as a support to this vibrating rod, the plane of oscillation is not carried with it."

This persistence of the plane of oscillation of a vibrating rod, notwithstanding the rotation of the point to which its end is fixed, does not appear to have hitherto been made the subject of philosophical observation. Ordinary notions even seem to have been opposed to this now recognized fact. Chladni, in his treatise on Acoustics, in the chapter "On the co-existence of Vibrations with other kinds of Motion," states as follows :—

"Vibratory motions may co-exist with all other kinds of motions in an infinity of different manners, as has been demonstrated by Dan. Bernoulli and L. Euler in vols. xv. and xix. of the *Nov. Comment. Acad. Petrop.*, and confirmed by experiment. These coexistences of different motions occur in all sonorous bodies without exception. We may, for example, produce the sound of a string stretched on a board, or that of a plate, a tuning-fork, a bell, &c., and while the vibrations still last, impress on this sonorous body a motion of rotation round its axis, and at the same time a progressive motion : thus all these motions may be performed in the same time, without one being hindered by the other; but the absolute motion of each point will be very complicated."

Now this is true only when the vibrating body is constrained to vibrate in one direction. When the rod or string is equally flexible in every direction, the plane of vibration given to it from any original impulse is constantly maintained whatever may be the velocity of rotation communicated to its point of support, provided the axis of vibration remains in the same position, or moves only parallel to itself.

This observed independence of the plane of oscillation on the point of attachment led M. Foucault to assume that were a flexible pendulum suspended from a fixed point in the prolongation of the axis of the earth, that is above one of the poles, and maintained constantly in vibration, the plane of oscillation maintaining an invariable position in space would appear to a spectator on the earth's surface and moving with it to make an entire revolution in twenty-four hours, but in the opposite direction to that of the rotation of the earth.

What takes place at other points of the earth's surface is more difficult to determine; but M. Foucault, from mechanical and geometrical considerations, was led to the conclusion that the angular displacement of the plane of oscillation is equal but opposite to the angular motion of the earth multiplied by the sine of the latitude. According to the theory of rotation, first established by Frisi and more fully developed by Euler and Poinsot, the velocity of rotation of the earth may be considered as the resultant of two angular velocities, one round the vertical of the point where the observer is placed, and the other round the meridian or horizontal line lying N. and S. The component of the angular velocity estimated round the vertical axis is $n \sin \gamma$, and the plane of oscillation not participating in this motion remains at rest with respect to it, and therefore appears to an observer moving with the point to rotate with the same velocity in the contrary direction.

The experiment made by M. Foucault is said, both in the

direction and magnitude of the motion of the plane of oscillation of the pendulum, fully to confirm the indications of theory. The difficulty, however, of the mathematical investigation of the subject, and the delicacy of the experiment, liable as it is to so many extraneous causes of error, have induced many persons to doubt either the reality of the phenomenon or the satisfactoriness of the explanation. Another experimental proof, therefore, not depending on the rotation of the earth, that the plane of oscillation of a vibrating line remains at rest with relation to the vertical component of the real axis of rotation, may not be unacceptable. With this view I have devised the apparatus I am about to describe.

A semicircular arch from one to two feet radius is fixed vertically on a horizontal wheel, and may thus be moved with any degree of rapidity from any one azimuth to another. A rider slides along the inner edge of the arch, which is graduated, and may be fixed at any degree marked thereon. A spiral spring wire, by means of which a slow vibration is obtained with comparatively a short length, is attached at the lower end to a pin fixed in the axis of the semicircle, so that the point of attachment may be in the axis of rotation, and at the upper end it is fixed to a similar pin in a parallel position fixed to the rider. The vertical semicircle is not placed in a diameter of the horizontal wheel, but parallel to it, at such distance as not to intercept, from the eye of the observer, the vertical plane passing through the diameter, and in which plane the wire in all its positions remains.

When the upper end of the wire is placed at 90° (that is, when it coincides with the axis of rotation), if the wire be caused to vibrate in any given plane, say from N. to S., it will continue to do so whatever rotation may be communicated to the wheel; so that with respect to the moving wheel, or the axis of the wire, the plane of vibration will move with the same velocity and in the opposite direction. When the rider is fixed at 30°, and the wire makes, therefore, an angle of 60° with the axis of rotation, so as to describe in its motion the surface of a cone having this inclination to the vertical, it will be observed that the plane of vibration makes one

x

complete rotation during two rotations of the wheel: this is best observed by fixing the eye so that its axis shall coincide with a line in the same vertical plane with the wire, while walking round with the wheel during its rotation. When the rider is fixed at $19\frac{1}{2}°$, the plane of vibration makes one rotation during three rotations of the wheel; when fixed at $14\frac{1}{2}°$, it makes one rotation during four of the wheel, &c.; and when it is fixed at $0°$, the wire lying horizontally, no rotation of the plane of vibration occurs. It is needless to observe that the sines of $90°$, $30°$, $19\frac{1}{2}°$, $14\frac{1}{2}°$, $0°$, correspond to the numbers 1, $\frac{1}{2}$, $\frac{1}{3}$, $\frac{1}{4}$, 0, the reciprocals of the numbers expressing the respective times of rotation*.

It is not necessary that the wire should have one of its ends fixed in the axis of rotation; if it be parallel to a wire so fixed, the rotation of the plane of vibration will be exactly similar; in such case the wire or axis of vibration will describe the surface of two cones having their common apex in the axis of rotation.

The axis of a flexible pendulum can only assume a position vertical to the point of the earth's surface over which it is placed. Were it possible to maintain the vibration of a stretched wire occasioned by an original impulse for a sufficient length of time, the apparent rotation of its plane of vibration would vary with the inclination of the wire to the axis of the earth: placed in this axis, it would make a rotation in 24 hours, it would become progressively slower according to the law above given, as it approaches the plane of the equator, and when anywhere in this plane the vibrations would always be performed in the same direction.

* When the dimensions of the apparatus are as above given, I find that hardened brass wire (No. 26), coiled so as to form a helix of one quarter of an inch in diameter, shows the effect well. The thickest spiral wire employed in the manufacture of artificial flowers, which can be procured of any wire-drawer, will also answer the purpose.

The best way of setting the wire in vibration is to press the finger upon it in the middle, so as to deflect it in the plane in which the vibrations are required to continue, and then suddenly to withdraw the finger in the direction of the vibrations. The deflection must not be too great, or the elasticity of the wire will be injured.

On Fessel's Gyroscope.

[From the 'Proceedings of the Royal Society,' vol. vii. 1854. pp. 43–48.]

SINCE the announcement of M. Foucault's beautiful experiment which has afforded us a new mechanical proof of the rotation of the earth on its axis, the phenomena of rotary motion have received renewed attention, and many ingenious instruments have been contrived to exhibit and to explain them. One of the most instructive of these is the Gyroscope invented by M. Fessel of Cologne, described in its earlier form in Poggendorff's 'Annalen' for September 1853, and which, with some improvements by Prof. Plücker and some further modifications suggested by myself, I take the present opportunity of bringing before the Royal Society.

It is thus constructed : a beam is capable of moving freely

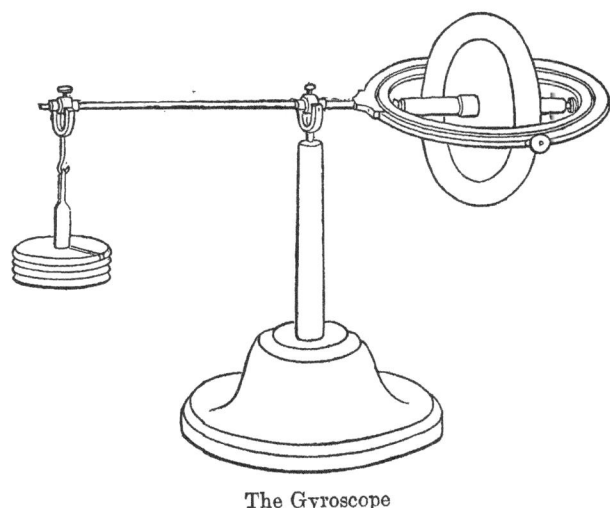

The Gyroscope

round a horizontal axis which is itself movable round a vertical axis, so that the beam may move in any direction round
a fixed point; at one end of the beam is fixed a horizontal
ring which carries a heavy disc, the axis of rotation of which
is in a line with the beam; at the opposite extremity is a
shifting weight by means of which the equilibrium of the
beam may be established or disturbed at pleasure.

If the beam be brought into equilibrium, and the disc be
rapidly rotated, by means of a thread unrolled from its axis,
it will be seen that the beam has no tendency to displace
itself in any direction.   Not so, however, if the equilibrium
be in any way disturbed; on moving the weight towards the
centre of the beam, thus causing the *disc* to preponderate, it
will be observed that if the disc rotates from right to left the
beam will move round the vertical axis also from right to
left; and if the motion of the disc be reversed the rotation
of the beam will be reversed also.   On causing the *equipoise*
to preponderate contrary effects will take place.   The velocity of the rotation of the beam round the vertical axis increases in proportion to the disturbance of the equilibrium.
It will also be observed that, notwithstanding the increased
or diminished action of gravity on the disc, its axis of rotation always preserves the same inclination to the vertical axis
at which it has been originally placed.   The effect produced
is a seeming paradox.   When the equilibrium is disturbed
while the disc is at rest, the beam being placed in any other
position than the vertical, gravity acts so as to turn it round
a horizontal axis; but when the disc is in motion the usual
effect of gravity disappears, and there is substituted for it a
continued rotation round a vertical axis, that is, round an
axis perpendicular to the plane which contains the axes of
the two original rotations.

A similar composition of forces takes place when the disc
is caused to rotate while the equilibrium of the beam is maintained, by impressing on the beam a rotation round the vertical axis.   When the disc rotates from right to left, the
slightest pressure tending to produce rotation round the ver-

tical axis in the same direction causes the end of the beam carrying the disc to ascend, and a pressure in the opposite direction causes it to descend—that is, the beam is constrained to move round a horizontal axis perpendicular to the vertical plane which contains the two axes of impressed rotation, a case exactly analogous to the preceding. The beam ascends and descends, in like manner, after rotation has spontaneously taken place round the vertical axis in consequence of the equilibrium being disturbed, whenever this rotation is anyhow accelerated or retarded : the disc rotating from right to left and its weight predominating, the rotation round the vertical axis is from left to right; accelerating the latter motion will cause the disc to descend, and retarding it will occasion it to ascend.

As the centre of gravity of the beam is below its point of suspension, even when equipoised it is in perfect equilibrium only when it is horizontal : consequently, if it be elevated above or depressed below this position it will endeavour to resume it, tending to produce in the two cases rotations in opposite directions round a horizontal axis ; the rotation of the disc combined with this tendency gives rise, as in the other cases I have mentioned, to a continued rotation round the vertical axis. If the disc rotate from right to left, and the end of the beam carrying it be elevated above the horizontal position, the rotation round the vertical axis will be from right to left; if, on the contrary, the same end of the beam be depressed below the horizontal position, that rotation will be from left to right.

In all the experiments above mentioned the axis of the rotating disc has remained in the prolongation of the beam, but, by means of an internal ring movable round a line perpendicular thereto, this axis may be placed at any inclination and at any azimuth with respect to it. Very obvious considerations show that the inclination of this axis should produce no difference in the character of the effects, but merely in their intensity, since in any inclined position of the disc its rotation is resolvable into two others, one perpendicular to the beam, and the other, which is incapable of producing any effect, in a plane containing it. When the axis of the rotating disc is vertical and at right angles to the beam, no rotation on the vertical axis ought to take place in any case; but, contrary to this expectation, although the beam be horizontal and in perfect equilibrium, a motion round the vertical axis results, which is in opposite directions according as one or the other end of the axis of the disc is uppermost. It is, however, easy to see that this rotation is not owing to the same cause which gives rise to the phenomena hitherto considered, for whether it be accelerated or retarded no change is produced in the horizontal position of the beam; it is, in fact, occasioned by the difference of leverage of the centrifugal forces acting in opposite directions at those points of the diameter of the disc which are at the greatest and least distances from the vertical axis of rotation of the beam. Attention to this extraneous cause of rotation will explain numerous anomalies which present themselves in many of the instruments contrived to exemplify the phenomena of combined rotary motions. It is one of the advantages of Fessel's apparatus that the phenomena may be exhibited in their more important phases without being affected by this source of error.

We may form a clearer conception of these phenomena by first considering some simpler facts which do not appear to me to have been hitherto sufficiently attended to. For this purpose let the system of rings carrying the disc be removed from the rest of the apparatus, and by unfastening the tightening screw let the inner ring be allowed to move freely within the outer. Having set the disc in rapid rotation, hold the outer ring at the extremities of the diameter which is in the plane in which the axis of motion of the disc is free to move; then giving to the outer ring a tendency to rotation round that diameter, it will be observed that, in whatever position the axis is, it will fly to place itself in the fixed axis thus determined, and rotation will take place round it in the same direction. Considerable resistance is felt so long as the movable axis is changing its position; but when once it coincides with the fixed axis, the rotation of the external ring round its diameter is effected with facility. A slight alternate motion of the outer ring, tending to give to it rotations in opposite directions, will occasion a continued rotation of the movable axis. The same result takes place when an endeavour is made to rotate the outer ring round an axis perpendicular to its plane. In all cases when the axis of the rotating disc is free to move in a plane, and the outer ring is constrained to rotate round a line in this plane, the movable axis will place itself so as to coincide with that line, and so that the disc shall rotate in the same direction as the ring; if the fixed axis be in a different plane, the movable axis will assume permanently that position in its plane which approaches nearest to the former. The movable axis is thus apparently attracted towards the fixed axis if the rotations are in the same direction, and repelled from it if the rotations are in opposite directions.

In the experiments just described the free and constrained axes of rotation intersect, but in Fessel's apparatus they are distant from each other. In the latter case the rule must be thus modified, that the free axis of rotation tends to place itself *parallel* to the constrained axis of rotation, or to as near a position thereto as possible. By this principle all the results manifested are easily explained. The beam being in equilibrium, a motion impressed on it round the vertical axis causes it to ascend or descend, because the axis of the rotating disc tends to place itself parallel to the vertical axis of rotation, and so that the disc rotates in the impressed direction. When the equilibrium of the beam is destroyed, gravity tends to make it rotate round a horizontal axis; the axis of the disc endeavours to place itself parallel with that axis; but both being unchangeably at right angles to each other, the tendency to place itself there gives rise to a continued rotation. Other results with this apparatus, to which I have not yet adverted, are similarly explained. Fix the outer ring horizontally and loosen the inner ring, keeping

them, both, however in the same plane; then, on moving the
beam round the vertical axis, the axis of the rotating disc
will immediately fly to place itself parallel thereto, with rotation of the disc in the impressed direction.   The rings being
placed in the vertical plane, the same result will take place
if the beam be moved in a vertical plane, *i. e.* round a horizontal axis.

The following additional experiments may be made with
the rings detached from the apparatus.   The results are
necessary consequences of what has been previously explained :—

1. Suspend, by means of a string, the outer ring at the
extremity of a diameter perpendicular to the axis of the inner
ring; and, having loosened the latter, place it at right angles
to the former.   On causing the disc to rotate, its axis will
retain its original position; but if the slightest effort be
made to turn the outer ring round the vertical line, the axis
of the rotating disc will instantly fly into this position, and
the disc will move in the same direction as that of the impressed rotation.

2. The horizontality of the loose inner ring being restored,
if a weight be suspended from either end of the axis of the
disc, that axis will, while it preserves its horizontal or any
inclined position, revolve round the vertical line; the direction of the motion will change if either the weight be
applied to the opposite end of the axis or the disc rotate
in the opposite direction.   If this rotation be arrested, gravity will immediately cause the weighted end of the axis to
descend.

3. Clamp the rings together either in the same plane or
at right angles to each other, and fasten a string, in the
first case, at the extremity of a diameter coinciding with
the axis of the inner ring, and in the latter case at the
extremity of a diameter perpendicular thereto.   Having set
the disc spinning, if a rotation round the vertical line be
given to the system the axis of the disc will ascend, carrying
with it the disc and rings notwithstanding their weight, and,

even when the impressed rotation has ceased to act, will continue to rotate in the same direction until the motion of the disc ceases.

In this note I have purposely avoided entering into the mathematical theory of the phenomena my intention having been solely to describe the apparatus exhibited and to give an intelligible account of its effects. Those who wish to investigate the subject more profoundly will find the best guide in the Astronomer Royal's essay on Precession and Nutation, published in his Mathematical Tracts.

On the Formation of Powers from Arithmetical Progressions.

[From the 'Proceedings of the Royal Society,' vol. vii. 1855, pp. 145–151.]

THE same sum n^a may be formed by the addition of an arithmetical progression of n terms in various ways. Hence we are enabled to construct a great variety of triangular arrangements of arithmetical progressions, the sums of which are the natural series of square, cube, and other powers of numbers. Among these there are several which render evident some remarkable relations.

Each of the following triangles is formed of a series of arithmetical progressions, the number of terms increasing successively by unity.

The first term of an arithmetical progression of n terms having a common difference δ, and whose sum is n^a, is equal to

$$n^{(a-1)} + \frac{\delta}{2}(1 - n).$$

§ 1. SQUARE NUMBERS.

If $S = n^2$, the first term $= n + \dfrac{\delta}{2}(1 - n)$.

A.

Every square n^2 is the sum of an arithmetical progression of n terms, the first term of which is unity, and the difference 2.

$$
\begin{aligned}
1 &\ \ldots\ldots\ldots\ldots = 1^2 \\
1+3 &\ \ldots\ldots\ldots\ldots = 2^2 \\
1+3+5 &\ \ldots\ldots\ldots\ldots = 3^2 \\
1+3+5+7 &\ \ldots\ldots\ldots\ldots = 4^2 \\
1+3+5+7+9 &\ \ldots\ldots\ldots\ldots = 5^2 \\
1+3+5+7+9+11 &\ \ldots\ldots\ldots\ldots = 6^2 \\
1+3+5+7+9+11+13 &\ \ldots\ldots\ldots\ldots = 7^2
\end{aligned}
$$

Thus every square number is formed by the addition of a series of odd numbers commencing with unity—a result universally known.

The difference of any two squares is either an odd number, or the sum of consecutive odd numbers.

Each series may be resolved into two others consisting of alternate odd numbers, the respective sums of which are two adjacent triangular numbers, the addition of which, it is well known, forms a square. *Ex.* :

$$1+5+9+13=28$$
$$3+7+11=21$$
$$\overline{}$$
$$49=7^2$$

B.

Every square n^2 is the sum of an arithmetical progression of n terms, the first term of which is $\dfrac{n+1}{2}$, and the common difference 1.

$$1 \ \ldots\ldots\ldots\ldots\ldots\ldots =1^2$$
$$1\tfrac{1}{2}+2\tfrac{1}{2} \ \ldots\ldots\ldots\ldots\ldots =2^2$$
$$2+3+4 \ \ldots\ldots\ldots\ldots =3^2$$
$$2\tfrac{1}{2}+3\tfrac{1}{2}+4\tfrac{1}{2}+5\tfrac{1}{2} \ \ldots\ldots\ldots\ldots =4^2$$
$$3+4+5+6+7 \ \ldots\ldots\ldots =5^2$$
$$3\tfrac{1}{2}+4\tfrac{1}{2}+5\tfrac{1}{2}+6\tfrac{1}{2}+7\tfrac{1}{2}+8\tfrac{1}{2} \ \ldots\ldots\ldots =6^2$$
$$4+5+6+7+8+9+10 \ \ldots\ldots =7^2$$

This arrangement renders evident that every square of an odd number is the sum of as many consecutive natural numbers as the root has units.

Every square of an odd number is the difference between two triangular numbers the bases of which are respectively $(3n+1)$ and n. For, the sum of any series of natural numbers is the difference of two series of natural numbers commencing with unity; and since, as it is shown above, every square of an odd number is the sum of a series of natural numbers, it is also the difference between two triangular numbers.

It is also evident that series, the sums of which are squares of odd numbers, may be so taken that, when placed in succession, they will form an uninterrupted progression of natural numbers commencing with unity, the sum of which is a triangular number:

$$(1)+(2+3+4)+(5+6+7+8+9+10+11+12+13)..\&c.=$$
$$(1^2+3^2+9^2+27^2\ldots\ldots+(3^n)^2)=$$

a triangular number the base of which is the series

$$(1+3+9+27\ldots\ldots+3^n).$$

§ 2. CUBE NUMBERS.

If $S = n^3$, the first term $= n^2 + \dfrac{\delta}{2}(1-n)$.

C.

Every cube n^3 is the sum of an arithmetical progression of n terms, the first term of which is unity, and the difference $2(n+1)$.

$$
\begin{aligned}
1 &\ldots\ldots\ldots\ldots =1^3\\
1+7 &\ldots\ldots\ldots\ldots =2^3\\
1+9+17 &\ldots\ldots\ldots\ldots =3^3\\
1+11+21+31 &\ldots\ldots\ldots\ldots =4^3\\
1+13+25+37+49 &\ldots\ldots\ldots\ldots =5^3\\
1+15+29+43+57+71 &\ldots\ldots\ldots\ldots =6^3\\
1+17+33+49+65+81+97 &\ldots\ldots\ldots =7^3
\end{aligned}
$$

D.

Every cube n^3 is the sum of an arithmetical progression of n terms, the first term of which is the root n, and the difference $2n$.

$$
\begin{aligned}
1 &\ldots\ldots\ldots\ldots =1^3\\
2+6 &\ldots\ldots\ldots\ldots =2^3\\
3+9+15 &\ldots\ldots\ldots\ldots =3^3\\
4+12+20+28 &\ldots\ldots\ldots\ldots =4^3\\
5+15+25+35+45 &\ldots\ldots\ldots\ldots =5^3\\
6+18+30+42+54+66 &\ldots\ldots\ldots\ldots =6^3\\
7+21+35+49+63+77+91 &\ldots\ldots\ldots =7^3
\end{aligned}
$$

The last terms of these series are the alternate triangular numbers. If they be respectively divided by the first terms, the quotients will be the series of odd numbers.

E.

Every cube n^3 is the sum of an arithmetical progression of n terms, the first term of which is $(n^2 - n + 1)$, and the difference 2.

$$
\begin{aligned}
1 &\quad\dots\dots\dots\dots\dots = 1^3 \\
3+5 &\quad\dots\dots\dots\dots\dots = 2^3 \\
7+9+11 &\quad\dots\dots\dots\dots = 3^3 \\
13+15+17+19 &\quad\dots\dots\dots\dots = 4^3 \\
21+23+25+27+29 &\quad\dots\dots\dots = 5^3 \\
31+33+35+37+39+41 &\quad\dots\dots\dots = 6^3 \\
43+45+47+49+51+53+55 &\quad\dots\dots = 7^3
\end{aligned}
$$

This, it will be observed, is a triangular arrangement of the uneven numbers in their regular order.

Every cube is the sum of as many consecutive odd numbers as there are units in the root[*].

The known theorem, that the sum of the cubes of any succession of the natural numbers commencing with unity is equal to the square of the sum of the roots, or, in other words, to the square of the corresponding triangular number, is an immediate consequence of the above.

$$
(1^3 + 2^3 + 3^3 + 4^3 \dots\dots + n^3) = (1+2+3+4 \dots + n)^2 = \left(\frac{n^2 + n}{2}\right)^2.
$$

The sum of any series of odd numbers commencing with unity being equal to the square of the number of terms (A.), the sum of the numbers in any triangle formed as above is necessarily equal to the square of a triangular number. It is also easy to see that each cube is the difference between the squares of two consecutive triangular numbers; and that the difference between the squares of any two triangular numbers whatever is the sum of consecutive cubes. The following equations have been found by ascertaining what

[*] Since the present note was communicated to the Royal Society, I have found that this relation has been already noticed by Count d'Adhémar (Comptes Rendus, tom. xxiii. p. 501). Cauchy observes, " quoiqu'elle puisse, comme on le voit, se déduire des principes déjà connus, toutefois, elle est assez curieuse et très-simple."

differences of the squares of two triangular numbers are equal to single cubes:—

$$3^3 + 4^3 + 5^3 = 6^3$$
$$11^3 + 12^3 + 13^3 + 14^3 = 20^3.$$

F.

Every cube n^3 is the sum of an arithmetical progression of n terms, the first term of which is a triangular number $\frac{n^2 + n}{2}$, and the difference $= n$.

$$
\begin{aligned}
1 &\ \ldots\ldots\ldots\ldots\ldots = 1^3 \\
3 + 5 &\ \ldots\ldots\ldots\ldots = 2^3 \\
6 + 9 + 12 &\ \ldots\ldots\ldots\ldots = 3^3 \\
10 + 14 + 18 + 22 &\ \ldots\ldots\ldots = 4^3 \\
15 + 20 + 25 + 30 + 35 &\ \ldots\ldots = 5^3 \\
21 + 27 + 33 + 39 + 45 + 51 &\ \ldots\ldots = 6^3 \\
28 + 35 + 42 + 49 + 56 + 63 + 70 &\ \ldots\ldots = 7^3
\end{aligned}
$$

Each number contained in this triangle is itself the sum of an arithmetical progression of n terms. Thus, taking the fifth row for example:—

$$
\begin{aligned}
1 + 2 + 3 + 4 + 5 &= 15 \\
2 + 3 + 4 + 5 + 6 &= 20 \\
3 + 4 + 5 + 6 + 7 &= 25 \\
4 + 5 + 6 + 7 + 8 &= 30 \\
5 + 6 + 7 + 8 + 9 &= 35 \\
\hline
125 &= 5^3
\end{aligned}
$$

The sum of all the numbers contained in a square thus formed is equal to the cube of the number which occupies the upper right-hand and lower left-hand corners. The sum of the numbers in either of the diagonals is the corresponding square, and in the case of the odd numbers the sum of the middle horizontal or vertical line is also the square.

This last-mentioned relation was pointed out by Lichtenberg[*], who stated the theorem thus:—If a be a whole

---

[*] G. C. Lichtenberg's Vermischte Schriften, Band ix. p. 359. Göttingen, 1806.

number, and A be the sum of all the natural numbers from 1 to a, then:

$$a^3 = A + (A+a) + (A+2a) + (A+3a)\ldots\ldots + (A+[a-1]a).$$

G.

Every cube n^3 above 1 is the sum of an arithmetical progression of n terms, the first term of which is $(n-2)^2$, and the difference $= 8$.

$$
\begin{aligned}
0+8 &\ldots\ldots\ldots\ldots\ldots = 2^3 \\
1+9+17 &\ldots\ldots\ldots\ldots = 3^3 \\
4+12+20+28 &\ldots\ldots\ldots = 4^3 \\
9+17+25+33+41 &\ldots\ldots\ldots = 5^3 \\
16+24+32+40+48+56 &\ldots\ldots = 6^3 \\
25+33+41+49+57+65+73 &\ldots\ldots = 7^3
\end{aligned}
$$

Each progression of this triangle, consisting of an uneven number of terms, contains two consecutive odd square numbers.

An *uninterrupted* arithmetical progression commencing with unity and proceeding by the constant addition of 8, arranged in a triangular form, presents some curious results. 1st. The first terms of each line are the squares of the odd numbers in their regular sequence. 2nd. The sum of all the numbers in any two adjacent lines is the cube of an odd number.

$$
\begin{aligned}
1 &\ldots\ldots\ldots \} \ldots\ldots = 3^3 \\
9+17 &\ldots\ldots\ldots \} \ldots\ldots = 5^3 \\
25+33+41 &\ldots\ldots \} \ldots\ldots = 7^3 \\
49+57+65+73 &\ldots \} \ldots\ldots = 9^3 \\
81+89+97+105+113 &\} \ldots\ldots = 9^3 \\
121+129+137+145+153+161 &\} \ldots\ldots = 11^3
\end{aligned}
$$

It is evident from the preceding arrangement that

$$(2n+1)^2 = 1 + 8\left(\frac{n^2+n}{2}\right).$$

Thus any triangular number multipled by 8 with 1 added is equal to the square of an odd number; or, any square of an uneven number minus 1 is divisible by 8, and the quotient is a triangular number.

§ 3.

Of the higher powers I will confine myself to one example.

H.

Every fourth power n^4 is the sum of an arithmetical progression of n terms, the first term of which is n^2, and the difference $2n^2$.

$$1 \quad \dots\dots\dots\dots\dots = 1^4$$
$$4+12 \quad \dots\dots\dots\dots\dots = 2^4$$
$$9+27+45 \quad \dots\dots\dots\dots = 3^4$$
$$16+48+80+112 \quad \dots\dots\dots = 4^4$$
$$25+75+125+150+225 \quad \dots\dots = 5^4$$
$$36+108+180+252+324+396 \quad \dots\dots = 6^4$$

This triangle consists of the progressions in (D.) multiplied respectively by n, or of those in (A.) multiplied by n^2.

Interpretation of an important Historical Document in Cipher*.

[From the 'Memoirs of the Philobiblon Society.']

7 Chester Terrace, Regent's Park,
Feb. 3, 1862.

MY DEAR SIR,—I herewith send you the translation of a document in cipher which I made two or three years ago, and which you suggested to me might interest the members of the Philobiblon Society. The accompanying letter from Sir Henry Ellis will sufficiently explain its history ; and, to assist any person desirous of verifying the accuracy of the translation, I have added the key of the cipher employed, with a few particulars respecting its construction.

Yours very truly,

C. WHEATSTONE.

R. M. Milnes, Esq., M.P.

---

* [The following note is appended to the original :—" The Editor has no other preface to make to this interesting communication, except to tender to Professor Wheatstone the thanks of the Philobiblon Society, and to express their admiration of this additional instance of that wonderful faculty of interpretation which seems to ordinary minds a special intuition, not unworthy of the great scientific discoverer and practical benefactor of his age."]

24 Bedford Square,
June 1, 1858.

MY DEAR SIR,—A good many years ago the Trustees of the British Museum purchased, at a large price, what appeared, and no doubt must be, a very important document in cipher; occupying seven folio pages closely filled with numerals; every page signed at top by King Charles the First, and countersigned below by Lord Digbye.

I was long ago in hope to have got it deciphered, and to have inserted it, with its explanation, in my second series of original letters illustrative of English history; but every effort on my part to procure the deciphering failed. Being unable to obtain its explanation, I gave up the thought of its publication.

A few evenings ago, at Earl Stanhope's, I met Lord Wrottesley, to whom I accidentally mentioned that I had just been talking on the subject of this letter to Mr. Babbage, and he immediately mentioned that no likelier person would be found to get at the secrets in this letter than Mr. Wheatstone. Will you, therefore, allow me to lay before you an exact copy of this letter, and to request your kind investigation of what may turn out to be its contents?

I enclose the only copy which I have, which was made for me at the time I have already mentioned. If length implies importance in such a document, this ought to be a very important letter.

Yours faithfully,

HENRY ELLIS.

Charles Wheatstone, Esq.

On an examination of this cipher it appeared that about 90 different numerals were employed to represent the letters of the alphabet, each letter being, consequently, represented by several distinct numerals. These figures representing letters were intermixed with higher numbers, referring evidently to a vocabulary. Besides the difficulty arising from this redundancy of characters, another was superadded in no divisions having been made between the words, so that no assistance could be obtained from comparing groups of characters representing the same words with each other. The task of translation was rendered more tedious from the original being in a different language to that in which I had first assumed it to be.

In translating the literal portion of the cipher I have preserved the irregularities of spelling; in this part there can be no uncertainty respecting the meaning, except only in two or three places, where it has no doubt arisen in the process of transferring the original into cipher. As to the references to the vocabulary, when the same numbers occur frequently there is no difficulty in ascertaining the proper corresponding word, as it must be one suited to complete the sense in every case : but when a number occurs only once a greater latitude in determining its meaning must necessarily be assumed; and it is probable that in several cases I have not selected the appropriate word. Those words respecting which I have any doubt are, however, marked in the key to the vocabulary with an asterisk. As the higher numbers evidently form part of a vocabulary arranged in alphabetical order, if I have in any instance mistaken a word it must necessarily be supplied by one commencing with the same or with an adjacent letter.

The words and portions of words placed between parentheses in my translation correspond with numbers in the vocabulary; the remainder is orthographical.

KEY TO THE LITERAL CIPHER.

A	12.	13.	14.	15.	16.	17.
B	18.	19.				
C	20.	21.	22.			
D	23.	24.	25.			
E	26.	27.	28.	29,	30.	31. 32.
F	33.	34.	35.			
G	36.	37.	38.			
H	39.	40.	41.			
IJ	42.	43.	44.	45.	46.	47.
L	48.	49.	50.	51.		
M	52.	53.	54.	55.		
N	56.	57.	58.	59.		
O	60.	61.	62.	63.	64.	
P	65.	66.	67.			
Q	80.	81.	82.			
R	83.	84.	85.	86.	87.	
S	88.	89.	90.	91.	92.	
T	93.	94.	95.	96.	97.	
UV	98.	99.	100.	101.	102.	
X	103.	104.	105.			
Y	108.	109.	110.			

KEY TO THE VOCABULARY.

113 ances 1	333 même 3
121 avoir 3	336 mauvais 1
126 alliance 2	347 nécessaire 1
130 affaire. 6	348 nous 6
160 advantage . . . 1	*361 opinion 1
194 bien 1	372 pouvoir 1
203 connaître. . . . 3	376 pour 29
213 comme 1	380 post 1
220 donner 2	384 proposé 1
232 dans 8	*385 preparé 1
241 être 4	391 quand 1
246 effet 3	392 qu'il 10
247 elle 4	393 qu'elle 3
*253 franchise 1	394 que 32
255 faire 10	395 qui 2
265 faite 1	396 quelque 2
266 fait 10	409 raison 4
273 guerre. 1	*420 support 1
299 intérêt 8	*421 satisfaction 1
*300 insister 1	*426 toutefois 1
305 jour 3	439 tems 4
*315 joindre 1	442 troubles 1
316 lui 12	448 trouver 2
318 lequel 2	462 tenir 1
319 laquelle 1	467 vous 1
320 les 32	*469 volonté ⎱ 2
324 ligue 2	vouloir ⎰

474 Angleterre 6
478 France 14
479 Espagne 3
483 Hollande 3
484 1
495 le roi d'Angleterre 12
496 la reine d'Angleterre 1
497 le prince de Galles 3
509 le prince d'Orange 5
512 la princesse Henrietta d'Orange 4
522 1
562 un Envoi 4

INSTRUCTIONS POUR LE SIEUR DE GOFFE.

(Vous) { (faire)s / ferez } (connaître) à son altesse (le prince d'Orange) que selon un discours (que) (lui) a esté (fait) par (522) durant le séjour de (la reine) à la Haye par son commandement touchant le marriage du (prince) et (la princesse) (qu'elle) a comuniqué au (roi) ce qui s'est passé et a continuellement fait sçavoir à son altesse (les) résolutions qu'ont esté prises par les (quelle)s son altesse a peut re (connaitre) la { (satisfaction) / soin } de (laquelle) ceste (affaire) a esté commencée et (pour) suivie.   (Le roi) voulant demeurer tou (jour)s dans la même (post)ure { asteur / à cette heure } (que) ses (affaire)s (lui) permettent de (tenir) à la dernier résolution, a pris celle aussi de (faire) part à son altesse avec la (franchise) qu'il ne (fait) point de doute de rencontrer auprès d'(elle).   Son altesse sçaura donc (que) (les) (mêmes) (faite)s qui ont donné mouvement à la première proposition sont ceux qui obligent présentement (le roi) d'en parler (comme) il fera ; c'est à dire, l'estime (qu'il) (fait) de la persone et maison de son altesse, joint avec l'utilité (qu'il) espère rencontrer (pour) restablisement. de ses (affaire)s et corone dans un (tems) moins embrouillé (que) cellui cy ; la première considération auroit pu (être) suffisante ; mais dans (les) (trouble)s (que) (lui) sont survenus (les) loix de Dieu et des hommes (lui) nécessitent à regarder davantage le seconde.   (Pour) cest (effet) faisant réflexion partout où il se povoit promettre un (support) considérable et convenable, autant à son estat qu'à ses inclinations, a (fait) ceste conclusion : (qu'il) n'en povoit point rencontrer (que) d'une des corones de (France) ou de (l'Espagne), et de plus de pas une d'icelles sans l'assistance et concurence de son altesse ; (le roi) manquant (les) moyens (pour) leur (donner) la { (volonté) / vouloir } et (elle)s le (pouvoir) de (lui) procurer (quand) (même)s (elle)s auroient le desire entre (les) deux.   Il n'a pas esté difficile au (roi) de resoudre par où il devoit commencer, regardant dans ses inclinations, aux (intérêt)s de son estat, et à ceux de ses anciennes alli (ance)s; et en ces considérations a premièrement jetté (les) y (eux)

sur la (France), se promettant (que) par l'image (qu'il) s'est
figuré des (intérêt)s de cest estat et des (intérêt)s des persones
(qui) en mannient (les) (affaire)s qu'en les povant assurer des
advantages (qui) (leur) pouront (être) procurés par son alteses
(que) ceux (qu'il) aura desiré d'(elle)s lui seront assurés. (Le
roi) a creu par les tesmoinages (que) son altesse a rendu de
son inclination (pour) une seconde (alliance) avec (Angleterre)
(qu'elle) feroit toutes sortes des choses (raison)ables (pour)
la (faire) réussir (bien) a aussy cru (que) ces choses là (pour)-
oient estre tellement (proposé)s et (preparé)s que (France) y
(trouver)oit ses (intérêt)s et par consequent (le roi) les siens
avec (France).   (Le roi) avoit (proposé) d'envoyer (562) en
(France) se croiant assez assuré des inclinations de son altesse
(pour) (faire) l'ouverture du traité, qui auroit porté qu'en
cas (que) (nous) eussions pu (le) porter aux résolutions (les)
plus advantageuses pour eux, (qu'il)s auroient pu (toutefois)
sçavoir si eux de leur costé seroient venus à celles (pour)
(nous) (que) (nous) leur aurions proposés.   (Les) choses que
(lui) seront desirés de (faire) en considération du (roi) et du
mariage sont celles de quoy les (intérêt)s d'(Angleterre) et de
(France) requièreront d'(être) (fait)es (pour) leur (advantage), et devant (avoir) fait l'ouverture en (France) et entré
(dans) (les) considérations qu'un traité feroit naistre, il ne se
peut dire exactement ce (qui) sera desiré de ce costé là, mais
en apparence ce seront (les) particularités suivantes, ou de
ceste nature :—

1. D'entrer et de (faire) entrer (Hollande) en (ligue)
offensive et défensive avec (Angleterre) et (France).

2. De rompre avec l'(522) en cas que (France) le propose.

3. D'assièger telle ou telle place (que) le tems (dans)
(lequel) la proposition sera (fait)e, (pour)ra permetre.

4. De demeurer en campagne aussy long (tems) qu'eux,
et s'il n'y auroit point de siège, au moins de marcher et
(joindre) le plus utilement (pour) leur dessein (qu'il) se
pourroit.

5. De fournir (pour) deux mois XV ou XX vaisseaux de
(guerre), et des vaisseaux (pour) passer deux milles chevaux
et quatre milles hommes de pied de (France) en (Angleterre).

Les choses (que) (nous) demanderons de (France) :—

I. (Les) hommes susdits et telle somme d'argent (que) (nous) (pour)rions procurer.

II. D'entrer (dans) une (alliance) offensive et défensive avec (nous) et (Hollande).

S' il arive {astur / à cette heure} que (France), par (quelque) (intérêt) caché ou (quelque) (mauvaise) (volonté) (que) la jalousie du (tems) passé auroit fait passer (dans) (les) humeurs (pour) (les) (affaire)s du roi, se trouvoit repugnante à toutes nos propositions, en ce cas là (le roi), se contentant de s'(être) acquité de touts (les) devoirs (que) (lui) pouvoient presser, propose nonobstant le mariage du (prince) et de (la princesse), et au lieu des advantages (qu'il) se promettoit du costé de (France) ne doute point (que) (le prince d'Orange) ne (lui) peut (faire) recevoir d'aussy utiles du costé de (l'Espagne) ; (pour) cest (effet) il lui faudroit s'engager de (faire) la trève avec ceste corone en considération de celle d'(Angleterre) faisant (connaître) aux (Espagn)ols (qu'il)s ne la (pour)ront point (avoir) qu'en y comprenant (les) (intérêt)s du (roi) et lui laisant l'advantage de ce traité, la preuve (fait)e.

(Le prince d'Orange) :—

I. Disposer (les) Anglois (dans) le service de (Hollande) de passer en (Angleterre) (dans) des régiments entiers, et leur (donner) (les) moyens de passer.

II. Il sera achever l'(affaire)d'Amboyn présentement.

III. Il payera l'argent de quoy il sera convenu (pour) la dote de (la princesse) présentement ; (les) conditions (pour) (les) advantages de (la princesse) seront selon ce qui a tou-(jour)s esté practiqué (pour) (les) femmes des (prince)s.

(Pour) conserter (lequel)es il sera immédiatement envoyé une persone instruite (pour) cest (effet) de tout ce (qui) sera (nécessaire) selon la réponse (que) son altesse fera à cest ou-verture. Il est dit ci dessus que (562) devoit aller en (France) (pour) ce traité ; il est vray (que) cela avoit esté résolu et qu'il iroit après (484) (trouver) (le prince d'Orange) sur

l'(opinion) (que) l'on avoit icy que (le prince d'Orange) sou-
haitoit assez ceste (alliance) (pour) (faire) toutes sortes
choses (raison)ables (pour) la (faire) réussir; mais après,
faisant un peu de réflection plus profonde il a esté plus selon
l'ordre de la (raison) d'en (avoir) un esclaircissement plus
assuré (que) l'on ne pouvoit tirer des conjectures, et (pour)
ceste raison (562) a esté arresté et le porteur choisy (pour)
raporter en toute dilligence (les) sentiments de son altessse
sur toute la matière, et ce porteur a esté choisy (pour) (les)
(raison)s que sont assez amplement dittes (dans) ses letres
de créance.

C. R.

George Digbye.

CHARLES R.

42. 56. 90. 93. 83. 99. 20. 95. 43. 60. 57.
91. 376. 48. 27. 90. 43. 28. 99. 83. 24.
28. 38. 60. 33. 34. 26.

467. 255. 88. 203. 17. 92. 61. 59. 15. 48.
94. 27. 90. 88. 30. 509. 394. 92. 27. 49. 60.
57. 99. 59. 25. 47. 89. 20. 61. 100. 83. 92.
394. 316. 12. 29. 90. 93. 31. 266. 65. 16. 84.
562. 23. 102. 85. 15. 57. 97. 48. 27. 89. 26.
305. 23. 31. 496. 13. 49. 14. 40. 15. 108. 27.
66. 16. 86. 90. 61. 56. 22. 57. 53. 55. 17.
58. 25. 27. 54. 31. 57. 95. 94. 65. 99. 20.
39. 15. 59. 97. 49. 28. 54. 12. 85. 44. 15.
38. 29. 24. 99. 497. 28. 93. 512. 393. 14.
2064. 54. 98. 56. 46. 80. 98. 26. 16. 101.
495. 21. 29. 81. 99. 13. 66. 17. 89. 90. 30.
28. 95. 15. 20. 60. 57. 96. 44. 56. 102. 27.
50. 48. 32. 54. 29. 59. 97. 266. 91. 22. 14.
102. 62. 43. 86. 13. 90. 64. 58. 17. 46. 94.
31. 90. 92. 28. 320. 87. 32. 88. 63. 48. 99.
93. 47. 60. 56. 90. 81. 102. 62. 58. 95. 29.
90. 97. 32. 66. 84. 43. 90. 29. 91. 69. 15.
84. 320. 393. 91. 92. 61. 57. 14. 49. 96. 27.
89. 91. 28. 16. 67. 30. 99. 86. 32. 203. 50.
13. 421. 23. 30. 319. 20. 31. 91. 95. 30. 130.
14. 26. 88. 94. 29. 21. 64. 52. 53. 30. 58.

George Digbye.

C. R.

21. 29. 30. 97. 376. 89. 98. 45. 100. 46. 28.
32. 495. 98. 60. 99. 48. 13. 56. 93. 23. 27.
54. 30. 100. 85. 28 86 95. 61. 100. 305. 88.
232. 49. 12. 333. 380. 99. 84. 26. 13. 89.
93. 27. 100. 84. 394. 89. 28. 90. 130. 91.
316. 65. 29. 85. 52. 30. 94. 95. 31. 56. 96.
24. 32. 462. 14. 49. 15. 25. 26. 85. 57. 42.
27. 86. 87. 28. 92. 61. 50. 102. 97. 43. 62.
58. 17. 66. 86. 44. 90. 20. 28. 50. 51. 26.
13. 98. 88. 89. 45. 23. 29. 255. 67. 14. 83.
93. 14. 89. 63. 56. 15. 49. 96. 30. 90. 91.
32. 17. 101. 26. 21. 50. 13. 253. 392. 29.
97. 58. 31. 266. 66. 68. 46. 59. 97. 25. 28.
24. 63. 99. 95. 29. 23. 30. 86. 30. 57. 22.
64. 56. 96. 85. 30. 86. 13. 98. 68. 86. 28.
92. 24. 247. 90. 63. 57. 16. 50. 96. 27. 88.
90. 28. 91. 20. 16. 99. 85. 13. 24. 62. 58.
21. 394. 320. 333. 90. 265. 91. 81. 99. 61.
56. 93. 23. 62. 57. 58. 28. 55. 62. 98. 99.
26. 52. 27. 56. 93. 13. 48. 14. 65. 83. 42.
53. 43. 84. 28. 66. 85. 60. 66. 61. 88. 44. 94.
30. 61. 57. 89. 62. 58. 95. 20. 30. 99. 103.
395. 63. 18. 49. 45. 36. 31. 59. 96. 66. 85.
32. 90. 26. 56. 97. 27. 55. 28. 56. 93. 495.
23. 29. 57. 67. 14. 86. 50. 30. 87. 213. 45.
51. 33. 31. 87. 15. 21. 32. 88. 96. 16. 24.
42. 83. 28. 50. 28. 88. 94. 43. 55. 27. 392.
266. 23. 29. 50. 13. 67. 30. 83. 90. 60. 57.
32. 31. 94. 53. 13. 46. 90. 63. 58. 25. 27.
90. 62. 58. 15. 50. 95. 26. 90. 91. 27. 44.
61. 45. 59. 93. 14. 99. 28. 22. 50. 99. 94.

George Digbye.

C. R.

43. 51. 44. 94. 27. 392. 28. 89. 65. 29. 86.
30. 83. 31. 57. 20. 60. 57. 95. 85. 29. 86.
376. 87. 30. 91. 93. 13. 18. 48. 42. 90. 27.
52. 28. 57. 96. 25. 31. 90. 30. 92. 130. 92.
31. 94. 22. 60. 84. 62. 58. 32. 24. 12. 59.
92. 102. 56. 439. 53. 62. 46. 58. 90. 32. 54.
19. 85. 62. 99. 45. 50. 49. 27. 394. 21. 30.
49. 50. 99. 46. 22. 107. 48. 15. 68. 86. 47.
55. 46. 27. 83. 30. 21. 60. 56. 90. 45. 24.
28. 87. 17. 94. 44. 63. 58. 15. 98. 86. 61.
43. 96. 66. 99. 241. 88. 98. 34. 35. 43. 92.
14. 57. 94. 30. 52. 13. 27. 92. 25. 14. 57.
92. 320. 442. 91. 394. 316. 90. 63. 57. 94.
92. 100. 83. 102. 27. 58. 101. 90. 320. 48.
62. 26. 104. 23. 28. 24. 46. 26. 99. 29. 95.
23. 27. 90. 41. 62. 54. 55. 28. 91. 316. 57.
28. 20. 27. 91. 92. 43. 93. 26. 56. 96. 14.
85. 27. 37. 17. 87. 23. 30. 83. 25. 14. 99.
15. 59. 96. 14. 36. 30. 49. 13. 90. 27. 21. 63.
58. 25. 32. 376. 20. 27. 91. 95. 246. 35. 15.
42. 92. 17. 57. 93. 85. 27. 34. 48. 26. 103.
43. 64. 59. 67. 16. 84. 93. 62. 99. 97. 64.
98. 44. 49. 90. 27. 67. 60. 100. 63. 45. 94.
68. 86. 62. 54. 30. 95. 86. 27. 102. 59. 420.
20. 60. 59. 90. 42. 24. 27. 86. 12. 19. 49.
28. 29. 96. 21. 63. 57. 101. 29. 59. 15. 19.
50. 30. 16. 101. 93. 13. 56. 96. 15. 91. 63.
59. 28. 91. 94. 15. 97. 81. 99. 12. 91. 27.
92. 43. 57. 21. 49. 44. 57. 13. 95. 45. 64.
56. 90. 16. 266. 22. 27. 90. 95. 28. 20. 61.
57. 21. 48. 100. 89. 45. 63. 58. 392. 57. 26.

George Digbye.

C. R.

56. 66. 64. 101. 61. 44. 96. 68. 60. 42. 58.
96. 87. 29. 57. 22. 62. 57. 93. 83. 27. 85.
394. 23. 100. 59. 27. 24. 31. 90. 21. 61. 86.
63. 58. 28. 92. 25. 27. $\underset{\wedge}{v}$. 478. 63. 99. 24.
26. 476. 27. 96. 24. 30. 66. 49. 100. 92. 24.
28. 67. 13. 90. 100. 59. 29. 23. 43. 22. 30.
49. 50. 27. 90. 91. 26. 56. 90. 48. 14. 88.
43. 90. 94. 15. 56. 20. 27. 29. 96. 21. 63.
58. 22. 102. 86. 28. 58. 20. 29. 24. 29. 92.
62. 57. 13. 48. 94. 28. 90. 91. 27. 495. 55.
16. 58. 80. 99. 14. 58. 96. 320. 52. 60. 42.
28. 56. 90. 376. 49. 27. 100. 84. 220. 48. 16.
469. 26. 93. 247. 88. 49. 30. 372. 23. 27. 50.
30. 316. 66. 86. 61. 20. 99. 83. 32. 87. 391.
333. 90. 247. 92. 13. 99. 86. 62. 45. 26. 57.
95. 48. 27. 23. 28. 90. 43. 85. 29. 27. 56. 93.
83. 28. 320. 24. 27. 99. 103. 42. 48. 56. 15.
67. 16. 92. 29. 88. 93. 28. 24. 42. 34. 33. 43.
21. 45. 48. 32. 13. 100. 495. 23. 26. 83. 27.
89. 60. 99. 25. 84. 30. 67. 13. 83. 61. 100.
43. 50. 24. 27. 101. 64. 42. 93. 20. 62. 54.
55. 28. 56. 22. 30. 83. 85. 28. 36. 16. 86.
24. 16. 59. 94. 23. 27. 57. 90. 91. 28. 92.
44. 56. 21. 48. 42. 57. 17. 93. 45. 62. 57.
92. 17. 99. 104. 299. 89. 23. 26. 91. 63. 59.
30. 90. 95. 13. 96. 29. 95. 15. 21. 28. 99.
105. 25. 29. 90. 31. 92. 14. 57. 20. 44. 29.
59. 90. 16. 48. 50. 47. 113. 32. 95. 31. 57.
89. 28. 90. 20. 60. 56. 92. 42. 23. 30. 85.
16. 96. 44. 62. 58. 92. 14. 68. 85. 46. 54.
45. 27. 83. 28. 55. 31. 58. 97. 43. 27. 93.

George Digbye.

C. R.

94.   26.   320.   110.   27.   99.   107.   89.   100.   83.
48.   13.   478.   90.   28.   69.   84.   64.   54.   30.   95.
96.   16.   58.   97.   394.   67.   16.   87.   50.   42.   55.
16.   37.   27.   392.   91.   28.   92.   95.   33.   43.   38.
99.   83.   29.   23.   29.   91.   299.   92.   24.   26.   21.
27.   91.   93.   27.   88.   95.   13.   96.   30.   96.   24.
28.   92.   299.   89.   23.   27.   88.   68.   28.   85.   91.
62.   56.   28.   92.   395.   28.   58.   55.   16.   56.   57.
42.   29.   56.   94.   320.   130.   90.   81.   98.   27.   57.
320.   65.   64.   99.   15.   57.   93.   13.   90.   92.   100.
87.   28.   87.   23.   26.   91.   13.   24.   99.   14.   56.   94.
16.   37.   26.   91.   394.   320.   66.   62.   99.   83.   64.
58.   95.   241.   67.   84.   62.   22.   99.   87.   29.   90.
68.   14.   86.   90.   61.   58.   15.   51.   95.   30.   89.   92.
32.   394.   21.   26.   98.   105.   392.   15.   99.   85.   17.
25.   29.   90.   42.   84,   28.   86.   24.   247.   90.   316.
90.   27.   86.   60.   58.   96.   15.   90.   92.   100.   85.
32.   92.   495.   13.   21.   83.   27.   101.   65.   15.   85.
320.   93.   26.   89.   54.   60.   42.   57.   15.   36.   30.
90.   394.   90.   61.   59.   17.   59.   93.   27.   89.   88.
30.   13.   84.   29.   58.   25.   98.   23.   28.   90.   60.   56.
43.   57.   20.   51.   43.   57.   14.   95.   46.   61.   58.   376.
99.   59.   27.   90.   27.   20.   62.   56.   24.   28.   126.
14.   100.   27.   20.   474.   393.   33.   28.   83.   63.   43.
93.   94.   64.   102.   95.   27.   92.   88.   64.   83.   93.   26.
88.   23.   28.   89.   21.   39.   62.   90.   28.   92.   409.
13.   19.   49.   29.   90.   376.   51.   15.   225.   84.   27.
99.   91.   92.   42.   85.   28.   194.   14.   15.   100.   90.
91.   110.   20.   85.   101.   394.   21.   30.   92.   22.
40.   60.   90.   27.   92.   48.   15,   376.   61.   43.   27.

George Digbye.

C. R.

57. 96. 29. 90. 94. 84. 27. 94. 28. 43. 49.
29. 55. 31. 58. 96. 384. 92. 28. 93. 385. 92.
394. 478. 110. 448. 61. 45. 95. 89. 29. 90.
299. 92. 27. 96. 67. 15. 83. 22. 62. 53. 92.
32. 82. 99. 29. 59. 93. 495. 320. 90. 44. 27.
59. 92. 14. 100. 28. 22. 478. 495. 14. 102.
60. 42. 93. 384. 23. 26. 56. 99. 62. 108. 30.
87. 562. 28. 58. 478. 90. 29. 20. 85. 63. 43.
13. 56. 96. 14. 90. 91. 29. 92. 16. 89. 90.
100. 85. 29. 24. 27. 92. 44. 58. 21. 49. 46.
58. 16. 96. 47. 60. 59. 88. 24. 26. 89. 61.
56. 15. 49. 93. 27. 90. 91. 28. 376. 255. 48.
62. 102. 28. 85. 96. 100. 87. 31. 24. 98. 93.
83. 13. 43. 95. 27. 80. 99. 16. 100. 85. 60.
46. 97. 66. 63. 86. 96. 32. 81. 102. 29. 58.
22. 12. 90. 394. 348. 28. 99. 90. 91. 42. 62.
58. 92. 66. 99. 316. 67. 60. 83. 95. 27. 87.
16. 98. 104. 86. 27. 90. 62. 50. 100. 93. 44.
63. 59. 91. 320. 68. 48. 102. 90. 13. 23. 98.
14. 56. 93. 15. 36. 42. 26. 99. 88. 27. 89.
376. 27. 100. 103. 392. 88. 16. 99. 83. 60.
43. 28. 57. 95. 65. 100. 426. 89. 20. 17. 101.
61. 44. 83. 90. 45. 30. 102. 104. 24. 31. 48.
32. 99. 83. 20. 61. 91. 93. 32. 90. 26. 84.
60. 42. 27. 59. 94. 101. 30. 56. 99. 88. 12.
21. 28. 48. 49. 29. 92. 376. 348. 394. 348.
320. 12. 98. 85. 43. 60. 56. 90. 66. 85. 61.
67. 62. 92. 29. 91. 320. 21. 41. 61. 91. 27.
92. 394. 316. 91. 27. 84. 63. 57. 95. 25. 28.
90. 46. 86. 29. 90. 24. 300. 255. 27. 57. 21.
61. 58. 89. 42. 24. 27. 83. 12. 93. 43. 61.

George Digbye.

C. R.

57. 25. 99. 495. 28. 96. 24. 99. 52. 13. 84.
44. 14. 36. 27. 90. 60. 59. 95. 21. 28. 48. 50.
31. 88. 23. 27. 80. 99. 60. 110. 320. 299. 91.
24. 474. 28. 94. 23. 478. 85. 28. 81. 101.
30. 85. 31. 83. 60. 56. 93. 25. 241. 266. 27.
21. 376. 49. 29. 99. 87. 160. 27. 94. 24. 26.
99. 12. 59. 97. 121. 266. 51. 61. 99. 27. 83.
95. 101. 84. 29. 28. 56. 478. 31. 95. 26. 57.
95. 84. 28. 232. 320. 22. 62. 58. 89. 42. 23.
27. 83. 12. 94. 42. 61. 57. 91. 80. 99. 100. 57.
93. 83. 13. 44. 95. 27. 33. 26. 87. 62. 42. 93.
59. 14. 47. 90. 94. 84. 27. 47. 48. 56. 28. 90.
29. 66. 26. 99. 94. 23. 46. 86. 26. 27. 105. 13.
20. 96. 27. 54. 30. 58. 96. 22. 27. 394. 90. 32.
87. 14. 24. 26. 90. 43. 86. 32. 25. 26. 88. 30.
20. 60. 90. 96. 30. 48. 16. 52. 15. 42. 92.
28. 57. 17. 67. 68. 16. 87. 29. 58. 21. 31.
89. 31. 90. 27. 83. 60. 59. 93. 320. 67. 13.
87. 94. 42. 21. 100. 50. 44. 32. 86. 92. 90.
99. 42. 101. 14. 59. 93. 27. 89. 60. 99. 23.
26. 20. 27. 88. 94. 28. 58. 15. 95. 101. 83.
30. 1. 24. 27. 57. 96. 86. 26. 87. 30. 93. 24.
26. 255. 30. 59. 93. 83. 27. 84. 483. 27. 59.
324. 60. 33. 34. 27. 56. 90. 42. 99. 28. 28.
96. 23. 31. 35. 28. 58. 92. 42. 99. 30. 13.
100. 26. 20. 474. 27. 93. 478. 11. 24. 27.
83. 63. 53. 67. 87. 27. 15. 99. 29. 21. 49.
522. 28. 59. 22. 17. 90. 394. 478. 48. 28.
65. 85. 64. 66. 63. 92. 27. 3. 23. 15. 91. 92.
42. 30. 37. 26. 83. 95. 29. 49. 50. 60. 99.
93. 27. 51. 48. 30. 65. 48. 15. 21. 27. 394.

George Digbye.

C. R.

51.   27.   439.   232.   318.   48.   13.   66.   86.   60.   65.
61.   90.   42.   93.   43.   62.   57.   92.   28.   83.   13.
266.   27.   376.   83.   13.   65.   28.   84.   54.   27.   94.
87.   28.   4.   25.   26.   24.   27.   52.   28.   99.   83.   27.
84.   30.   57.   20.   14.   54.   65.   15.   37.   58.   30.
16.   101.   90.   91.   110.   50.   60.   56.   38.   439.
80.   99.   28.   98.   105.   26.   93.   88.   42.   48.   56.
108.   12.   98.   83.   60.   43.   94.   69.   60.   46.   58.
94.   24.   27.   92.   44.   26.   37.   32.   16.   100.   55.
64.   44.   58.   90.   23.   26.   54.   14.   83.   21.   39.
29.   87.   30.   94.   315.   50.   31.   65.   49.   102.   90.
98.   93.   47.   48.   49.   29.   52.   27.   57.   55.   376.
50.   30.   99.   83.   23.   30.   90.   91.   26.   45.   36.
92.   392.   92.   31.   376.   86.   64.   44.   93.   5.   23.
27.   34.   61.   99.   83.   58.   44.   86.   376.   24.   26.
99.   103.   52.   60.   44.   88.   XV.   60.   99.   XX.
99.   13.   43.   92.   27.   16.   102.   103.   23.   26.   273.
27.   93.   24.   28.   99.   15.   42.   92.   28.   12.   102.
105.   376.   65.   17.   91.   89.   30.   83.   II.   54.   44.
48.   49.   29.   21.   39.   26.   98.   14.   99.   107.   27.
93.   IIII.   52.   47.   50.   51.   30.   41.   61.   54.   55.
27.   90.   25.   28.   67.   44.   32.   23.   24.   26.   478.
27.   58.   474.   320.   20.   41.   61.   91.   27.   92.
394.   348.   24.   26.   52.   12.   57.   23.   30.   83.   60.
57.   90.   24.   32.   478.   I.   320.   40.   60.   52.   53.
26.   90.   91.   99.   92.   23.   43.   93.   27.   39.   30.
94.   93.   31.   49.   50.   32.   88.   60.   52.   55.   28.
23.   13.   85.   37.   30.   57.   95.   394.   343.   376.
86.   42.   62.   57.   92.   65.   83.   60.   20.   99.   86.
26.   87.   II.   23.   27.   57.   93.   85.   30.   86.   232.
99.   57.   28.   324.   62.   33.   34.   32.   57.   92.   42.

George Digbye.

C. R.

99. 27. 28. 94. 24. 26. 34. 30. 56. 90. 42.
99. 32. 13. 102. 30. 21. 348. 29. 93. 483.
90. 42. 43. 48. 12. 83. 44. 99. 27. 17. 91.
94. 98. 86. 394. 478. 65. 16. 83. 396. 299.
20. 15. 21. 40. 27. 60. 99. 396. 336. 469.
394. 50. 17. 42. 13. 50. 61. 100. 91. 27. 23.
99. 439. 67. 17. 90. 92. 28. 12. 102. 87. 63.
45. 95. 266. 65. 13. 90. 92. 27. 83. 232.
320. 40. 99. 52. 27. 98. 87. 92. 376. 320.
130. 88. 24. 102. 495. 90. 26. 93. 83. 62.
99. 61. 42. 94. 85. 26. 65. 99. 36. 56. 13.
58. 96. 30. 13. 93. 60. 99. 94. 27. 90. 58. 60.
88. 66. 83. 62. 65. 61. 90. 42. 95. 45. 60. 59.
90. 26. 56. 20. 27. 21. 12. 92. 48. 16. 495.
88. 28. 21. 61. 59. 93. 29. 56. 94. 13. 59.
97. 25. 27. 92. 241. 15. 20. 80. 99. 43. 93.
29. 13. 96. 60. 102. 97. 89. 320. 24. 28. 100.
60. 43. 83. 92. 394. 316. 65. 60. 99. 98. 61.
45. 27. 56. 95. 67. 83. 28. 91. 88. 30. 87. 68.
85. 60. 69. 64. 89. 27. 56. 61. 58. 63. 18. 89.
94. 12. 57. 93. 49. 27. 52. 14. 84. 44. 13. 36.
27. 24. 99. 497. 28. 93. 23. 30. 512. 27. 94.
15. 99. 51. 43. 29. 98. 24. 28. 90. 15. 23. 99.
16. 58. 95. 13. 37. 29. 88. 392. 90. 27. 65.
83. 60. 52. 26. 93. 94. 60. 42. 95. 23. 99. 20.
61. 90. 95. 27. 23. 26. 478. 57. 29. 25. 60.
98. 95. 29. 65. 61. 44. 58. 95. 394. 509. 59.
28. 316. 67. 28. 100. 96. 255. 84. 26. 21. 27.
100. 61. 43. 86. 24. 12. 99. 90. 91. 108. 102.
96. 43. 49. 27. 92. 25. 102. 21. 61. 91. 93.
27. 23. 479. 376. 21. 27. 90. 93. 246. 42. 50.

George Digbye.

C. R.

316. 34. 16. 98. 23. 84. 62. 42. 94. 90. 29.
59. 36. 13. 38. 29. 83. 23. 28. 255. 48. 14.
93. 83. 99. 102. 30. 12. 102. 28. 21. 22. 26.
90. 93. 27. 22. 62. 85. 60. 59. 27. 28. 58.
22. 61. 56. 96. 42. 23. 26. 84. 12. 93. 45.
64. 59. 24. 27. 22. 27. 49. 50. 29. 24. 474.
34. 13. 42. 88. 17. 56. 93. 203. 15. 99. 104.
479. 60. 48. 90. 392. 91. 56. 27. 51. 13. 376.
83. 61. 57. 93. 65. 60. 42. 57. 94. 121. 80.
99. 27. 57. 108. 21. 62. 54. 65. 83. 27. 56.
12. 57. 93. 320. 299. 90. 23. 99. 495. 28.
94. 316. 49. 13. 43. 89. 15. 59. 93. 50. 15.
23. 99. 16. 56. 94. 17. 36. 27. 24. 28. 20.
27. 95. 83. 12. 42. 96. 28. 50. 13. 94. 85.
27. 100. 102. 30. 266. 29. 509. I. 24. 42. 92.
67. 60. 91. 27. 83. 13. 320. 14. 56. 36. 48.
60. 44. 90. 232. 50. 27. 90. 27. 84. 99. 44.
20. 30. 24. 28. 483. 24. 30. 65. 13. 89. 90.
28. 86. 28. 56. 474. 232. 23. 26. 88. 85. 28.
37. 44. 54. 30. 57. 93. 90. 28. 57. 96. 42.
29. 83. 90. 27. 94. 49. 27. 91. 220. 12. 320.
54. 60. 42. 27. 57. 23. 28. 66. 13. 92. 91.
27. 83. II. 44. 51. 35. 31. 86. 16. 13. 20.
40. 27. 99. 30. 86. 50. 130. 25. 13. 54. 19.
60. 43. 58. 66. 85. 28. 92. 26. 56. 94. 30.
52. 31. 59. 93. III. 44. 48. 67. 12. 109. 28.
83. 13. 50. 16. 85. 38. 29. 59. 94. 24. 32.
80. 99. 60. 109. 44. 51. 90. 28. 83. 13. 20.
61. 59. 99. 27. 58. 99. 376. 51. 30. 25. 62.
95. 26. 24. 28. 512. 66. 83. 26. 90. 28. 57.
94. 27. 53. 26. 59. 96. 320. 20. 60. 56. 23.

George Digbye.

C. R.

42. 93. 45. 61. 56. 90. 376. 320. 12. 23. 99.
13. 56. 94. 15. 38. 27. 89. 24. 28. 512. 90.
27. 83. 60. 57. 93. 90. 26. 48. 60. 56. 21.
27. 82. 100. 12. 94. 61. 99. 305. 90. 27. 91.
93. 28. 65. 83. 14. 20. 94. 43. 80. 99. 28.
376. 320. 34. 28. 52. 53. 29. 90. 24. 27. 90.
497. 90. 376. 22. 62. 56. 90. 28. 84. 94. 27.
86. 318. 28. 90. 42. 48. 90. 27. 87. 13. 42.
52. 53. 26. 25. 42. 12. 93. 27. 54. 30. 56.
95. 28. 56. 99. 60. 109. 27. 102. 59. 28. 65.
27. 83. 90. 60. 56. 26. 42. 57. 91. 94. 84.
102. 43. 95. 29. 376. 20. 29. 90. 93. 246.
24. 26. 96. 60. 99. 93. 22. 29. 394. 90. 27.
83. 12. 347. 90. 28. 49. 61. 56. 48. 13. 84.
26. 65. 60. 56. 22. 30. 394. 90. 60. 57. 14.
49. 94. 27. 90. 91. 28. 34. 26. 84. 12. 14.
20. 27. 88. 93. 29. 60. 102. 27. 83. 94. 100.
86. 32. 42. 48. 27. 90. 93. 23. 43. 95. 20.
109. 25. 26. 89. 88. 100. 92. 394. 562. 24.
28. 99. 60. 42. 95. 14. 49. 48. 26. 84. 28.
57. 478. 376. 20. 27. 94. 84. 14. 42. 93. 32.
43. 48. 26. 90. 95. 99. 87. 12. 109. 394. 20.
27. 48. 13. 13. 102. 60. 45. 96. 29. 90. 94.
27. 84. 27. 88. 60. 50. 102. 28. 93. 81. 99.
42. 58. 45. 87. 61. 46. 96. 16. 65. 85. 27.
90. 25. 57. 484. 448. 509. 90. 99. 83. 50.
361. 394. 51. 61. 56. 13. 98. 61. 46. 96. 43.
22. 109. 394. 509. 90. 60. 100. 40. 14. 42.
94. 60. 44. 95. 15. 89. 90. 27. 92. 22. 28.
91. 93. 32. 126. 376. 255. 94. 61. 98. 93. 26.
90. 92. 61. 83. 96. 26. 92. 22. 40. 62. 91.

George Digbye.

C. R.

30. 92. 409. 13. 18. 48. 27. 92. 376. ∠9. 12.
255. 83. 102. 89. 90. 44. 83. 52. 12. 42. 90.
12. 65. 83. 27. 90. 33. 13. 43. 92. 15. 56.
93. 99. 57. 65. 27. 99. 23. 27. 87. 30. 33.
48. 27. 104. 42. 61. 57. 65. 48. 99. 88. 66.
85. 61. 34. 60. 56. 24. 28. 42. 48. 12. 27.
90. 93. 27. 66. 50. 99. 91. 92. 28. 48. 60.
56. 51. 61. 83. 23. 84. 29. 24. 30. 51. 17.
409. 24. 26. 56. 121. 99. 56. 27. 90. 20. 48.
12. 42. 83. 21. 43. 89. 27. 52. 29. 56. 93.
65. 48. 99. 90. 12. 90. 91. 99. 85. 28. 394.
49. 61. 56. 56. 29. 66. 60. 99. 102. 60. 42.
93. 95. 43. 87. 29. 86. 24. 27. 92. 20. 60.
56. 45. 27. 22. 94. 102. 87. 32. 88. 27. 95.
66. 60. 102. 87. 22. 29. 90. 93. 32. 86. 13.
43. 92. 60. 59. 562. 17. 27. 88. 93. 28. 16.
84. 86. 30. 92. 93. 26. 26. 95. 48. 30. 66.
62. 85. 95. 27. 99. 87. 20. 40. 60. 12. 92.
110. 376. 85. 16. 67. 62. 84. 95. 28. 83. 28.
57. 95. 60. 99. 96. 27. 25. 43. 49. 48. 42.
38. 30. 56. 22. 29. 320. 90. 30. 56. 56. 43.
54. 28. 59. 95. 89. 25. 27. 90. 61. 57. 13.
49. 95. 29. 90. 91. 30. 92. 102. 83. 95. 60.
102. 94. 27. 48. 13. 54. 17. 95. 45. 27. 86.
28. 26. 93. 22. 27. 66. 60. 83. 94. 27. 102.
83. 13. 28. 90. 95. 27. 22. 40. 60. 42. 92.
109. 65. 62. 99. 83. 320. 409. 90. 80. 102.
28. 92. 60. 56. 93. 12. 90. 92. 28. 90. 23. 54.
65. 48. 30. 52. 26. 57. 95. 24. 42. 95. 96. 32.
90. 232. 92. 28. 91. 48. 32. 95. 83. 28. 88.
24. 29. 22. 87. 28. 17. 57. 22. 28.

George Digbye.

Instructions for the Employment of Wheatstone's Cryptograph.

[From a Pamphlet published to accompany an instrument called
"The Cryptograph."]

A CIPHER which at the same time should be perfectly secure and easy in its application is a desideratum; and these combined advantages can only be obtained by means of an instrument in which all the complexity necessary to ensure security shall be effected by mechanical arrangements, whilst its manipulation shall be subjected to the simplest rules. Such an instrument is now offered to the public, after its utility has been proved by extensive employment in various departments of the public service and by telegraph companies.

One thing is yet wanting to render the benefits of the Electric Telegraph complete. Letters by post are sent sealed, and their contents are conveyed, with a secrecy seldom violated in free countries, to their destination; but telegraphic messages are in general transmitted so that their contents are understood by all the officials concerned in their conveyance. This arises from several causes : some persons are totally ignorant of the principles on which any ciphers are constructed; others acquainted with particular methods are afraid to employ them in their uncertainty regarding their security, and with good reason, seeing the facility with which those in ordinary use have been detected; and others who possess what they consider to be secure methods are deterred from using them on account of the great difficulties attending their translation and re-translation.

With the aid of this instrument an extensive secret correspondence can be carried on with several persons, and a separate cipher can be employed for each correspondent. The despatches prepared by it are indecipherable by any person unacquainted with the word that may have been selected for the base of the cipher alphabet; moreover, so long as the key-word remains undivulged, even the possession of the instrument employed, or of one similar to it, would not in the least assist any endeavour to discover the translations.

The number of telegraphic messages relating to domestic occurrences are very much limited by the disinclination of parties to let their family affairs be known to officials in their neighbourhood ; and there can be no doubt were this difficulty removed, this class of messages would be considerably augmented, to the benefit of the telegraphic department as well as to that of the public.

The advantages of communication by cipher for military purposes are too obvious to be insisted upon.

INSTRUCTIONS.

Formation of the Permutated Alphabet.

Choose any name or word, in which the same letter does not recur, as " France," " Carlton," " Palmerston."

Write down beneath it the omitted letters of the alphabet in their regular order, placing them in columns thus :—

C	a	r	l	t	o	n
b	d	e	f	g	h	i
j	k	m	p	q	s	u
v	w	x	y	z		

Then write the letters as they appear in the successive vertical columns, in a single row, thus :—

c b j v a d k w r e m x l f p y t g q z o h s n i u

The letters of the alphabet of the Cryptograph are then to be placed in the above order round the inner ring, the initial letter C in this case being placed exactly below the blank of the permanent alphabet.

It is not *necessary* to adopt a word in which the same

letters do not recur. Any word whatever will answer the purpose by erasing the redundant letters. For example :—

$$S \quad e \quad b \quad a \quad s \quad t \quad o \quad p \quad a \quad l$$
$$c \quad d \quad f \quad g \quad h \quad i \quad j \quad k \quad m \quad n$$
$$q \quad r \quad u \quad v \quad w \quad x \quad y \quad z$$

This gives

$$s \ c \ q \ e \ d \ r \ b \ f \ u \ a \ g \ v \ h \ w \ t \ i \ x \ o \ j \ y \ p \ k \ z \ m \ l \ n$$

If a different key-word be adopted for each correspondent, it will be impossible for any one of them to decipher a despatch addressed to another.

If the divided letters be employed, their subsequent distribution, after the despatch is prepared, will render the discovery of the key-word impossible even on possession of the instrument. This is of importance when the cryptograph is employed for military purposes ; but when no fear is entertained of its falling into other hands, it will be more convenient to make use of the entire card circles having the cipher alphabets written upon them. A distinct cipher may thus be always ready for each correspondent, one circle being instantly changeable for another.

The Cryptograph can be secured to a desk or table by means of the pins on the lower surface. As by these means it can be made perfectly steady, the hands can be worked by one hand. It will be found preferable to use the left hand for this purpose, as thereby the right hand is left free for writing.

Rendering a Despatch into Cipher.

At the commencement, the long hand must correspond with the blank of the outer circle and the short hand be directly under it.

The long hand must be brought successively to the letters of the despatch (outer circle), and the letters indicated on the inner circle by the short hand must be written down.

At the termination of each word the long hand must be brought to the blank, and the letter indicated by the short hand also written down. By this arrangement the cipher is continuous, no intimation being given of the separation of the words.

Whenever a double letter occurs, some unused letter (as, for instance, *q*) must always be substituted for the repeated letter; or the latter may be omitted.

It will be best to divide a despatch into short sentences, and to commence each sentence with the instrument adjusted as at the beginning of the despatch. By this arrangement an error in one sentence cannot affect the other sentences.

The full-stop at the end of each sentence should be represented by a dash following the letter that is used to conceal the blank or termination of the last word.

To translate a Cipher Despatch back to the original.

Make the long and the short hands coincide at the blank as before, and then by the movement of the long hand place the short hand opposite each of the letters of the cipher despatch in succession, writing down the letters indicated by the long hand.

The rule respecting the double letters is not to be observed in turning a cipher despatch back to the original. A slight error in copying the cipher can be retrieved in the translation by trying the letters beyond, omitting previously a turn, or adding an extra one according to circumstances.

After using the instrument, if there be any occasion to make it impossible for persons about one to discover the key-word of the cipher, the letters of the cipher or inner circle should be removed. This can be effected by applying the ivory punch to the apertures at the back. The order of the cipher can be reproduced by remembering the key-word.

In cases where no fear of loss or abstraction of the instrument is apprehended, the card circles may be employed. The letters of the cipher must be written in their proper order in the divided spaces. Ciphers with different keys appropriated to each correspondent may be thus always kept in readiness, and one may be promptly substituted for another.

The chief circumstances which render this cipher so secure are these :—

1. The same letter of the despatch is represented indifferently by any letter of the cipher.

2. No indication of the number of letters that there may be in any word is afforded.

3. There is no clue to the separation of the words, as the blank is itself represented indifferently by all the letters of the alphabet.

4. The changes in the signification of a letter depend on a *regular law,* the accessory hand making in some cases a complete revolution after one letter, and in others after two, three, or more letters.

5. The permutations of the cipher alphabets are practically infinite.

Before despatching a letter in cipher, the sender should translate the different sentences in order to ascertain that no mistakes exist.

In practice only, the most important parts of a despatch need be rendered into cipher.

————

The same instrument can be employed in corresponding with several persons, as the order of the letters in the cipher on the inner circle is capable of being changed in an almost infinite number of ways.

As the secret of the cipher is the particular order or succession of the letters, it is not necessary to keep the instrument under lock and key, provided that the cipher alphabet has been removed after use.

The following despatch of the Duke of Wellington, translated into a cipher constructed by means of this instrument, will afford an exercise both for translating into cipher and re-translating into ordinary language. The key-word is *France.* The dash indicates that the long and short hands of the cryptograph are both to be brought back to the blank, so that the following sentence may be translated without running on from the preceding.

Lieut.-Gen. Viscount Wellington, K.B., to Lieut.-Gen. Hill.

SIR,   Arruda, 8th October, 1810.

P Z L S P Q R E Q A J D I T F B U F Z O H Q O S U Q U
D I K I T O R T W E Z A C M T P L E R A U E S E G S O F G
F D K H L S J I R K H F H M F A D A Y I V U O H A O B L N O
G R E J A I B K M P J Z T M J A B Q C N F P O M Y H Y R C
Z D C W B X U B Z——

Z B I L I J T E J Y S P F D L C X E T K Q A S O X O U N
N O D Q J C W E C L X P U Y I E M M C M S Y V C F P O K
W C D E D V D A G L P E E K N A G V K M N U U L S H X Y
X Y V G F Q P U Y I O R Q K L P T C Z H H K——

Z B K U P V S W Z W X A Q X D R E K T K Q A S O X O U
I R S K O M F S T I I X G W T Q J J V D Y F N A H L S I I X I
A G Q L Z X V O G N H G R B U O H Y Z O O P W V Y D D M
Q J K F M O B J P D Y V R B A W K G W S J I R J G I T O W
T V E Z B H S O S L V U N B C H Q S O T E I E B D Q M G W
H G J A M I S X F I F B B P A V P E S V C J U T A D——

P Z L P T Y V X Q X D T G L T T A F C V M H O M B I N J
K W V Y A Z O C Q L A I U K F E G F N C N F I Z H H K V Z Y
Q U G L I V E N K A H T R V F V E B W H W L R Y C M L X
W S Y S Y J H U F S F P O K E G Z R X L U B F T——

(Signed)   WELLINGTON.

---

*The * Quarter * Master * General * sends * orders *
to * Major * General * Fane * to * withdraw * the *
Cavalry * under * his * command * to * Tojal * and *
Loures *——*
I * request * you * also * to * send * a * Brigade * of *
six * pounders *.to * Sobral * de * Monte * Agraca *
to * join * the * Sixth * Division * I * also * request *
you * to * send * from * Villa * Franca * the * nine *
pounder * Brigade * to * Copua * de * Montecheque *
where * it * is * to * remain * in * reserve * and * in *
readiness * to * move * at * a * short * notice *——*
There * must * be * a * Brigade * of * Infantry * for *
the * occupation * of * the * lines * extending * from *
the * high * road * to * the * Tagus.——*

N.B. The above despatch is given nearly at length in order
to illustrate the system of cipher; but in practice only the
most important parts requiring concealment need be rendered
in cipher.

(1.) *On the Vowel Sounds, and on Reed Organ-Pipes.
By* ROBERT WILLIS, *M.A., Fellow of Caius College, and of the Cambridge Philosophical Society.*
(From the Transactions of the Cambridge Philosophical Society. Cambridge, 1829.)

(2.) *Le Mécanisme de la Parole, suivi de la Description d'une Machine Parlante. Par* M. DE KEM-
PELEN, Conseiller Aulique Actuel de sa Majesté
l'Empereur Roi. Vienna, 1791.

(3.) C. G. KRATZENSTEIN. *Tentamen Coronatum de
Voce.* Petrop., 1780.

[From the 'London and Westminster Review,' No. xi. & liv.,
Oct. 1837.]

(The article is signed C[HARLES] W[HEATSTONE].)

WE propose in this article to give an account of the various
attempts which have been made to imitate the articulations
of speech by mechanical means, and to show how far the
united labours of the philosopher and the mechanician have
advanced the inquiry respecting the physical causes upon
which these articulations depend.

Before we proceed, it may not be uninteresting to take a
glance at the various accounts recorded of the speaking statues
of the Ancients, and of those pretended speaking machines
which at different times were exhibited prior to the present
century.

The voice, as is well known, is easily and audibly transmitted to distant places by means of communicating tubes;
this simple means of conveying words to the lips of a statue

did not escape the observation of the priests of antiquity, and we have evidence that some of their oracles spoke in this manner. It is true that in general the priests considered it more safe to deliver the answers themselves, or to cause them to be delivered by women instructed for the purpose. But we read frequently that their idols spoke; and the same thing is stated with regard to the images of saints.

Oracular responses were delivered by a speaking head at Lesbos; it predicted, though in equivocal terms, the violent death of Cyrus the Great, which terminated his expedition against the Scythians. This was the head of Orpheus.

The celebrated head of Memnon, though in general it emitted only a *musical* sound when the rising sun fell upon its lips, yet it is proved, by inscriptions engraven on the colossus, that the priests, proportioning the miracle to the credulity of the votary, caused the statue *sometimes* to speak.

The speaking heads of the Middle Ages had more of a philosophic character. They were not brought forward for the purpose of imposing on a superstitious multitude the belief that they were miracles emanating from supernatural intelligence; they professed merely to be the products of human ingenuity—the inventions of men distinguished from the rest of mankind only by their superior knowledge.

Gerbert, who, under the name of Sylvester II., occupied the Papal chair from 999 to 1003, constructed a speaking head of brass. For this exertion of his ingenuity he was, of course, accused of magic, the common charge against mechanical inventors in those times.

In the thirteenth century Albertus Magnus, the extent of whose knowledge was astonishing for the age in which he lived, made a head of earthenware. It is said that his disciple, Thomas Aquinas, was so terrified when he heard it speak that he broke it to pieces, upon which the mechanist exclaimed, "There goes the labour of forty years." In the same century lived the celebrated Roger Bacon, who is reported to have made a similar automaton.

It is both amusing and instructive to recur sometimes to those crude hypotheses respecting physical phenomena which were advanced in the early dawn of science. Walchius thought it possible so to contrive a trunk or hollow pipe that it should preserve the voice entirely for certain hours or days, so that a man might send his words to a friend instead of his writing. There being always a certain space of inter-mission, for the passage of the voice, betwixt its going into these cavities and its coming out, he conceives that if both ends were seasonably stopped, whilst the sound was in the midst, it would continue there till it had some vent. " Huic tubo verba nostra insusurremus, et cum probe munitur tabel-lario committamus," &c. When the friend to whom it is sent shall receive and open it, the words shall come out di-stinctly, and in the same order in which they were spoken. From such a contrivance as this (saith the author) did Albertus Magnus make his image, and Friar Bacon his brazen head, to utter certain words. This scientific hy-pothesis powerfully and worthily supports the credit of that Popish tradition concerning Joseph's " hah ! " or the noise that he made (as carpenters usually do) in giving a blow— which is said still to be preserved in a glass bottle among other ancient relics.

The celebrated Jesuit, Athanasius Kircher, who lived in the seventeenth century, and was, if not the inventor, the first describer of the Æolian harp, the speaking trumpet, and many other important acoustical inventions, assures us, in the ninth book of his ' Musurgia Universalis,' &c., that a statue might be made, perfectly isolated, the eyes, lips, and tongue of which shall have a motion at will, which shall pro-nounce articulate sounds, and which shall appear to be alive. He had an intention of making one of this kind for Queen Christina, but he was prevented, it is said, either by want of time or by the expense.

After this time speaking machines of this kind were fre-quently exhibited in different parts of Europe, and the means of communication were sometimes so ingeniously concealed as to excite the astonishment, and impose upon the credulity, of even men of learning. A Thomas Irson exhibited a speak-ing head, which excited the wonder of Charles II. and his court. It answered in several languages to questions whispered in its ear. When the astonishment was at its height, one of the pages discovered in the adjoining room a Popish priest, who answered through a pipe.

In the last century a figure of Bacchus seated on a barrel was exhibited at Versailles. It pronounced, in a loud and intelligible voice, all the days of the week, and wished the company good day. Many persons were deceived by it, because the owner of the machine allowed them to inspect the inside of the figure and the barrel, where nothing was perceived but organ-pipes, bellows, wind-chests, wheels, cylinders, &c. But the deception did not last long. A person more inquisitive than the rest discovered a false wind-chest, in which a dwarf was concealed, who articulated the words, the sound of which was conveyed by means of a tube to the mouth of the figure.

Professor Beckmann describes a similar pretended automaton which he saw. The figure was that of a Turk placed on a box, which was filled, as in the former case, with pipes, bellows, &c.; but the voice was communicated from a person in the adjoining room.

In consequence of the prevalence of these deceptions, suspicion was so much excited whenever a speaking machine was announced, that when De Kempelen first made public his invention in 1783, a pamphlet was published, entitled 'The Speaking Machine and Chess Player Exposed and Detected.' De Kempelen, we may remark, was the inventor of the latter curious figure; the sole merit of which, as he himself has acknowledged, consisted in the ingenuity of the deception.

As a specimen of the speculations respecting the mechanism of speech which were prevalent about a century and a half ago, and which then supplied the place of observation and experiment, we may mention a small work on this subject, written by Van Helmont, the celebrated alchemist. Its translated title is, 'A short Explanation of the true natural Hebrew Alphabet, which at the same time supplies a method by which those who are born deaf can instruct themselves, not only how to understand others when they speak, but like-wise how to speak themselves.' Van Helmont in this tract endeavours to establish that all the letters of the Hebrew alphabet are written in exact imitation of the position which the tongue assumes in pronouncing them. He then goes further, and proves to his own satisfaction that the letters of the alphabet ought to follow each other in the order they do, and in no other; because the tongue, whilst finishing the pronunciation of one letter, is already in the position for commencing the following. His imagination has repre-sented the tongue in the most extravagant contortions. His work is illustrated by thirty-four plates.

When absurd speculations like this supplied the place of accurate knowledge, it cannot be supposed that any progress was made towards constructing a speaking machine pro-perly deserving the name. Such an invention requires, above all, for its accomplishment, a full and exact knowledge of the mechanism of speech.

The first experimental investigation of the articulations of speech dates from 1779, when the Imperial Academy of St. Petersburg proposed for their annual prize the following two questions :—

First. What is the nature and character of the sounds of the vowels A E I O U, so different from each other?

Secondly. Can an instrument be constructed like the *vox humana* pipes of the organ, which shall accurately express the sounds of the vowels?

These questions were answered and the prize gained by Professor C. G. Kratzenstein. After examining the positions of the various organs, and measuring the apertures of the lips, teeth, &c. for the different sounds, he constructed a series of tubes, which, when applied to an organ-bellows, imitated with tolerable accuracy the five vowel sounds re-quired. The forms of the tubes which he employed cannot readily be rendered intelligible without the aid of figures; four were rendered vocal by the application of freely vibrating reeds, the other in the manner of an ordinary organ-pipe.

Though Kratzenstein has the undoubted merit of having

been the first who imitated any of the articulations of speech by mechanical means, yet we cannot give him the credit of having thrown any philosophic light on the subject. It has since been ascertained, as we shall presently see, that the forms of the tubes have not that relation to the different vowel sounds which he supposed to exist; but that tubes exactly the same in form will articulate different vowels when their dimensions are altered.

About the same time that Kratzenstein was engaged in this inquiry, M. de Kempelen of Vienna was occupied in a more extensive investigation of the same subject. He attempted not only to imitate the vowel sounds, but the consonants also : at present we will confine ourselves to his experiments with the vowels. De Kempelen found that by placing a conical tube, like the bell of a clarionet, before the reed he employed, different vowels were obtained according as the bell was more or less covered. On such a tube, U is heard when the tube is nearly closed, O when it is about half closed, and A when it is entirely open. It is only when a rapid transition is made from one of these sounds to another that each can be distinctly recognized; when the cover is kept upon the tube for some time in the same position, the sound appears always to be A. De Kempelen infers from this that the sounds of speech become distinct only from the contrast which exists between them, and that they obtain their perfect clearness only when connected in entire words and phrases.

Before proceeding to the researches of Mr. Willis on the vowel sounds, it will be necessary to explain, first, in what the vowels differ from the other articulations of speech, and, secondly, in what they differ from each other.

The vowels are formed by the voice, modified, but not interrupted, by the varied positions of the tongue and lips. Their differences depend on the proportions between the aperture of the lips and the internal cavity of the mouth, which is altered by the different elevations of the tongue. The vowel sound *aw*, as pronounced long in "fall" and

2 A

short in "folly," is formed by augmenting the internal cavity by the greatest possible depression of the tongue, and at the same time enlarging the separation of the lips. Departing from this sound there are two series, the one represented in the vertical row of the following Table, the other in the horizontal row. In the first, the external aperture remains open, and the internal cavity gradually diminishes by the successive alterations of the tongue. In the second, the tongue remains depressed, but the aperture of the lips is gradually diminished. There is also an intermediate series of vowel sounds, obtained by different elevations of the tongue when the lips are partially closed; these, though abounding in many foreign languages, are not used in our own; they are shown in the second vertical column.

| | | | 4·7 | |
			C²	Indefinite.
3·05	G²			
3·8	Eb²	Aw	O	Oo
2·2	Db²	Ah	Ou	
1·8	F³			
1·0	D⁴	Ae	Eù	
0·6	C⁵	A	Eú	
0·38	G⁵	E	Ü	

This table indicates all the most usually pronounced vowel sounds, but practised ears might distinguish others intermediate in each series; for each vowel may pass to the next in order, either above or below it, by imperceptible gradations. Each of these vowels may be long or short, according to the duration of its sound in a syllable.

When these vowels are sounded, the soft palate is raised so as to prevent the sound from issuing through the nasal channels; when, on the contrary, the soft palate is depressed, the partial escape of the breath through the nostrils modifies all the preceding sounds in a very evident manner. These modifications, however, are never employed in the English language.

The extreme inadequacy of our written language in its representation of the articulations of speech is very obvious. We have six characters which are called vowels, each of which represents a *variety* of sounds quite distinct from each other, and while each encroaches on the powers of the rest, many simple vowels are represented by combinations of two letters. On the other hand, some simple vowel letters represent true diphthongs, consisting of two distinct simple vowels pronounced in rapid succession; thus, *a* consists of *a* and *e*, *i* of *ah* and *e*, and *u* of *e* and *u*. Again, the literal diphthongs are mostly simple vowel sounds, as *ea* in " bleak," *ie* in " thief," &c.

This want of correspondence between the characters of our written and the sounds of our spoken language has been a great obstacle to the proper understanding of the real elements of speech. A child is taught that the letters W, H, Y, make the syllable "why"; now, if we examine the sound of this word, we shall find it to be formed by the rapid succession of the vowel sounds *U, ah, E*. In attempting, therefore, to imitate by artificial means the sound of this word, we should pay no regard to the letters of which it is formed; the elementary *sounds* alone are the objects of our attention. The same observation is generally applicable to the words of our language.

Turn we now to the recent and interesting researches of Mr. Willis on the vowel sounds. His first object was to verify De Kempelen's observations. Having obtained the vowel sounds U, O, A very distinctly, he found that on using a shallower cavity than De Kempelen had employed, and sliding a flat board on the top of the funnel, he could obtain the entire series in the order U, O, A, E, I. The success of this experiment induced Mr. Willis to try the effect of cylindrical tubes of different lengths, and, for the more complete investigation of this case, he constructed an apparatus consisting of a tube or *port-vent*, bent at right angles to connect

it with the wind-chest of the bellows. At the extremity of
this tube is a plug, in which is fitted a freely vibrating reed.
Another tube of equal length is fitted to this, so ·that, on
being gradually drawn out, it will show the effect of applying
to this reed a cylindrical tube of any length, from nothing to
about two feet; then if the tube be pushed back and another
joint equal to it be fitted on, a fresh drawing out of the tube
will show the effect of any length from that of the first tube
to double the length, and in this way with different joints
we may go on to any length we please. The result of experiments with this apparatus we will describe in general
terms.

Let the line *a b c d* represent the length of the pipe measured from *a*, and take *a b, b c, c d*, &c. respectively equal to
the length of the stopped pipe in unison with the reed employed; that is, equal to half the length of the sonorous wave
occasioned by the reed.

No. I.

IE A O U	U O A EI	IE A O U	
•	•	•	•
a	*b*	*c*	*d*

No. II.

IE A O U	U O A EI	IE A O U	U O A EI	IE A O U
•	•	•	•	•
a	*b*	*c*	*d*	*e*

No. III.

	IE A O	U	
•	U • O	A EI •	
a	*b*	*c*	

The lines in these diagrams must be considered as measuring-
rods placed at the side of the tube, with their extremity *a*
opposite to the point at which the reed is placed; the letters
and other marks show the effects produced when the open
extremity of the pipe reaches the points so marked, and the
distance, therefore, from these points to *a* respectively is the
length of the pipe producing the effects in question.

Now, if the pipe be drawn out gradually, the tone of the reed, retaining its pitch, first puts on in succession the vowel qualities IE A O U ; on approaching *c* the same series makes its appearance in *inverse order*, as represented in the diagram; then in direct order again, and so on in cycles—each cycle being merely the repetition of *b d*, but the vowels becoming less distinct in each successive cycle. The distance of any given vowel from its respective centre points is invariably the same in all.

If another reed be tried which gives a higher sound, and whose wave, represented by the second figure, is consequently shorter, the centres of the cycles *a, c, e*, &c., will be at the distance of the sonorous wave of the *new* reed from each other, but the vowel distances (and this is an important point to be remarked) will be exactly the same as before. It may be stated as a general rule, that if a certain length added to the reed gives to its sound a determinate vowel character, the same vowel is heard when the same length is added to or subtracted from a length equal to any simple multiple of the original wave $(2\,n\,a + v)$.

When the pitch of the reed is high, some of the vowels become impossible. For instance, let half the wave of the reed (No. III.) be less than the length producing U. In this case it is found that the series never ascends higher than O, and that on passing *b*, O of the reverse series commences. In like manner, if reeds of higher notes be taken, more vowels are cut off. This is exactly, Mr. Willis remarks, the case in the human voice; female singers are unable to pronounce U and O in the higher notes of their voice; for example, the proper length of the pipe for O is that which corresponds to the note C on the third space in the treble, and beyond this note in singing it becomes impossible to pronounce a distinct O.

We have added to the table of vowel sounds at page 354 the lengths of a stopped pipe which Mr. Willis has ascertained to correspond with the different vowel-sounds, and also the musical notes which these pipes would give were they sounded themselves.

Mr. Willis further states that cylinders of the same length give the same vowel, whatever be their diameter or figure; and that, so far as he has tried, he has always found that any two cavities yielding the identical note when applied to the embouchure of an organ-pipe, will impart the same vowel-quality to a given reed, or, indeed, to any reed provided its note be flatter than that of the cavity.

From these experiments it is evident that the forms stated by Kratzenstein, as producing the different vowels, are perfectly arbitrary. The entire series of vowels can be produced from tubes of either of his forms by merely changing its dimensions.

Mr. Willis finally concludes, from his experiments, that the vowel quality, added to any sound, is merely the coexistence of its peculiar note with that sound; this accompanying note being excited by the successive reflections of the original wave of the reed at the extremities of the added tube.

This view of the matter naturally associates the phenomena of vowel-sounds with those of multiple resonance, a subject first investigated by Professor Wheatstone.

The phenomena of simple or unisonant resonance are so well known that we need only call attention to one or two of the most striking facts. If a vibrating body be brought near a column or volume of air, which would be capable of producing the same sound were it immediately caused to sound as an organ-pipe or otherwise, then the sound of the vibrating body is greatly reinforced, as when an harmonica-glass is brought before an unisonant cavity, or when a tuning-fork is placed at the embouchure of a flute, the apertures of which are stopped, so that, if blown into, the flute would sound the same note; in the latter case the experiment is more remarkable, as the sound of the tuning-fork is scarcely itself audible. The same effect takes place when the cavity of the mouth is adjusted so as to be in unison with the tuning-fork.

We now come to the new facts of resonance. A column of air will not only enter into vibration when it is capable of producing the *same* sound as the vibrating body which causes the resonance, but also when the number of the vibrations which it is capable of making is any simple multiple of that of the original sounding body, or, in other words, if the sound to which the tube is fitted is any harmonic of the original sound.

For instance, if a tube closed at one end by a movable piston is taken, and its length adjusted to six inches, it will resound as an unison to a C tuning-fork; and if we shorten the length of the tube to three inches, the unison will no longer be reciprocated, but its octave will be heard. The same effect is produced by altering the cavity of the mouth.

By placing a vibrating lamina which produces a lower sound than can be obtained from a tuning-fork (the tongue of a Jew's harp, for instance), and successively adjusting the column of air so as to be one half, one fourth, one fifth, &c. of the column reciprocating the fundamental sound, the octave, twelfth, double octave, seventeenth, &c. will be produced. The relative numbers, considering the vibrations of the tongue as unity, are 1, 2, 3, 4, 5, &c. The mouth produces precisely the same effect as this changeable tube does, and all the beautiful sounds which Mr. Eulenstein manages with so much skill [on the Jew's harp] are produced by this means; they are multiple resonances of the column of air, and not the vibrations of the tongue itself, as was formerly supposed.

Similar results are obtained when the vibrating tongue of an Æolina is brought before a tube, and its length is altered; and this case resembles Mr. Willis's arrangement. The same multiple resonances are produced also when the cavity of the mouth is substituted for the tube. In these cases the fundamental sound is louder than when the Jew's harp is employed. The sound of the larynx itself may be substituted for that of the vibrating tongue, and similar harmonic sounds will be heard; this experiment may easily be made by placing a piston-tube before the mouth while the voice continues to sound rather a low note.

About two years ago a young man named Richmond exhibited a novel kind of musical performance with the voice.

On examining the circumstances under which the sounds were produced, it was ascertained that the continued sound or drone was produced by the larynx, and that he had acquired the art of adjusting the cavity of the mouth so as to fit it for resounding to any multiple. In this way he was able to command these subordinate sounds in any succession, and even to dwell upon them; and he could thus perform a great number of airs.

Some kinds of sounds are better suited to produce these multiple resonances than others; and it is an universal fact that wherever these subordinate sounds can be distinguished, there also the vowel qualities are heard; and, reciprocally, when a sound puts on successively different vowel qualities, these multiple resonances are audible. The tongue of a Jew's harp, which so readily gives rise to these subordinate sounds, is obedient not only to the vowel sounds, but to almost all the articulations of speech. The free reed, or Æolina tongue, when it is such as can enter readily into vibration, is affected in a similar manner; but when it is too rigid, though it may produce as clear a musical sound as before, the multiple resonances and the vowel qualities are equally lost, not, perhaps, because they do not exist, but because they are overpowered by the original sound of the reed.

We do not mean to assert that each multiple resonance is a distinct vowel sound: but we infer that, when a tube is added to a reed or vibrating tongue, whatever may be its length, a quality is added to the original sound, which depends on the feeble vibrations of the air in the added tube; these increase in number in proportion to the shortness of the tube; and when the number of vibrations thus excited is any multiple of the original vibrations of the reed, the energy of the resonance is so greatly augmented as to produce the effect of a superadded musical sound.

Thus it is evident that the vowel qualities and multiple resonances are different forms of the same phenomena.

In considering the consonants to which we may now proceed, we may apply to them the same remark which has been already applied to the vowels. In attempting to class them, for the purpose of ascertaining in what respect these sounds agree with or differ from each other, we must pay no attention to the alphabetic characters which so inefficiently represent them, or to the names by which these characters are designated. For we find that a simple character sometimes represents a compound sound, as G and X; and a simple sound is as frequently denoted by two letters, as Sh, Th, Dh, Ng, in "song," &c. The same character often represents more than one sound; for instance, S has four different powers in *sea*, hi*s*, vi*s*ion, and *s*ure. Many other redundancies and deficiencies might be noticed, but these will suffice to show the danger of classifying letters instead of sounds. The letters *f, v, r, l, p, t, k, d, m*, and *n* are, however, generally constant in their signification.

CONSONANTS.

		Mutes.	Sonants.	Narisonants.
Explosive.		P	B	M
		T	D	N .
		K	G	Ng
Continuous.		F	V	
			Y	
		Sh	J	
		S	Z	
		Th	Dh	
			R	
		Ll	L	
		Kh	Gh	

In this Table all the elementary consonant sounds are

arranged, indicated by those characters by which they may be most readily recognized. These sounds will be best compared by articulating them all uniformly, followed by the same vowel, as *Pe, Be,* &c.

It would occupy too much space to explain the particular positions of the different organs during the pronunciation of all these sounds; it must suffice at present to state that for the six sounds characterized in the above table by the term explosive, the breath or voice is entirely obstructed, while for all the others it is allowed to pass, but with partial interruptions.

But there is one comprehensive and important division of these articulations which we cannot pass over in silence. It is found that when the organs are in the proper position for uttering any one of the articulations of the first vertical column, without the slightest change being made in that position, a different quality may be given to it, according as it is accompanied or not by the sound of the larynx, or the voice properly so called. It is in this respect alone that F differs from V, &c.

Every sonant has thus its corresponding mute. This rule seems to be violated, however, with regard to the sounds L and R. But though the corresponding mutes of these sounds are not in the English language, they are to be found in other tongues. The sound denoted in Welsh by Ll is the mute corresponding to our L; and though it is extremely difficult to catch this sound from merely hearing it pronounced, attention to the rule will enable any one to articulate it properly in a few minutes. The organs must be in the proper position for articulating L, but the voice must not sound; or, in other words, the L must be whispered, and then followed by the vowel; as, for example, Llewellyn, Llangollen.

The importance of this rule is obvious; for if in the construction of a speaking machine the voiceless articulations indicated in the first vertical column of the preceding Table were perfectly obtained, all the sounds in the second column could be immediately produced by adding the vocal sound, or what is substituted for it, to those of the first column. A single motion or key would therefore be sufficient to add eleven new articulations to the utterance of the instrument.

The positions of the organs for the three sounds M, N, Ng, are exactly the same as for B, D, G, only, in the former, the soft palate is depressed so as to allow the voice to escape through the nostrils; by this the articulation is rendered continuous. If in a speaking machine, therefore, these three sounds were perfect, the addition of a single key would convert them into six elementary sounds. To imitate all the usual articulations of speech, a machine would not require such complicated motions as might at first be supposed: thirteen keys would be sufficient to command the twenty-six consonant sounds of the above table. The difficulty in the construction of such a machine is therefore reduced to the perfect imitation of these eleven sounds and the vowels, which we have already seen are not beyond our command. How far these difficulties have been overcome we shall now proceed to show.

De Kempelen's machine consisted externally of a square box, and a bellows, of the ordinary construction, placed on a board. The bellows were pressed down by the right arm, and it expanded itself, by means of a weight or spring, when the pressure was removed. There were two openings in the box, for the purpose of introducing the hands to act on the keys &c. within, and numerous small round apertures concealed by silk, to prevent stifling the sound. When this cover was removed, a small box or wind-chest was seen communicating with the pipe of the bellows at one end, and with an India-rubber bell or funnel, from which the sound ultimately issued, on the other; on this wind-chest were various keys to be touched by the fingers of the right hand for the sounds S, Sh, R, &c. The sound was produced by an ivory or brass reed, covered, on the side which vibrated against the edges of the aperture, with very thin leather; this reed, representing the larynx of the vocal organs, was placed within, between the wind-chest and the narrow part of the India-rubber funnel or mouth. The sound P was produced by suddenly removing the left hand from the front of the mouth, which it had previously completely stopped; the sound B by the same action, but, instead of closing the mouth completely, a very minute aperture was left, so that the sound of the reed might not be entirely stifled; M was heard on opening two small tubes, representing the nostrils, placed between the wind-chest and the mouth, while the front of the mouth was stopped, as for P. A few vowel modifications, and the sounds F and V, were produced by modifying the form and size of the aperture of the mouth by the left hand; the continuous consonants S and Sh were obtained by causing the wind to pass through small tubes of particular forms, the passage of the wind being governed by keys; and R was imitated by occasioning a vibration or trembling of the reed when an appropriate lever was depressed*.

"In the space of three weeks," says De Kempelen, "any one may acquire wonderful skilfulness in performing on the speaking machine, especially if he applies himself to the Latin, French, and Italian languages; for the German language is much more difficult, on account of the consonants which so frequently occur in it, the sibilant sounds, and the mute letters which so often terminate the words. I can pronounce immediately every French or Italian word I am asked; a long German word, on the contrary, costs me much trouble, and it is rare that I perfectly succeed. As to entire phrases I can produce but few; and these must be short, because the bellows is not sufficiently large to furnish the necessary quantity of wind. For example : *Vous êtes mon ami—Je vous aime de tout mon cœur* ; or, in Latin : *Leopoldus secundus—Romanorum imperator—semper Augustus*, &c. I am perfectly convinced," says the author, "that without

* A speaking machine, made by Professor Wheatstone from De Kempelen's description, with some improvements, is at present in the collection of Philosophical Instruments of King's College, London.

much skill the machine may be arranged with keys like the harpsichord, or organ, so that it would be much more easy to play it than in the present manner; but this is another step towards perfection which I leave to those who may be inclined to give some attention to this new invention in its infancy, and who, by their meditations and labours, are willing to carry it further."

De Kempelen's is not the only speaking machine which has been made; the Abbé Mical, a celebrated French mechanician, constructed two colossal brazen heads, which are said to have uttered not only words, but entire phrases; he submitted these master-pieces of his ingenuity to the Academy of Sciences in July 1783, in the same year which De Kempelen was at Paris with his machine. No details of the construction of these were ever published; and the inventor is said to have broken them to pieces on being disappointed of the reward which, on the recommendation of the Academy, he had expected from the Government. He died shortly after, in 1789. We can, therefore, only judge of the merits of these automata from the accounts of writers who saw them. From the description which Rivarol gives, it appears that Mical applied to one of the heads a set of keys which were acted upon by pins fixed on a cylinder, so that a determinate number of phrases were produced, as tunes are upon a barrel-organ; the intervals between the words, and the accentuation, Rivarol states to have been marked correctly. The other head had a key-board, like that of a pianoforte, the keys corresponding to the different sounds and tones of the French language, which were reduced to a small number by an ingenious method peculiar to the inventor. "With a little practice and skill," says Rivarol, "any person may talk with the fingers as with the tongue; and he may give to the language of the heads, the rapidity, the stops, and in fine all the characters which a language can possess which is not animated by the passions. Foreigners may take the 'Henriade,' or 'Telemachus,' and cause them to be recited from one end to the other, by placing the volumes on the vocal instruments, as we place the book of an opera on a pianoforte."

No doubt Rivarol's enthusiasm has led him greatly to overrate the performance of these automata. It would be a very easy matter to add either a key-board or a pinned cylinder to De Kempelen's instrument, so as to make the syllables which it utters follow, each with their proper accentuations and rests; but unless the articulations were themselves more perfect, it would not be worth the trouble and expense. According to a more sober writer, Mical's, in this respect, was not superior to De Kempelen's invention. Vicq d'Azyr, the celebrated anatomist, gives the following account of the Abbé Mical's heads, in his report upon them to the Academy :—

"The heads," says this writer, "covered a hollow box, the different parts of which were connected together by hinges, and in the interior of which the inventor had disposed artificial glottises of different forms over stretched membranes. The air passing through these glottises, was directed on these membranes, which gave sounds of different pitches; and from their combination there resulted a very imperfect imitation of the human voice."

It is a curious circumstance that Kratzenstein, De Kempelen, and the Abbé Mical were all employed in their researches at the very same time, though in very different parts of Europe. Since that period no progress has been made in imitating the articulations of speech, and even the experiments of these mechanicians have been almost forgotten. The advantages that would result from the completion of a speaking machine render the subject worthy the attention of the philosopher and the mechanician. Dr. Darwin supposed that they might be made hereafter with voices so loud and sonorous as to command an army or instruct a crowd; but the most important uses to which they will perhaps in future times be applied, will be to fix and perpetuate the pronunciation of different languages.

There are no doubt a great many difficulties yet to be

contended with before we can succeed in perfectly imitating the articulations of speech; but the partial success of the attempts of which we have laid an account before our readers ought to encourage further trials. It is not too much to say, in the words of Sir David Brewster, " We have no doubt that, before another century is completed, a talking and a singing machine will be numbered among the conquests of science."

On the Vibrations of Columns of Air in Cylindrical and Conical Tubes.

(ROYAL INSTITUTION.)

[From the 'Athenæum,' March 24, 1832, p. 194.]

March 15.—Mr. C. Wheatstone gave a lecture " On the Vibrations of Columns of Air in Cylindrical and Conical Tubes." After enumerating the various modes by which columns of air may be put into sonorous vibration, and which constitute so many classes of wind instruments of music, the lecturer proceeded to detail the principal results of Bernouilli's "Theoretical Investigations." When a column of air in a cylindrical tube, open at both its ends, produces the lowest sound it is capable of rendering, according to this theory, the motions of the particles of air are made in opposite directions, alternately to and from the central point or node, where the variations of density are greatest. Mr. Wheatstone gave the following new and decisive experimental proof of this theoretical deduction. He took a tube bent nearly to a circle, so that its ends were opposite to each other, with a small space between them; he then took a glass plate capable of making the same number of vibrations as the air contained within the tube, and, causing it to sound by drawing a violin bow across it, placed it at equal distances between the two orifices, so that the impulses of the vibrating surface were made at the same instant of time, *towards* one, and *from* the other end of the tube; as might be expected from the theory, these effects neutralizing each other, no resonance took place, and the air in the tube remained at rest. But when (the two halves of the tube moving round each other by means of a joint) the orifices were brought opposite to different vibrating parts of the plate, so that the impulses were made at the same instant towards or from both the orifices, the column of air powerfully resounded.

He then proceeded to show that when a column of air sounded any other than its fundamental note, it did so in consequence of a division of the column into parts of equal length separately vibrating, in the same manner as the harmonic sounds of a string have been explained; that the air may vibrate when divided into any number of aliquot parts, and the corresponding sounds are as the series of natural numbers, 1, 2, 3, 4, 5, 6, &c.; that, at the limits of each vibrating part, a communication may be made with the atmosphere by an aperture, or even by entirely separating the tube, without any injury to the sound; that, in each mode of division in which there is a node in the centre (*i. e.* in each alternate mode), a solid partition may be placed at the centre of the tube, dividing it into two equal parts, each giving the same sound as the entire tube when the partition was removed; and that, consequently, a tube stopped at one end gives a series of sounds corresponding to the progression 1, 3, 5, 7, &c. of a pipe double its length and open at both ends.

After verifying these established results, the lecturer proceeded to show the erroneousness of the prevailing opinion, stated by Chladni and others, "that the end at which a tube is excited into vibration must always be considered as an open end, even if it be placed immediately to the mouth, as in the horn and trumpet." He showed that a cylindrical tube gave the same fundamental sound and the same series of harmonics, when it was excited, as a horn, or with a reed at one end, the other end being open, as when it was excited, like a flute or flageolet, at one end, the other end being shut. In proof of this he adduced the cremona pipe of the organ, which is a cylindrical tube, one half the length of the open diapason pipe, which gives the same note; and the clarionet, which is also a cylindrical tube (the conical bell which terminates it being merely a useless appendage), giving a fundamental sound, and an octave below that of a flute of equal

2 B

length, and the series of harmonics of a tube closed at one end.   He then adverted to the circumstance that, in all cases of the production of sound at the closed end of the tube, the tone is invariably more powerful than when the sound is produced at the open end of the same tube; and explained that in the one case the impulses are made at that part of the air where the condensations and dilatations are greatest, and in the other case where these variations of density are least.   This point was illustrated by some experiments with the flame of hydrogen gas, by which means a column of air can be excited into vibration at any point between the open end and the node, with a corresponding alteration of intensity.   At the orifice of the tube, the smallest possible flame is sufficient to excite the sound, which, however, ceases if the flame be made to move towards the node (*i. e.* the centre of a tube open at both ends, or the closed end of a tube stopped at one end); but if, at the same time that the flame is advanced in the tube, it be also enlarged in volume, the sound continues, and with increased intensity; by continuing to move the flame towards the node, and at the same time to proportionally enlarge the volume, the sound progressively increases in loudness until it attains its maximum at the node.

By analogous experiments on the sounds produced by the flame of hydrogen gas, in tubes of different diameters, Mr. Wheatstone showed that the loudest tone is produced in tubes of the smallest diameter (when a certain limit is not exceeded), which is exactly the reverse of the generally adopted opinion; and he stated the following to be the general results of numerous experiments—that the flame is required to be larger as the length of the tube is greater, as its diameter is less, and as the point of excitation is nearer the node.

The lecturer went on to give an exposition of the laws of the vibrations of the air in conical tubes, and explained that the air in a tube of this form, excited into vibration at its closed end or the summit of the cone, gave the same fundamental sound and the same series of harmonics as a cylindrical tube open at both ends.   To this similarity of effect

he ascribed the general error of considering all wind instruments as tubes open at both ends. To illustrate this subject,
he showed that the trumpet, French-horn, and hautbois pipes
of the organ, all being conical tubes, gave the same sound as
the cremona pipe (a cylindrical tube, excited precisely in the
same way), which is only one half their length. He compared also the hautbois, which is a conical tube, with a
clarionet, which is a cylindrical tube of the same length, and
proved that in the former the fundamental sounds were the
same, absolutely and relatively, as in the flute (a tube open
at both ends, of the same length), and that in the latter
they were the same with those of a stopped pipe of the same
length.

The lecture concluded with a variety of experiments on
the sounds of isolated portions of conical tubes, the situations
of their nodes, &c., with reference to their practical applications, which we cannot spare space to detail.

If a tube be cylindrical it will have only the odd numbers
as harmonics; the cone must be as 4 to 1 of the narrow end
to have all the harmonics.

INDEX.

Entries in Italics refer to complete Essays or Papers on the subject of reference.

A COUSTIC Figures: On the Figures obtained by Strewing Sand on Vibrating Surfaces, commonly called, 64–83; first discovered by Chladni, 64; table of Chladni's results, 65, 66; produced by simplest vibrations of rectangular surfaces, 67; produced by superposition of two similar modes of vibration, 68; resultant figures of two superpositions, 70; of four superpositions, 71; modes of superposition, 73; perfect resultant figures of square surfaces, 74; comparison of calculated figures with those obtained experimentally, 75; imperfect resultants, 76; modified by irregularity of plates, 77; rearrangement of Chladni's table, 78, 79; theories of Bernouilli, 80, of Chladni, *ib*., of Dr. Young, 81, of the Webers, *ib*.; how far obtainable on plates of wood, 81, 82.

Adhémar, Count d', on a relation in cube numbers, noticed, 317.

Aguilonius, on law of direction for monocular vision, 250.

Air, Column of, on the Resonances or Reciprocated Vibrations of, 36–46; resonant to tuning-forks, 37; to wind instruments, 37; its harmonic subdivisions, 38; law of resonance of, 42; simultaneous reciprocation of several sounds, 45; beats, how produced, *ib*.; vibrates when its sound is a harmonic of original sound, 349; experiments on, in cylindrical tubes, 368; vibrations of, in tubes with open and closed ends, 369, excited by hydrogen, 370; laws of vibration of, in conical tubes, 371.

Alembert, D', on law of direction for monocular vision, 249.

Alphabet, inadequate for expression of of sounds of the language, 355, 361.

Aluminum, voltaic position of, 158, 159.

Antinori, Cav., his experiments on the thermo-electric spark, 134.

Arcy, Chevalier d', on the duration of the sensation of sight, alluded to, 28, *n*.

Audibility, limits of, 34.

Audition, experiments on, 30–35; augmentation of sounds produced internally, 30, 31; augmentation of sounds produced externally—the microphone, 32, 33; consonance of sounds, 34; acute sounds rendered inaudible, *ib*.; limits of audibility, *ib*.; acute sounds increased in intensity, *ib*.; intensity of notes varied, 35.

Bacon, Lord, his experiments on resonance, 38, *n*.; on the transmission of sound, 47; on apparent superiority of monocular vision, 239, *n*.

Beats, how produced, 45.

Bernouilli, James, his investigations of acoustic figures, 80; *experimental verification of his theory of wind-instruments,* 220.

Biot, on resonance, alluded to, 37, *n*.; on the transmission of sound, noticed, 48.

Botto, Prof. G. D., his experiments on the chemical action of the thermo-electric pile, 136.

Bouvard, on the transmission of sound, noticed, 48.

Breguet, claims invention of electro-magnetic chronoscope, 143; is consulted by de Konstantinoff, 149; constructs a chronoscope, 150.

Brewster, Sir D., his explanation of Purkinje's experiment rejected, 221; on conversion of relief, 243; on law of direction for monocular vision, 249.

Chapman, Capt., introduces Wheatstone's chronoscope into Woolwich, 145.

Charge, instrument for effecting, 170; how influenced by temperature, 184; time of charging augmented by interposed resistances, 186; means of accumulation of, 196–205.

Charles I., translation of document written by him in cipher, 326–329.

Chladni, on the oscillations of phonics, 2, 7; on resonance, alluded to, 37, n.; on the velocity of sound, 48; his discoveries of acoustic figures, 64, 65, 80; on vibratory motions, criticised, 303.

Christie, on the measurement of resistance in voltaic circuit, 129, n.

Chronoscope électromagnétique, note sur le, 143–151; claimed by Breguet and de Konstantinoff, 143; invented by Wheatstone, ib.; description of, ib.; publication of invention, 144; communicated to de Konstantinoff, 146; modifications of, 147, 149, 150; instrument invented by Breguet, 150; principle already applied by Wheatstone, 151.

Cipher, Interpretation of historical document in, 321–341; history of, 321; principle of cipher, 323; key to literal cipher, 324; key to vocabulary, 325; translation of documents, 326–329; text of document in cipher, 330–341.

Circuit, Voltaic, an Account of several new Instruments and Processes for determining the Constants of, 97–133; Ohm's theory of, 98–102; resistance of, how determined, 103, 104; instruments for measuring resistance, 105–107; rheostat for measuring small resistance, 108; coils for measuring great resistance, 110; resistance of interposed body determined, 111; of wire of galvanometer, 112; sum of electromotive forces in, 112, 113; resistance of rheomotor determined, 118–121; resistance of liquids, how measured, 122–124; small differences of resistance, how measured, 126–128; comparative changes of rheomotors, how observed, 129; comparison of intensities of currents, 130–132; relations between degrees of force and of galvanometric scale determined, 132; influence of the earth, 154, 155; degree of tension in wire of insulated pole, 157.

Clock, description of the electro-magnetic, 138, 139.

——, polar, construction of, 287; is superior to sun-dial, 288; various forms of, 289.

Colours, Singular Effect of the Juxtaposition of, under particular circumstances, 284; two different, presented one to each eye, are not compounded, 248; apparent motion of certain patterns, 284: chiefly conspicuous in green and red, ib.; not apparent in daylight, ib.; how accounted for, ib.

Conductors of sound unaffected by length, 52; how to be connected with instruments, 53, 54; when bent, will transmit sound, 59; will transmit transverse vibrations, 61; possibility of transmitting human speech, 62.

Consonants, table of, 361; inadequately represented in alphabet, 361; sonant, have corresponding mutes, 362.

Copper, resistance to voltaic current less than iron, 175: best material for telegraphic wires, ib.

Cryptograph, Wheatstone's, Instructions for the Employment of, 342–347; benefits of, 342; mode of forming alphabet, 343; rendering into cipher, 344; retranslation, 345; advantages of system, 346; specimen in cipher, 347.

Crystals, plates of, polarize light, 290; action of uniaxal, 292; biaxal, 293; doubly refracting, 294; thickness of two films determined, 297.

Current, electric, its force determined, 99; force of, in voltaic circuit, 100; distribution of its force in circuit when a portion is diverted from limited extent of circuit, 101; comparison of intensities of, how effected, 130–132; strength can be increased without augmentation of effects of induction, 180; induced in a magnet by its own action, 212, 213.

Daniell, Prof., a modification of his battery, 107.

Darwin, Dr. R., on alteration of perceived magnitude of objects, noticed, 263.

De la Rive, Prof., on voltaic combinations, noticed, 117.

Dipolarization, plane, circular, and elliptic, defined, 291.

Discharge, inductive, instrument for effecting, 170; its amount proportionate to the electromotive force, 172; varies directly as length of wire, 173; effected simultaneously from number of wires, 174; its amount not dependent on conductivity of wire, 175; how influenced by diameter of wire and thickness of insulating coating, 175–180 ; how influenced by insulating material, 180–184 ; how influenced by temperature, 184 ; not influenced by pressure, 185 ; time of, augmented by interposed resistance, 186 ; from wire connected with earth, 187–190 ; true and apparent, considered, 191–194 ; in short wires, how measured, 191–196 ; means of accumulation of, 196–205.

Distance, inversion of, produced by pseudoscope, 279.

Du Tour, his experiments in optics, 248, 254.

Ear, experiments on the. See *Audition*.

Earth, its influence upon galvanometers in electric circuit, 154, 155.

——, *Rotation of, Note relating to M. Foucault's new Mechanical Proof of*, 303–306 ; theory of rotation of, 301 ; Foucault's experiment, 303 ; proved by further experiment, 303, 305.

Electricity, apparatus to determine velocity of, 84 ; velocity of, determined by means of revolving mirror, 85–88, 96 ; velocity of, through conductors, not previously known, 89 ; hypothesis of transmission of, 89–96 ; experiments on velocity of transmission of, 90–95 ; greatest velocity detected, 93 ; results of experiments, 93, 96 ; spark obtained from thermo-electric pile, 134, 135 ; chemical action of thermo-electric pile, 136 ; applied to clocks, 138, 139 ; applied to telegraphic thermometer, 206, 207, 209 ; of operator a source of error in experiment, 216–219.

Electro-magnetic clock, 138, 139.

Electrometer, sources of error in indications of, 216 ; affected by electrical condition of operator, 217.

Electro-motive forces, sum of, in voltaic circuit ascertained, 112, 113 ; various measures of, described, 113–118 ; their influence on amount of discharge, 172.

Electroscopic Experiments, on a Source of Error in, 216–219.

Element, voltaic, used by Wheatstone, described, 107.

Ellis, Sir Henry, letter to Wheatstone, 322.

"Enchanted Lyre," the, 8.

Enregistreur électromagnétique described, 141, 142.

Eülenstein, his performance on the jew's harp explained, 44, 45, 359.

Euler, his investigations of acoustic figures, 78.

Eyes, single object projects dissimilar pictures on retinæ, 225, 226 ; projection of dissimilar figures not previously observed, 227 ; modes of facilitating convergence of, 228 ; perspective projections of solid object, how perceived, 229 ; assisted by the stereoscope, 230, 231 ; binocular appearance of two drawings, presented one to each eye, identical with that of two real objects having projections the same as those of the drawings, 235 ; binocular perspective, 236 ; correct vision by one eye explained, 238 ; conversion of relief, 242 ; double appearance of similar pictures falling on corresponding points of the retinæ, 245 ; binocular vision of images of different magnitudes, 246 ; effect of presenting different pictures to both, 247 ; law of visible direction for monocular vision, 249 ; circle of binocular vision, 252, 253 ; theory of vision with one eye only, 254 ; abnormal adjustment of, how effected, 262, 265 ; table of inclination of optic axes, 270.

Faraday, on inductive action in telegraph cables, referred to, 168.

Fechner, on measurement of force of electric current, referred to, 104.

Fessel, his gyroscope explained, 307–313.

Foucault, his researches on the rotation of the earth, 303–306.

Fraunhofer, on the spectrum of the electric spark, alluded to, 224.

Fresnel, on the polarization of light, 11, *n.* ; his explanation of the successive polarization of light, 290 ; experiment on polarized light, 299.

Fusinieri, his experiments on electricity alluded to, 224.

Galileo, on acoustic figures, alluded to, 64.

Galvanometer, resistance of wire of, in voltaic circuit, 112 ; use of same instrument to measure forces in circuits of different kinds, 124, 125 ; applied to measurement of small resistances, 126–128 ; relations between its degrees and the degrees of force, 132 ; action of, in telegraph thermometers, 206.

Galvanometers, used in experiments on telegraph-wires, described, 171.

Gay-Lussac, on the transmission of sound, noticed, 48.

Geometry, descriptive, its processes, 233.

Génder, the, its principles explained, 40, 41.

Gerbert, his speaking head, 349.

Gmelin, Prof., on conversion of relief, referred to, 243.

Goffe, instructions pour le sieur de, historical document translated, 326–329.
Gyroscope, Fessel's, description of, 307; phenomena exhibited by, 308–312.

Harmonic Diagram, Explanation of, 14–20; facilitates acquisition of musical theory, 14; method of using, 16.
Hassenfratz, on the transmission of sound, noticed, 48.
Herhold, on the transmission of sound, noticed, 48, 49.
Hook, Dr., on acoustic figures, alluded to, 64.
Hooke, Dr., his experiments on transmission of sounds noticed, 47.

Insulation affected by pressure, 185.
Insulator, influence of coating of wire upon amount of discharge, 175–180; insulating material, 180–184.
Intervals. See *Melody, Elements of.*
Iron, its resistance to voltaic current greater than copper, 175.

Jacobi, Prof., his instruments for measuring force of electric current, 104, 105.
Jew's harp, theory of, 43, 44.

Kaleidophone, the, Description of, 21–29; renders visible the paths described by the points of greatest excursion in vibrating rods, 21; construction of apparatus, 22, 23; experiments with one rod, 24, 25; experiments with a number of rods, 26, 27; multiplication of images, 28; duration of impressions of light, how illustrated, 29.
Kempelen, de, *Le Mécanisme de la Parole*, reviewed, 348–367.
Keys. See *Melody, Elements of.*
Konstantinoff, de, claims invention of electro-magnetic chronoscope, 143; his acquaintance with Wheatstone, 146; receives a specimen of the chronoscope, *ib.*; communicates with Breguet, 149.
Kratzenstein, C. G., *Tentamen coronatum de Voce*, reviewed, 348–367.

Lehot, his theory of monocular vision, 255.
Light, Electrical, on the Prismatic Decomposition of, 223, 224; spectra of various metals, 223; duration of, determined, 87, 88, 93, 94, 96; hypotheses of origin of, 89; elongation of spark of, 93; applications of instantaneousness of, 95; obtained from thermo-electric pile, 134.
——, *Experiments on the Successive Pola-*

rization of, with description of a new Polarizing Apparatus, 290–302.
Light, polarized by reflection from sky, 285; successive polarization, explained by Fresnel, 290; plane dipolarization defined, 291; elliptic dipolarization defined, *ib.*; polarizing apparatus, *ib.*; polarization of uniaxal crystals, 292; biaxal crystals, 293; successive polarization of doubly-refracting laminæ, 294–296; explanation of successive polarization, 296; relative thickness of two films of crystal determined, 297; results of changing plane of reflection, 298; Fresnel's experiment, 299; rectilinear polarization, how obtained, 300; circular, how obtained, 301; elliptical, how obtained, *ib.*
Linari, Prof., his experiments on the thermo-electric spark, 134.
Liquids, resistance of, in voltaic circuit, measured, 122–124.
Loomis, Prof., his observation of remarkable electrical phenomena, 219.

Magnet, On the Augmentation of the Power of, by the reaction thereon of Currents induced by the Magnet itself, 211–215.
——, magnetized by its own currents, 211; description of new magnet, *ib.*; effect of excitation by rheomotor, 211, 212; effects after removal of rheomotor, 212; effects after junction of two circuits, *ib.*; explanation of phenomena, *ib.*; residual magnetism of magnet, 213; effect strongest at moment of completing circuit, *ib.*; result of connexion with inductorium, *ib.*; effect of diverting portion of current from magnet, 214; resistances necessary to produce effects, 214, 215.
Martin, on the transmission of sound, noticed, 48.
Melody, elements of, 15; scales—diatonic, chromatic, enharmonic, 15, 16; modes, 17; keys, *ib.*; progression of key in major mode, how found, *ib.*; progression of key in minor mode, how found, 18; relative minor to major key, how found, *ib.*; tonic minor of major key, how found, *ib.*; progression of major keys, *ib.*; progression of minor keys, 19; transposition of keys, *ib.*; intervals, 19, 20, when reciprocal, 20, simple and compound, *ib.*
Mical, Abbé, his speaking machine, 365.
Microphone, the, experiments with, 32, 33.
Mirror, revolving, vertical form of, described, 86; appearances presented by, 87; inclined form of, described, *ib.*;

appearances presented by, *ib.*; various applications of, 88; angular velocity of axle, how determined, 92; results of experiments with, 93; polygonal form of, proposed to be used, 95.

Modes. See *Melody, Elements of.*

Musical instruments: laws of construction of stringed instruments, 51, 52; transmission of their sounds to a distance, 53, 54; reciprocation by two stringed instruments, how effected, 56; sounds transmissible simultaneously to several places, 57; transmission of sounds of reed wind-instruments, *ib.*; Bernouilli's theory of wind-instruments verified, 220.

Necker, Prof., on apparent changes of figures of geometrical solids, 240.

Newton, Sir Isaac, an illustration of his theory of the reflection of colours, 39.

Nobili, on films of peroxide of lead, referred to, 117.

Numbers, square, formed from arithmetical progression, 314–316; cube, 316–319; higher powers, 320.

Oersted, Prof., his experiments on the vibrations of phonics, 7.

Ohm, his theory of the electric circuit, 98–100.

Perault, on the existence of more minute motions than sensible oscillations, 6.

Perolle, on the transmission of sound, noticed, 49.

Perspective, binocular, in plate of metal, 236; in compasses, 237; effect of, increased by viewing with single eye, 239; illusion of unusual projections, *ib.*

Petrina, Prof., on measuring electric currents, referred to, 125.

Phonics, general view of, 1, 2; oscillations of, 2; vibrating particles do not vary in different media, 3; phenomena, how observed, *ib.*; shown by experiment with fluids, *ib.*; with metallic plate, *ib.*; absolute magnitude of particles, how ascertained, 4; modify various properties of sound, 6.

——, oscillations of: are not immediate causes of sound, 2; proved by experiment, *ib.*; manner in which they induce vibrating particles, 5.

——: vibrations rendered visible by columns of air, 5; their excursions control loudness of sound, 6; rectilineal transmission of, 7; transmission influenced by thickness of conductors, 8, 9; how affected by passage through bent conductors, 9; nature of, 10. 11; how pola-

rized, 11, 12; law of polarization, 12; double refraction of, *ib.*; diffraction of, 12, 13; law of transmission when their source is in progressive motion, 13; to what extent audible, 50; plane of vibration affects intensity of sound, 51.

Pictures, cannot faithfully represent a near solid object, 226; Leonardo da Vinci on, 227; two, may be drawn to present identity with object, 233; perspective effect increased by viewing with single eye, 239.

Polarization, of light reflected from sky, 285; plane of, for any point of sky, defined, 286; plane of, for zenith, *ib.*; for north pole, *ib.*

Polarizing Apparatus, described, 291; experiments with, 292, 294, 296, 297, 298, 300; conversions of, 301.

Pouillet, on measures of electro-motive forces, referred to, 118; is allowed to copy the electromagnetic chronoscope, 145.

Progressions, Arithmetical, on the Formation of Powers from, 314–320; triangular arrangement of, 314; triangles which produce square numbers, 314–316, cube numbers, 316–319, higher powers, 320.

Pseudoscope, construction of, 275; its effect on perception of an object, 275, 276; illusions produced by, 277; involuntary changes of figures produced by, 278; illusions of distance produced by, 279; its effect on the perception of objects in motion, 281; modifications of, 282.

Purkinje's Experiments, Remarks on, 221; experiments described, 221; explanation of, 221; variation of, 222.

Quetelet is informed of the electromagnetic chronoscope, 144.

Rafn, on the transmission of sound, noticed, 48, 49.

Raspail, on conversion of relief, 244; on place of formation of visual images, 255.

Reduced length of wire in electric circuit, defined, 99.

Reid, Dr., on optics, alluded to, 248, 249, 252.

Relief, conversion of, Brewster's explanation of, 243; explanation of, 244; how effected, 273; cause of, *ib.*; not observed with reflexion or conversion of object, 274; appears when one of the pictures remains constant, 275; produced by the pseudoscope, 277; sometimes observable in pictures viewed directly with both eyes, 283.

Relief, inversion of, 245.

Resistance, in electric circuit, 98, 101; interposed, weakens electric force, 99; how determined, 103, 104; standard of, 110; of interposed body, how determined, 111; of rheomotor, ascertained by various processes, 118–121; of liquids, instrument for measuring, described, 122; method of measurement, 122, 123; small differences of, how measured, 126–128; interposed, augments time of charging or discharging wire, 186.

Resistance-coils, described, 110.

Resonance, how effected, 36; of column of air to vibrating tuning-fork, 37; to wind instruments, 37; of subdivisions of columns of air, 38; principle employed in musical instruments, 40, 43; law of, 42, 50; application of law to jew's harp, 43, 44; of imperfect unisons produces beats, 45; simultaneous, of several sounds, ib.; of sounding-boards, how influenced, 50; peculiarities of, explained, 51; takes place when sound of column of air is harmonic of original sound, 359; multiple resonances and vowel qualities heard together, 360; are different forms of same phenomena, ib.

Rheometer, magnetic, defined, 103; mode of using, 195, 196; construction of, 195.

Rheomotor, term defined, 102; resistance of, ascertained, 118–121; determined by comparison with standard rheomotor, 121; comparative changes of two instruments, how observed, 129.

Rheoscope, defined, 103.

Rheostat, instrument described, 105–107; used for small resistances, described, 108; its use as a regulator of the current, 109; in volta-typing, ib.

Rheotome, defined, 103.

Rheotrope, defined, 103.

Richmond, his performance with the voice explained, 360.

Rotation, phenomena of, shown by gyroscope, 307–313.

Savart, Dr., on modes of division in surfaces put in motion by communicated vibrations, 6; his experiments on audition, alluded to, 30; his experiments on resonance, 37, n.; his researches in acoustics noticed, 82.

Scales See Melody, Elements of.

Schönbein, on voltaic combinations, noticed, 117.

Siemens, Werner, on inductive action in telegraphic cables, referred to, 168.

Sky, polarizes light, 285; planes of polarization of, 286.

Smith, Dr., his system of optics referred to, 226, 237; on law of direction for monocular vision, 249, 251; on alteration of perceived magnitude of objects, noticed, 263.

Sound, New Experiments on, 1–13; its proximate causes, 1; its influence on vibrating particles, 2, 3; its properties attributable to modifications of vibrating particles, 6; its transmission through linear conductors, 7, 8; through other communicating mediums, 8; influenced by thickness of conductor, 8, 9; polarization of, 9, 10; effect of bent conductors on transmission, 9, 10; polarization observable in corded instruments, 10; phenomena analogous to those of polarized light, 11, 12; refraction and diffraction of, 12; motions of vibrating rods rendered visible, 21, 28; figures produced by motion of vibrating rods, 26, 27; see Kaleidophone; augmentation of sounds produced internally, 30, 31; of sounds produced externally, 32, 33; production of grave harmonic, 34; acute sounds rendered inaudible, ib.; rendered intense, ib.; varying intensity of certain notes, 35; augmentation by resonance of column of air, 37, 38; what sounds can be reciprocated by a column of air, 38; one of several simultaneous sounds rendered separately audible, 39; grave harmonic, cannot be heard separately in resonance, ib.; resonance of, employed in musical instruments, 40; law of resonance of, 42; simultaneous reciprocation of several sounds, 45; alternations in intensity of, produces beats, ib.

——. Transmission of Musical Sounds through Solid Linear Conductors, and on their subsequent Reciprocation, 47–63; its velocity in solid substances, 48, 49; augmentation of, 50, 51; its transmission, unaffected by length of conductor, 52; vibrations of sounding-boards, how transmissible, 53; of pianoforte transmitted upwards, ib., downwards, 54, horizontally, 56; of an instrument transmissible simultaneously to several places, 57; of stringed instruments transmissible, ib.; of reed wind-instruments, ib.; of certain instruments not transmissible, 58; transmission of, through bent conductors, 59; diffusion of, 60; longitudinal undulations preferable to transversal, 61; Bernouilli's theory of wind-instruments verified, 220; of columns of air in cylindrical tubes, 368, 369; in conical tubes, 370.

Sounding-boards, peculiarities of, explained, 50, 51.

Speaking-machines used in ancient oracles, 348 ; in Middle Ages, 349 ; principles of construction, 363.

Spectrum of electric spark from various metals described, 223.

Speech, human, possibility of transmitting, 62, 63.

Stereoscope, the, described, 230–232 ; outline figures for use in, 232 ; effect of relief produced by, 233 ; effect of transposition of pictures in, 234 ; conversion of relief, 235 ; effect produced by varied positions of skeleton figures, *ib.* ; projections of a symmetrical object, 236 ; conversion of plane outline into figure of three dimensions, *ib.* ; experiments on double appearance of two similar pictures projected on corresponding points of the retinæ, 245 ; binocular vision of images of different magnitudes, 246 ; experiments on theory of corresponding points, 256 ; portable reflecting, described, 266 ; refracting, 267 ; application to drawings of figures of three dimensions, *ib.* ; assistance afforded by photography, 268 ; how to obtain photographs for, 269 ; combinations of magnitude and solidity, 271, 272.

Strehlke, Prof., on acoustic figures, criticised, 78.

Sylvester II., Pope. See *Gerbert.*

Talbot, his talbotypes noticed, 268.

Télégraphe automatique écrivant, description of system, 160 ; the *perforateur,* 161 ; the *transmetteur,* 162 ; the *recepteur,* 164 ; the *traducteur,* 165 ; advantages of the system, 165.

Telegraph cables : *Experiments on Mediterranean Cable,* 152–157 ; first series of experiments, 153 ; phenomena shown by iron envelope of conductor, 153 ; second series, 154 ; third series, 155.

—— ——. *On the Circumstances which influence the Inductive Discharges of Submarine Telegraphic Cables,* 168–235 ; inductive action in, 168 ; effect of coating wire with insulating substances, *ib.* ; charging and discharging wires, how effected, 170 ; description of galvanometers used in experiments, 171 ; influence of electro-motive forces on amount of discharge, 172 ; influence of length of wire on inductive discharge, 173 ; simultaneous discharges from number of wires, 174 ; influence of conductivity of wire, 175 ; of diameter of wire, 175–180 ; of thickness of insular coating,

175–180 ; of insulating material, 180–184 ; of temperature, 184 ; of pressure, 185 ; of interposed resistance on time of charging and discharging, 187 ; current produced by discharging electricity to the earth, *ib.* ; true and apparent discharges of, 191–194 ; measurement of charges in short wires, 194–196 ; accumulation of charges and discharges, 196–205.

Telegraph Thermometer, Description of the, 206, 207 ; construction of instrument, 206 ; method of using, 206, 207 ; defects of, 208.

——. *On a new Telegraphic Thermometer, and on the application of the Principle of its construction to other Meteorological Indicators,* 208–210 ; improved instrument, 208 ; construction of, 209 ; questioner, *ib.* ; responder, *ib.* ; connection by telegraphic wires, *ib.* ; mode of using, 210 ; accuracy of, how affected, *ib.*

Temperature, its influence on amount of inductive discharge, 184.

Thermo-electric pile, spark obtained from, 134 ; chemical action of, 136.

Thermo-electric Spark, on the, 134–137.

Time, apparent Solar, on a means of determining, by the Diurnal Changes of the Plane of Polarization at the North Pole of the Sky, 285–289. See *Clock, Polar.*

Vallée, his theory of monocular vision, noticed, 255.

Vibration of rods, plane of oscillation persistent, 303, at the poles, 304 ; persistence of, shown by experiment, 305.

——, of surfaces, first investigated by Chladni, 64 ; figures of, how represented, 65 ; Chladni's table of, 66 ; modes of vibration of rectangular surfaces, 67, 70 ; states of vibration, 68 ; superposition of two similar modes of, *ib.* ; modes of superposition, 69 ; resultant figures of superposition, 70 ; superposition of four similar modes, 71 ; imperfect resultants, 76 ; figures of irregular surfaces, 77 ; Chladni's table rearranged, 78, 79 ; different theories of, 80, 81 ; plates of wood, figures produced by, 81.

Vinci, Leonardo da, on inferiority of paintings to natural objects, 226.

Vision, Contributions to the Physiology of, part the first, 225–259 ; *ib.,* part the second, 260–283 ; of objects when optic axes of both eyes are parallel, 225 ; when optic axes converge, *ib.* ; two similar objects, how made to fall on similar parts of retinæ. 228 ; effect of substituting two perspective projections of one object, 229 ; appearances of two vertical lines, *ib.* ; of a straight wire, *ib.* ; the stereoscope, 230–232 ; binocular perspective, 236 ; correctness of vision of single eye, 238 ; illusions of perspective, 239 ; changes in appearance of solid geometrical figures, 240 ; Necker's explanation, 241 ; phenomenon dependent on action of mind, *ib.* ; conversion of relief, 242 ; double appearance of similar pictures falling on corresponding points of retinæ, 245 ; binocular vision of images of different magnitudes, 246 ; effect of presenting different pictures to both eyes, 247 ; monocular, law of visible direction for, 249 ; views of Reid, Smith, D'Alembert, Brewster, *ib.* ; of Aguilonius, Wells, 250 ; theory of, that pictures fall on corresponding points of the retinæ, 251 ; theory inconsistent with law of Aguilonius, 252 ; theory of corresponding points disproved by facts, 253, 256 ; theory of vision with one eye only, 238, 254 ; appearance of relief independent of motion of eye, 256, 257 ; necessity of abnormally combining the elements of vision, 261 ; instrument for effecting combinations, 262 ; variation of perceived magnitude of object, 263 ; definitions of terms, *ib.* ; variation of apparent distance of object, 264 ; modifications of the stereoscope, 266, 267 ; combination of various inclinations of optic axes with various dissimilarities of pictures, 270 ; combination of increased magnitude with perception of solidity, 271 ; conversion of relief, 273 ; the pseudoscope, 275 ; gradual transition from normal to convex perception, 280.

Voice, human, cause of defective pronunciation of vowels in high notes, 357.

Vowels, formation of, explained, 353 ; table of, 354 ; inadequately represented by written signs, 355 ; researches of R. Willis upon, *ib.* ; production of, from cylindrical tube, 356, 357.

Weber, brothers, on theory of undulating surfaces, 81.

Wells, Dr., his "new theory of visible direction," 250.

Willis, Robert, *on the Vowel Sounds*, reviewed, 348–367.

Wollaston, Dr., his experiments on audition alluded to, 30, 31 ; on the diversity of limits of audibility, alluded to, 32.

Wood, plates of, acoustic figures produced by vibration of, 81, 82 ; Savart's experiments on circular plates of, 82.

Wunsch, Prof., on the transmission of sound, noticed, 47.

Young, Dr. Thos., his discoveries on light alluded to, 11, *n.* ; his experiments on excursions of vibrating chords, 22, *n.* ; on the origin of acoustic figures, 81 ; his instrument the basis of the electromagnetic chronoscope, 151.

Printed by TAYLOR AND FRANCIS, Red Lion Court, Fleet Street.

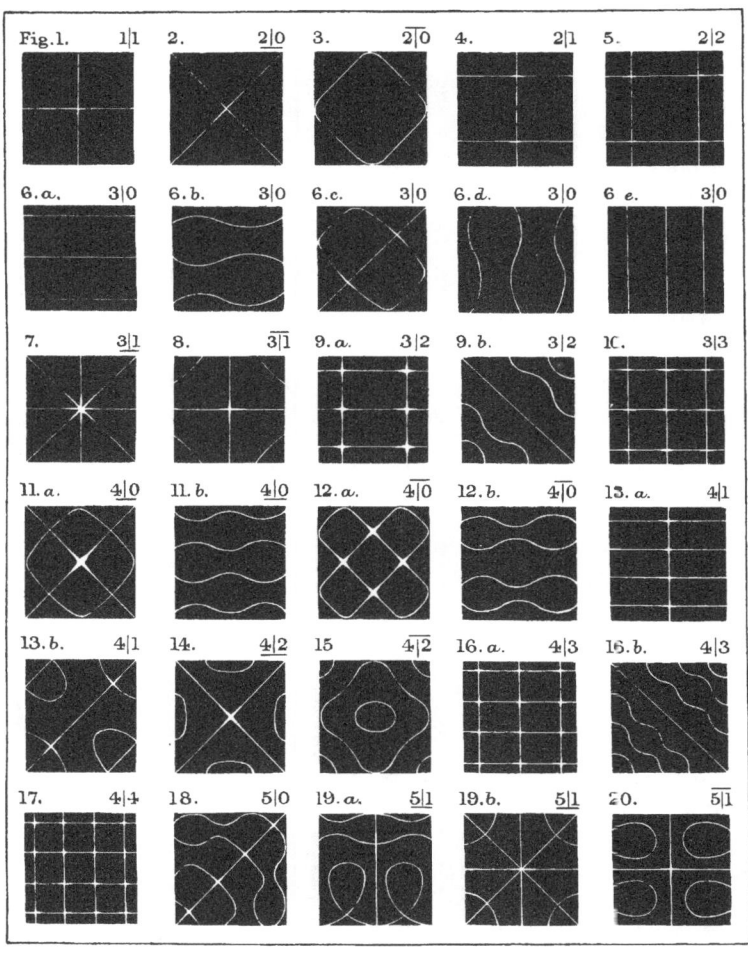

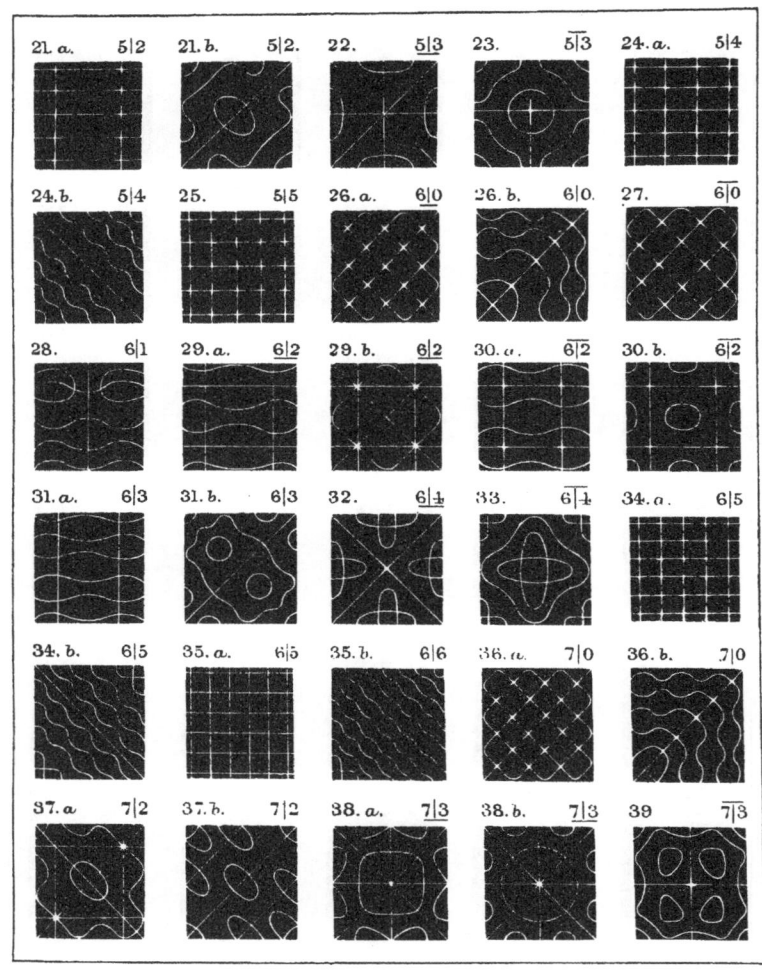

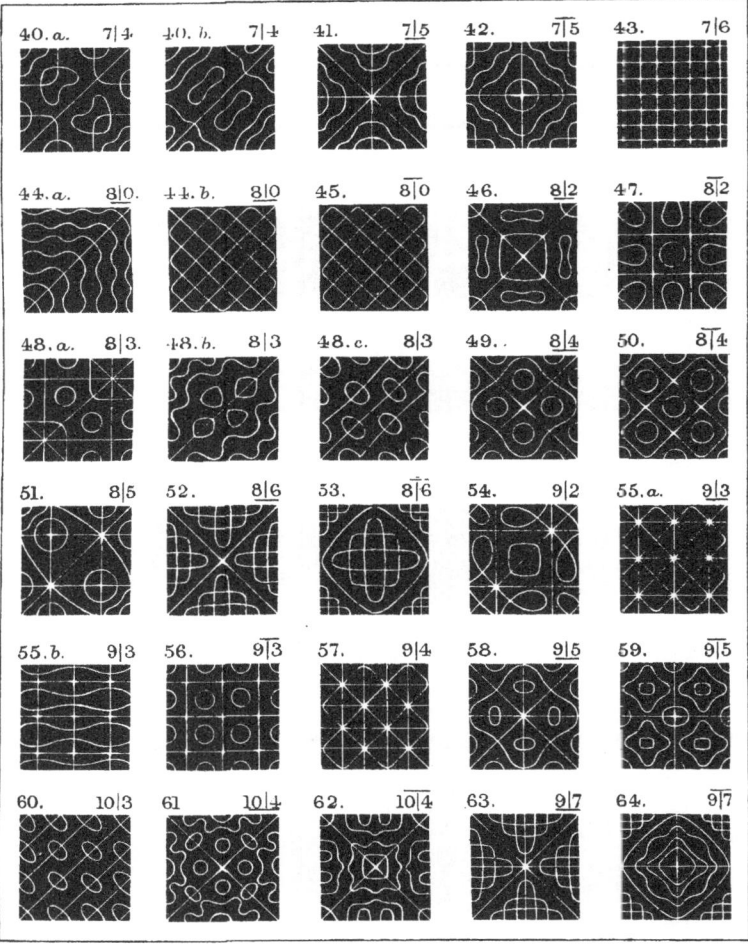

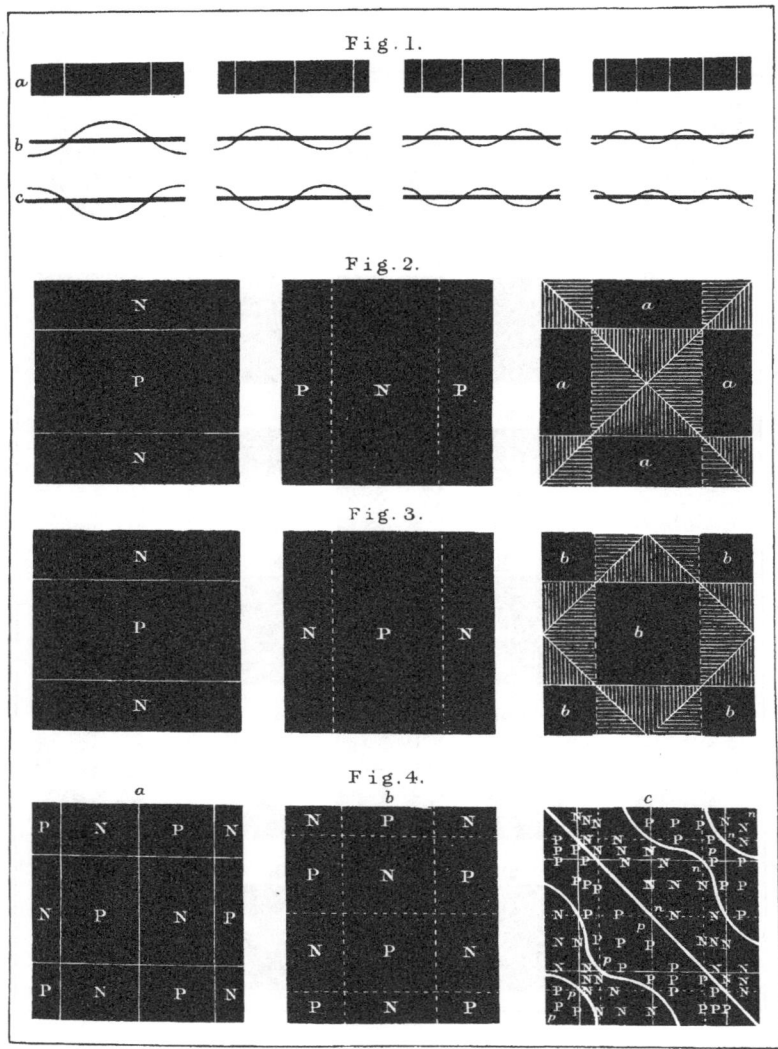

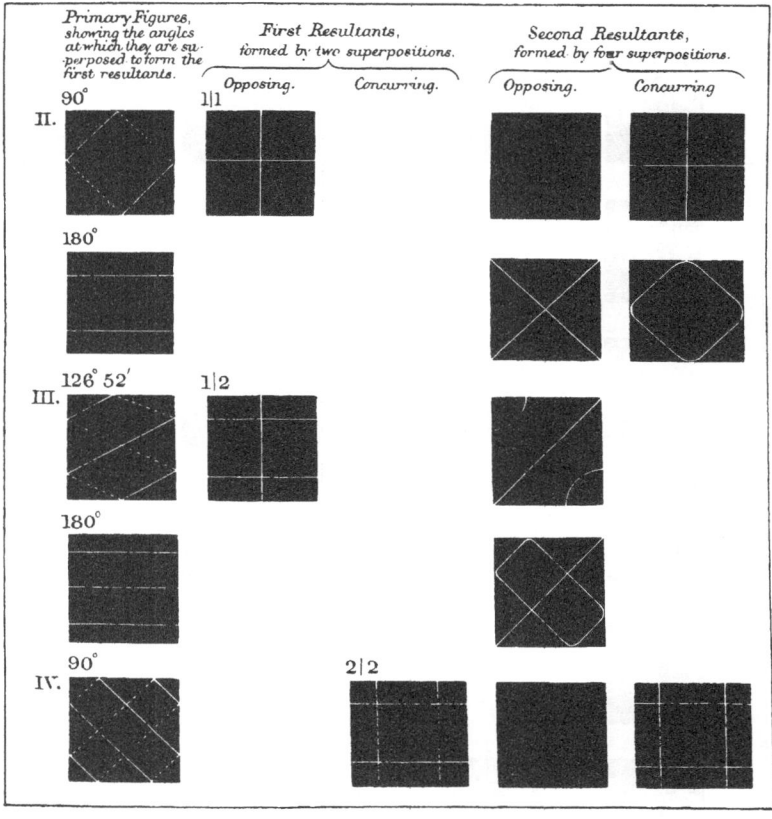

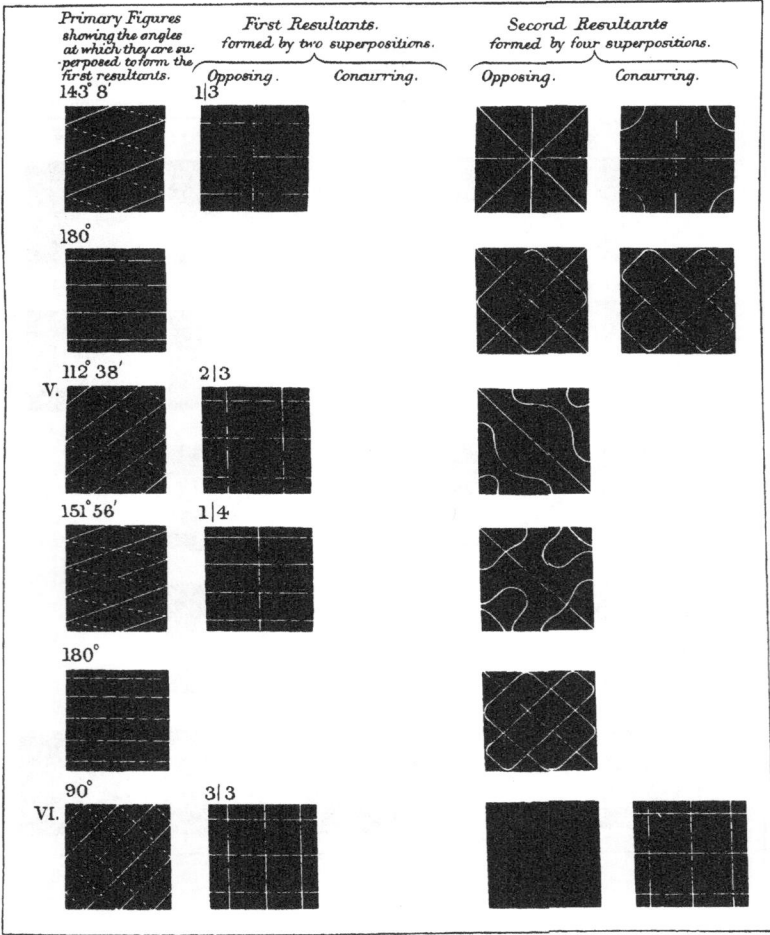

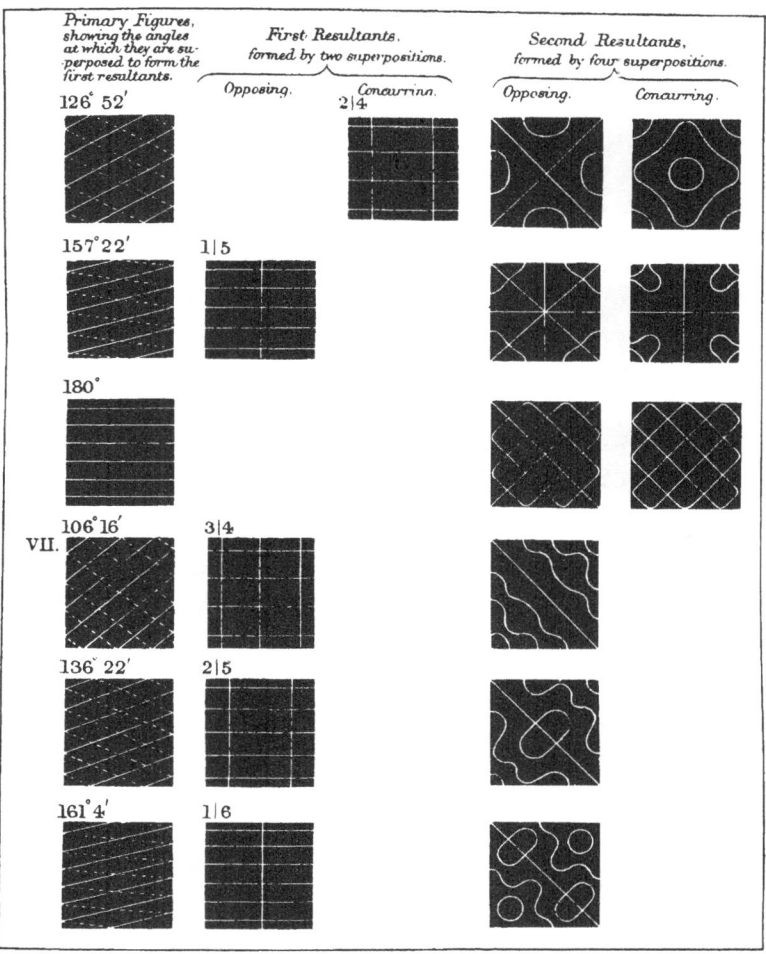

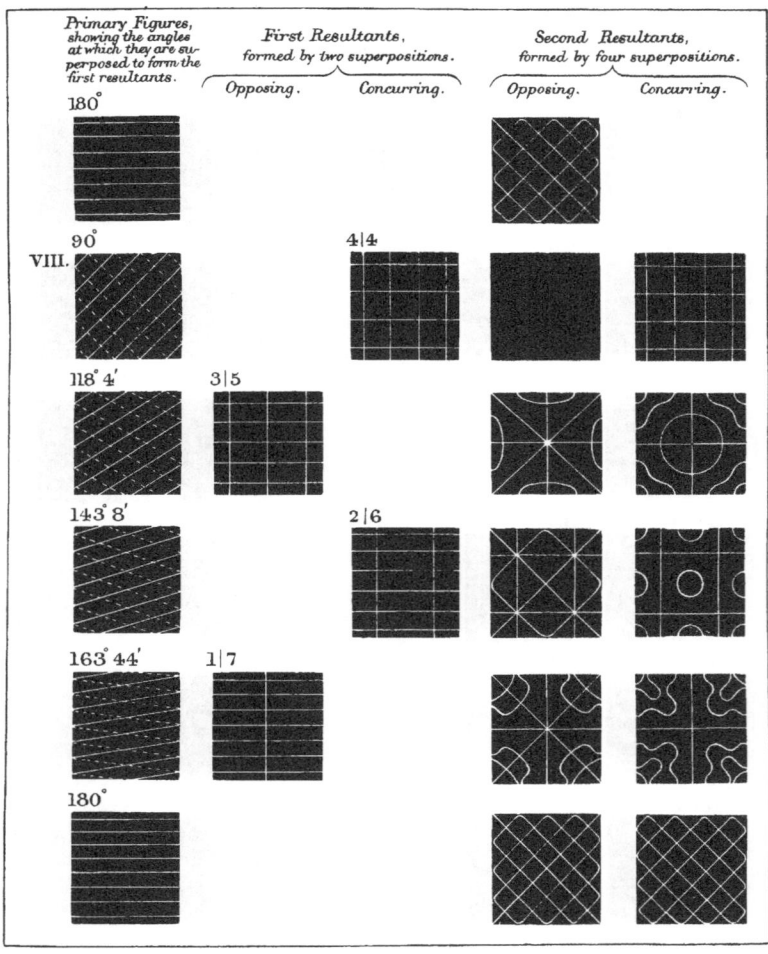

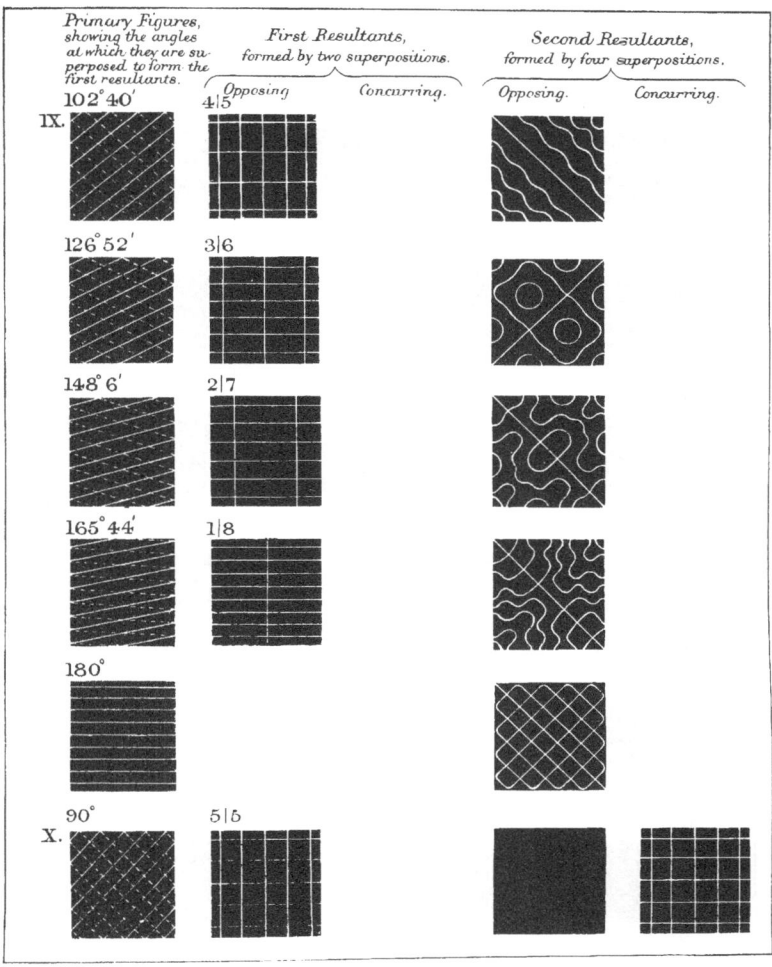

Mintern Bros. lith.

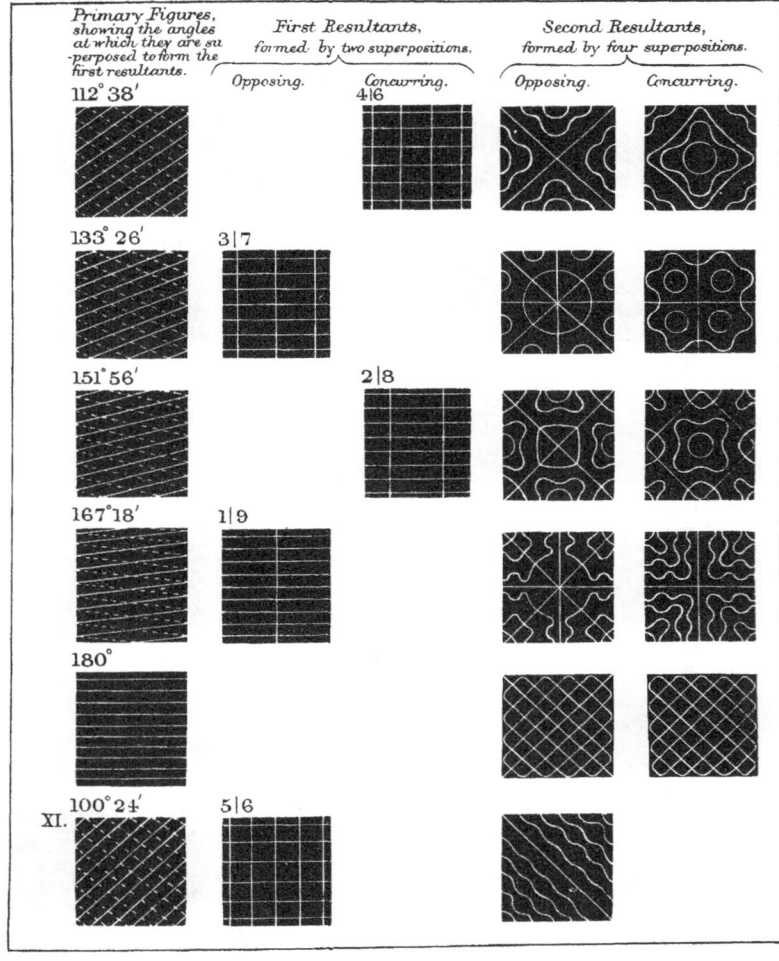

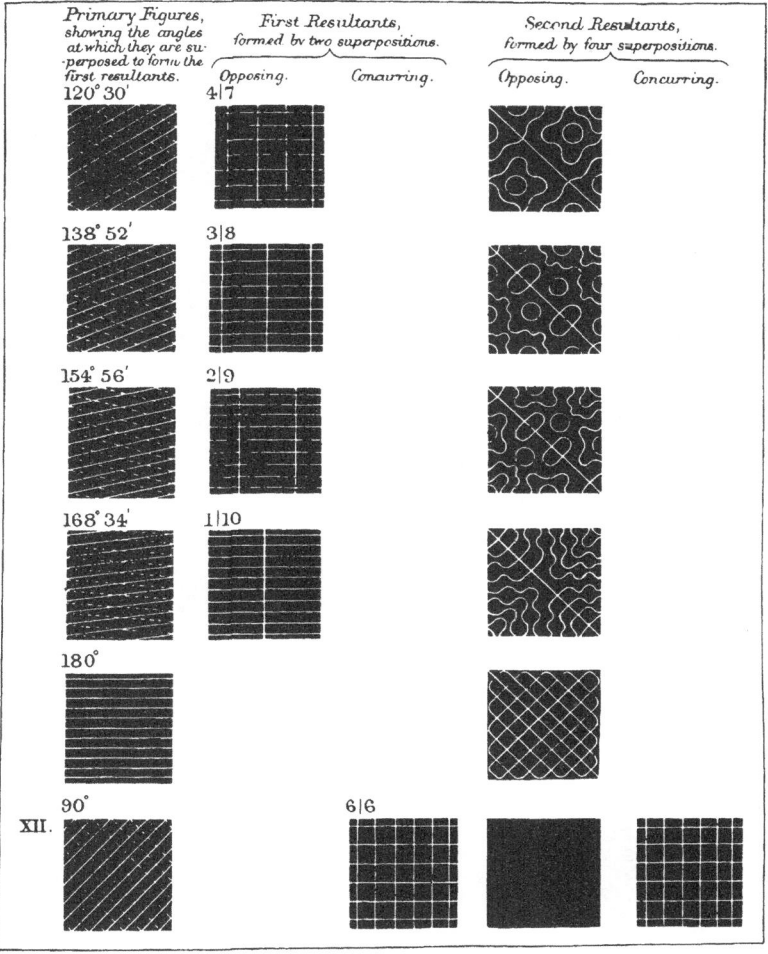

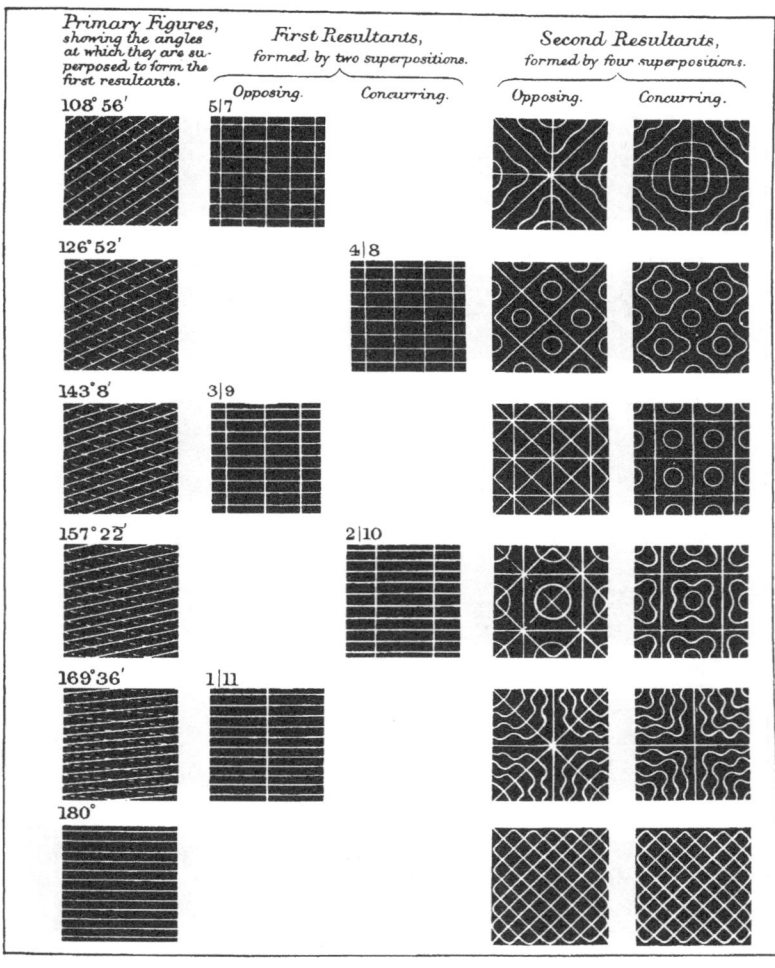

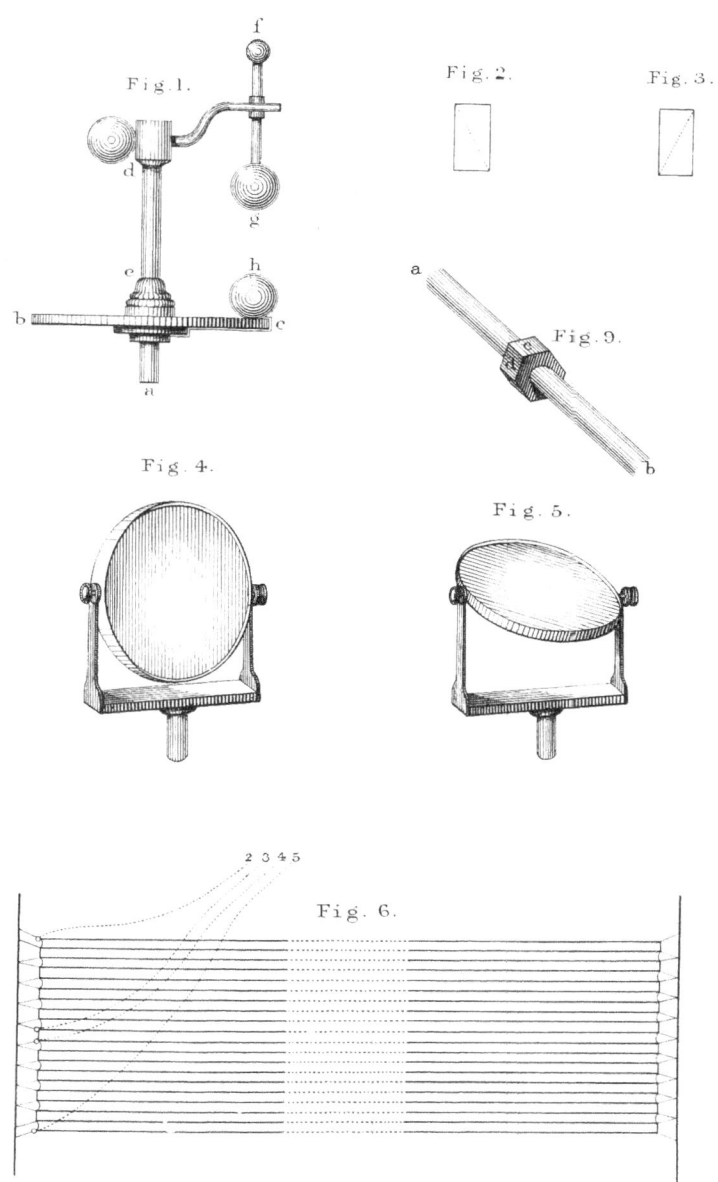

Fig.1.

Fig.2.

Fig.3.

Fig.9.

Fig.4.

Fig.5.

Fig.6.

Mintern Bros. lith.

Plate XIV.

Wheatstone.

Fig. 7.

Fig. 10.

Fig. 8.

Mintern Bros imp.

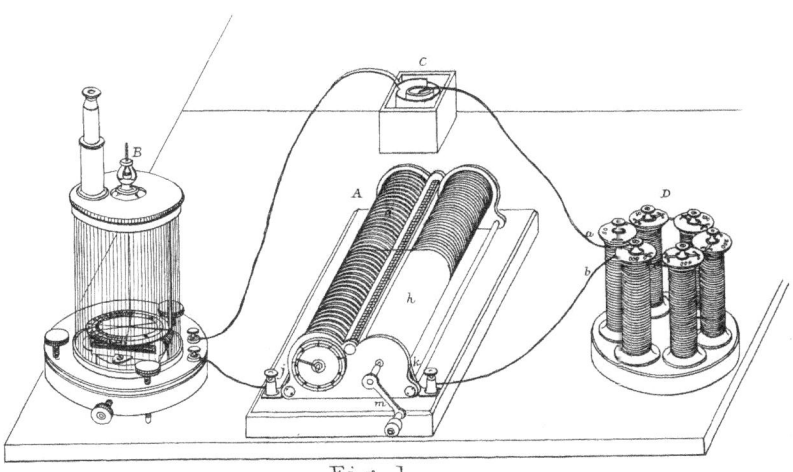

Fig. 1.

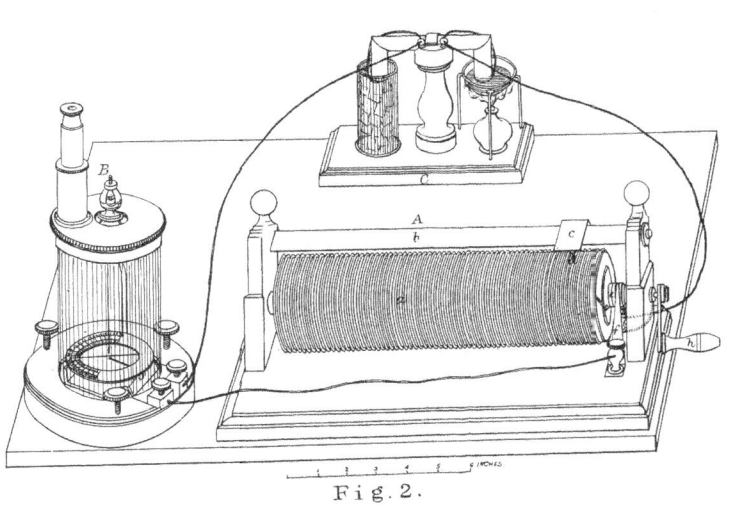

Fig. 2.

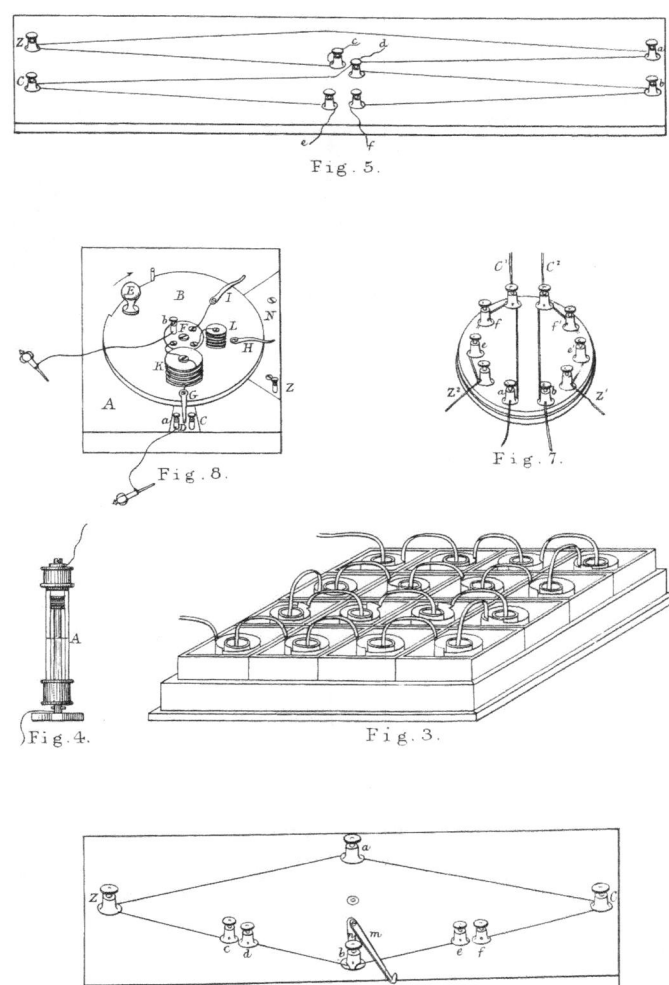

Fig. 5.

Fig. 8.

Fig. 7.

Fig. 4.

Fig. 3.

Fig. 6

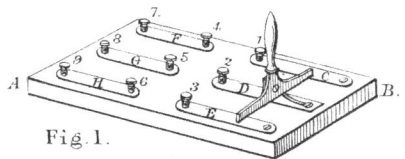

Fig. 1.

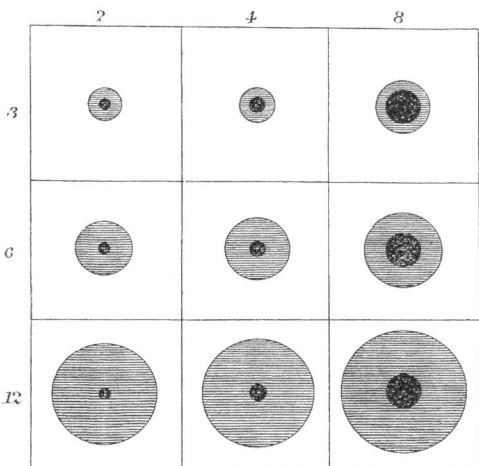

Fig. 2.

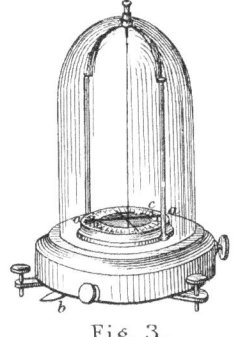

Fig. 3.

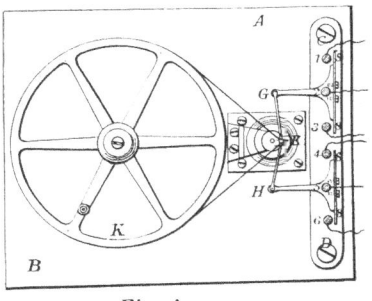

Fig. 4.

Plate. XVIII.

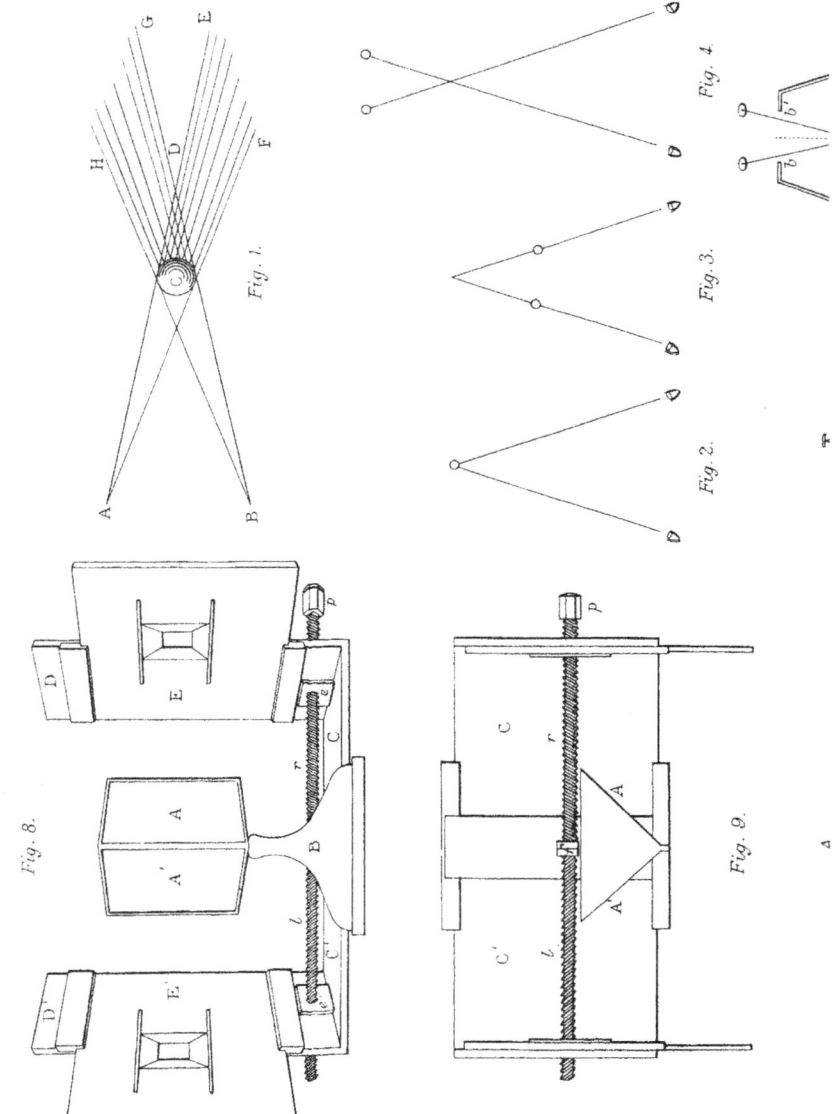

Fig. 1.

Fig. 2.

Fig. 3.

Fig. 4.

Fig. 8.

Fig. 9.

Fig. 6.

Fig. 5.

Fig. 25.

Fig. 22.

Fig. 7.

Fig. 24.

Fig. 26.

Fig. 21.

Wheatstone. Pl. 19.

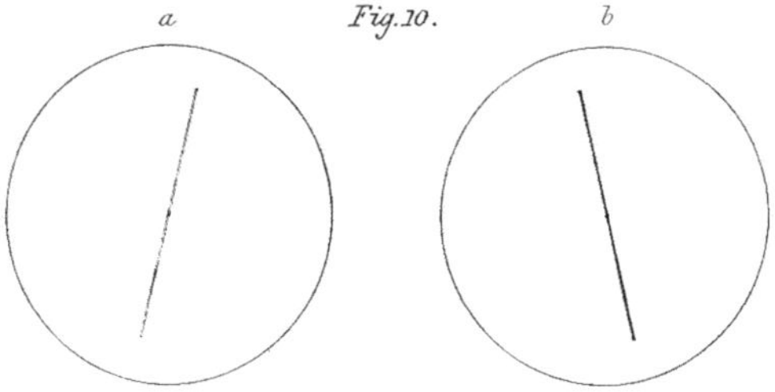

Fig. 23.

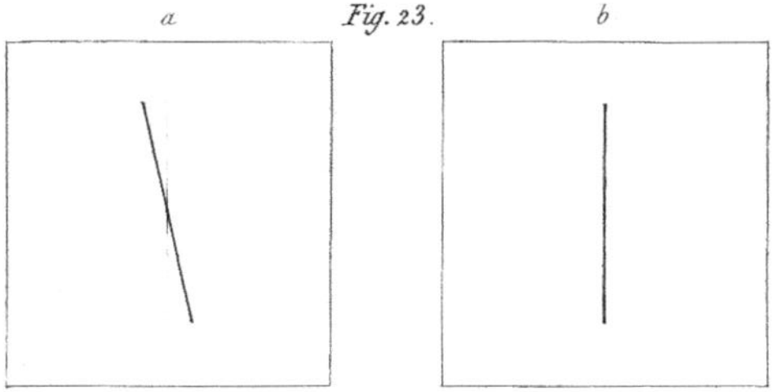

b

a

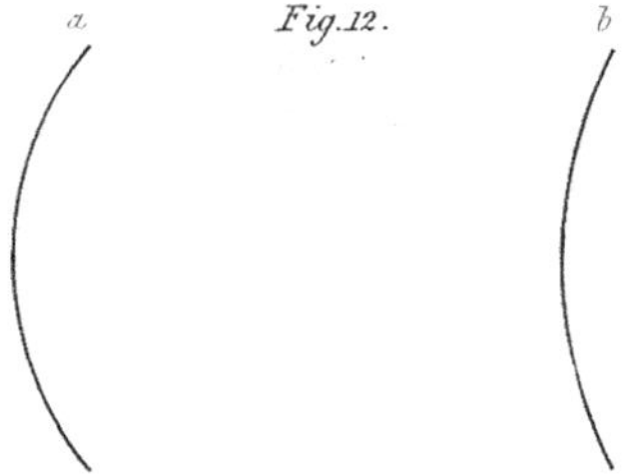

Fig. 11.

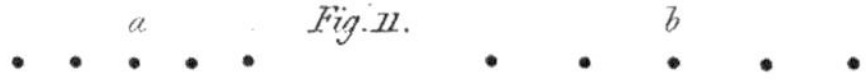

b

a

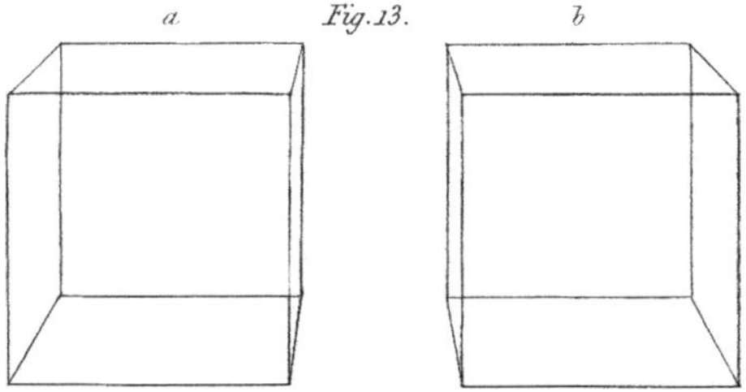

Fig. 14.

b

a

Fig. 10.

b

a

Fig. 12.

b

a

Fig. 13.

b

a

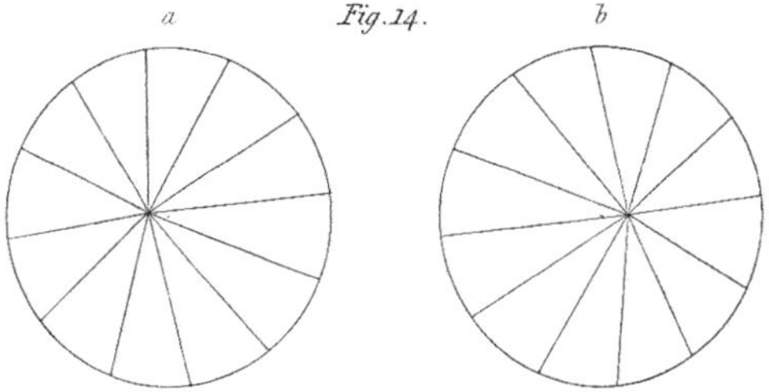

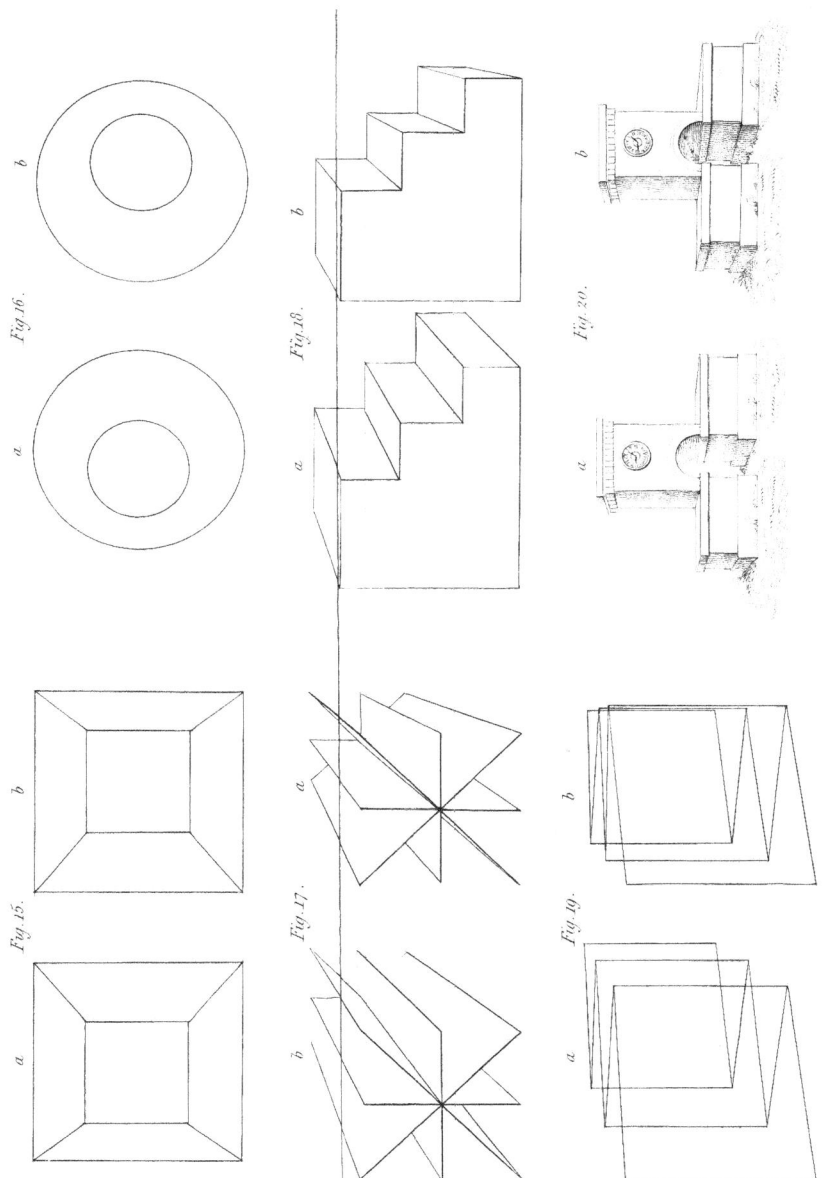

Fig.16.

Fig.15.

Fig.18.

Fig.17.

Fig.20.

Fig.19.

J. Basire sc.

Phil. Trans. MDCCCLII. Plate I. p. 3

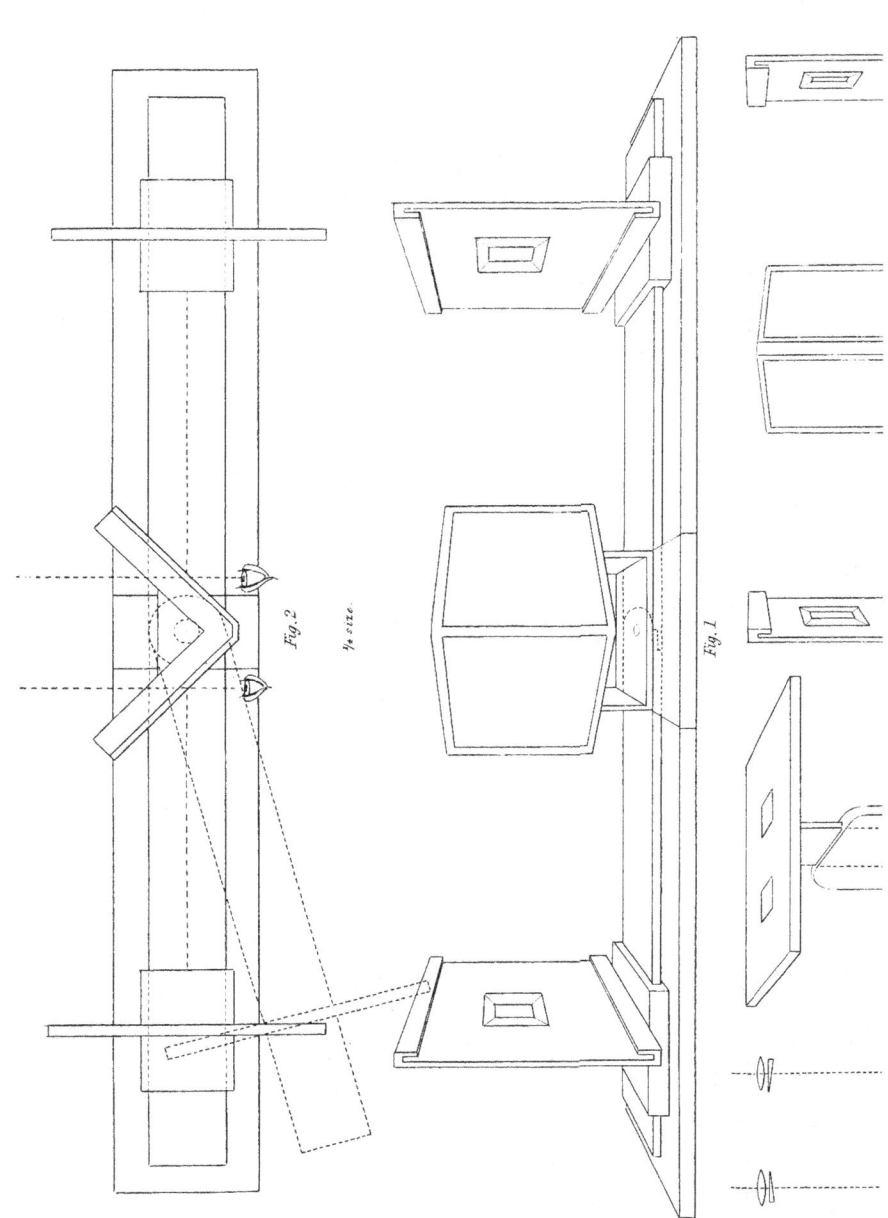

Fig. 2

¼ size.

Fig. 1

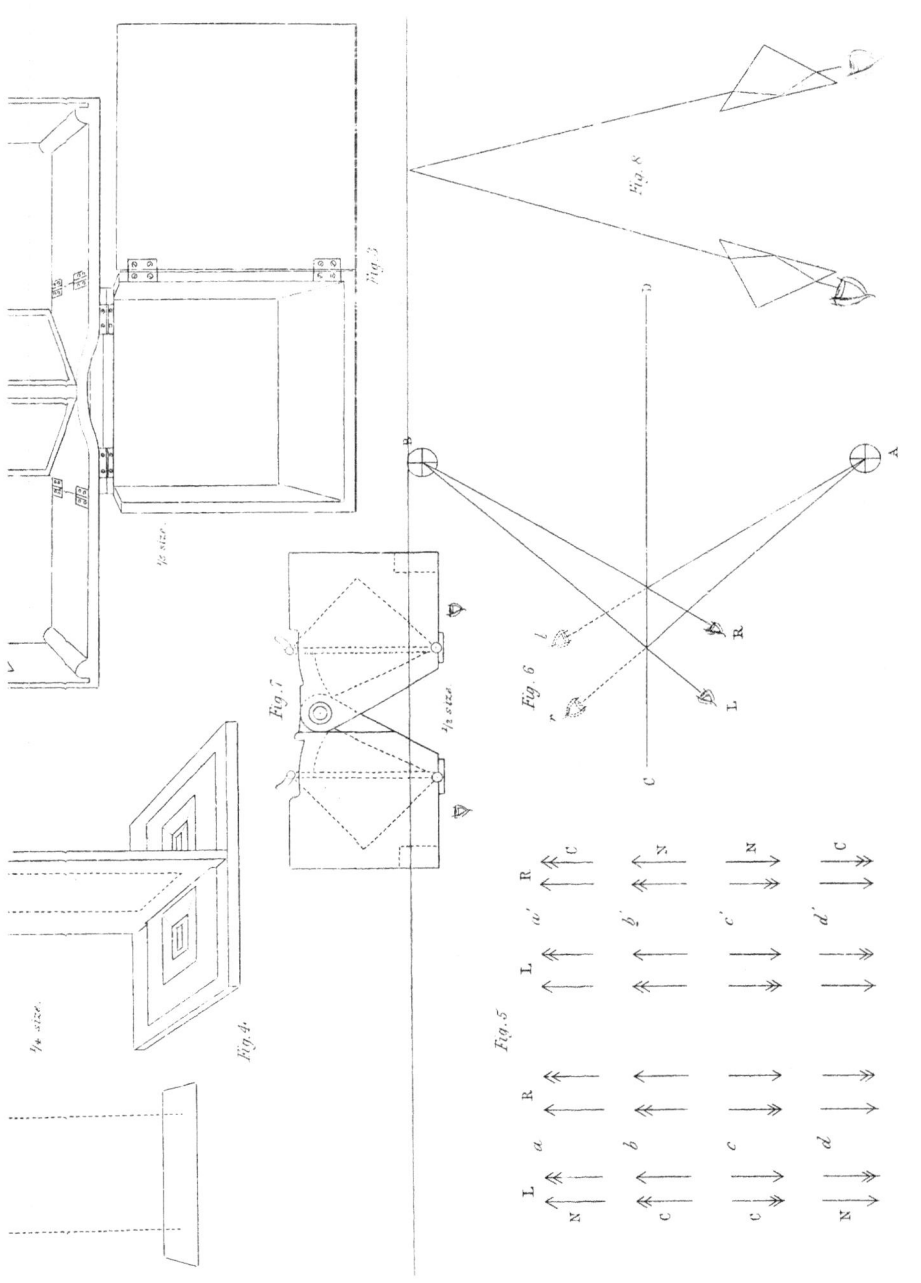

Fig. 1

Fig. 2

Fig. 3

Fig. 4

Fig. 5

Fig. 6

Fig. 8

½ size.

¼ size.

½ size.

¼ size.

J. Basire sc.

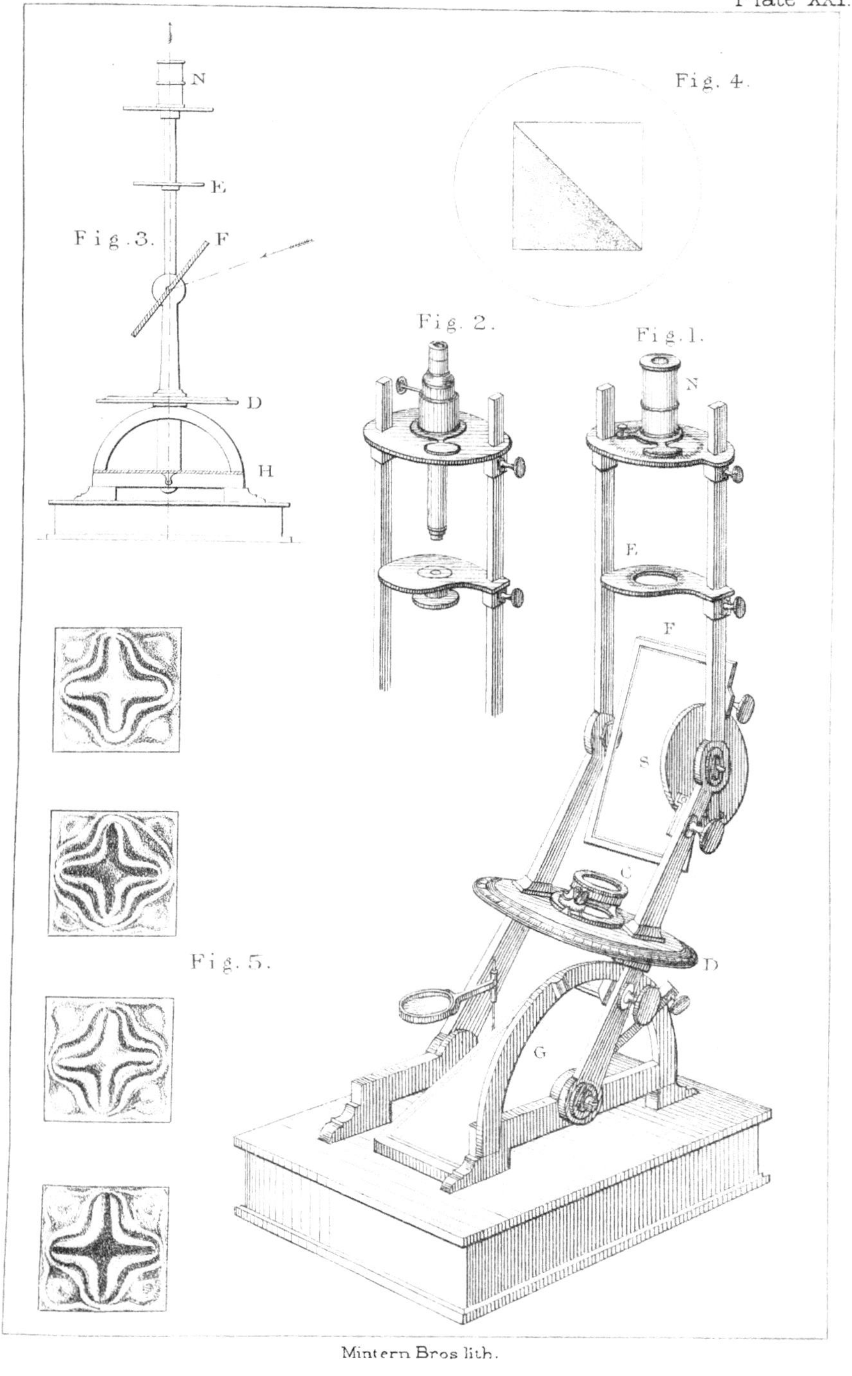
Fig. 4.
Fig. 3.
N
E
F
D
H
Fig. 2.
Fig. 1.
N
E
F
S
C
D
G
Fig. 5.

Mintern Bros lith.

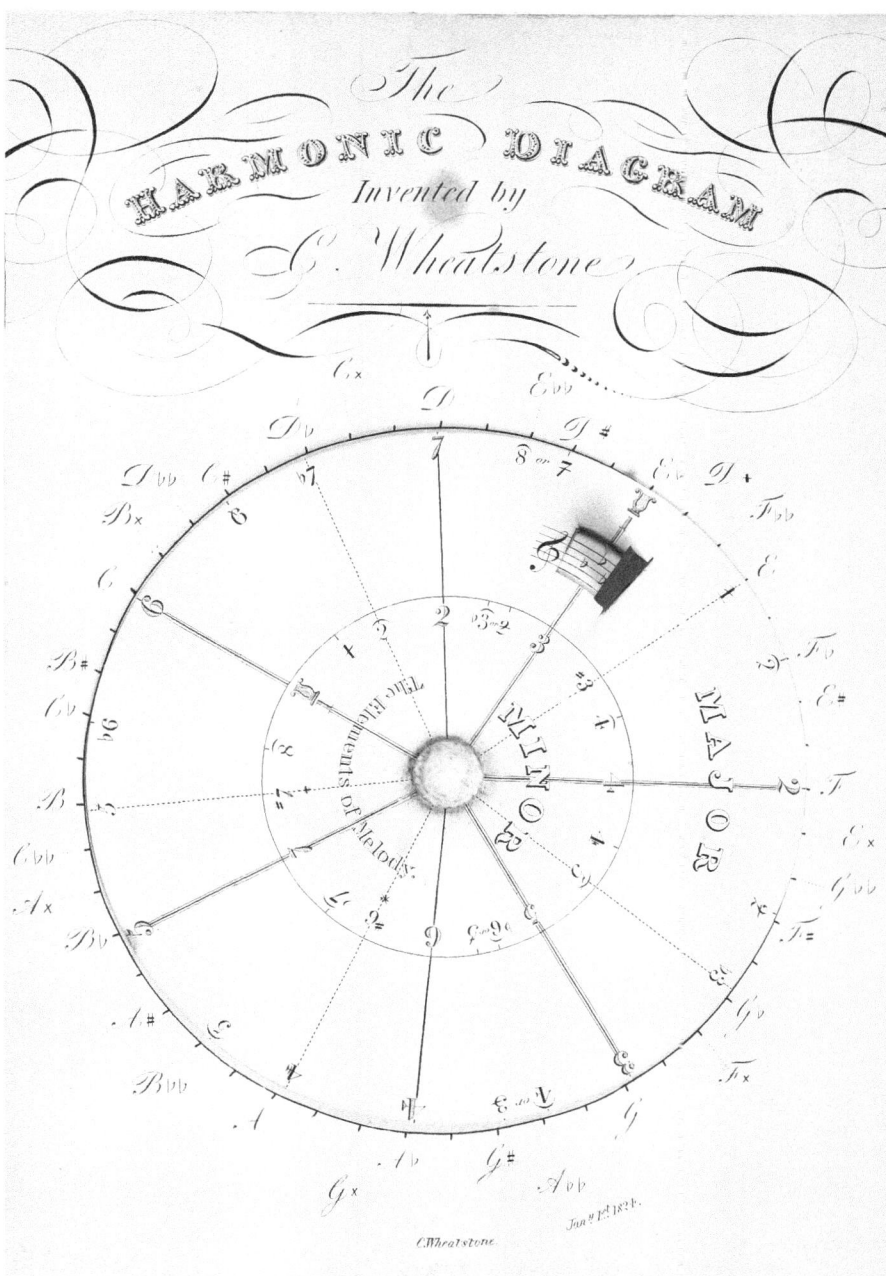

The
HARMONIC DIAGRAM
Invented by
C. Wheatstone